AF755736

Springer Tracts in Modern Physics

Volume 276

Series editors

Yan Chen, Department of Physics, Fudan University, Shanghai, China
Atsushi Fujimori, Department of Physics, University of Tokyo, Tokyo, Japan
Thomas Müller, Inst für Experimentelle Kernphysik, Universität Karlsruhe, Karlsruhe, Germany
William C. Stwalley, Storrs, CT, USA

Springer Tracts in Modern Physics provides comprehensive and critical reviews of topics of current interest in physics. The following fields are emphasized:

- Elementary Particle Physics
- Condensed Matter Physics
- Light Matter Interaction
- Atomic and Molecular Physics
- Complex Systems
- Fundamental Astrophysics

Suitable reviews of other fields can also be accepted. The Editors encourage prospective authors to correspond with them in advance of submitting a manuscript. For reviews of topics belonging to the above mentioned fields, they should address the responsible Editor as listed in "Contact the Editors".

More information about this series at http://www.springer.com/series/426

Russell E. Walstedt

The NMR Probe of High-T_c Materials and Correlated Electron Systems

Second Edition

Springer

Russell E. Walstedt
Department of Physics
University of Michigan
Ann Arbor, MI
USA

ISSN 0081-3869 ISSN 1615-0430 (electronic)
Springer Tracts in Modern Physics
ISBN 978-3-662-55580-4 ISBN 978-3-662-55582-8 (eBook)
https://doi.org/10.1007/978-3-662-55582-8

Library of Congress Control Number: 2017948210

1st edition: © Springer-Verlag Berlin Heidelberg 2008
2nd edition: © Springer-Verlag GmbH Germany 2018
This work is subject to copyright. All rights are reserved by the Publisher, whether the whole or part of the material is concerned, specifically the rights of translation, reprinting, reuse of illustrations, recitation, broadcasting, reproduction on microfilms or in any other physical way, and transmission or information storage and retrieval, electronic adaptation, computer software, or by similar or dissimilar methodology now known or hereafter developed.
The use of general descriptive names, registered names, trademarks, service marks, etc. in this publication does not imply, even in the absence of a specific statement, that such names are exempt from the relevant protective laws and regulations and therefore free for general use.
The publisher, the authors and the editors are safe to assume that the advice and information in this book are believed to be true and accurate at the date of publication. Neither the publisher nor the authors or the editors give a warranty, express or implied, with respect to the material contained herein or for any errors or omissions that may have been made. The publisher remains neutral with regard to jurisdictional claims in published maps and institutional affiliations.

Printed on acid-free paper

This Springer imprint is published by Springer Nature
The registered company is Springer-Verlag GmbH Germany
The registered company address is: Heidelberger Platz 3, 14197 Berlin, Germany

To Erwin, Vince, and Hiroshi

Preface

To the old saying that if you want to learn something about a subject, teach a course on it, I would add the much more arduous option of writing a book about it. The upside, of course, is that you may gain a perspective on a longtime favorite subject which can be achieved in no other way. The fact is that the opportunity to write such a book as this is, indeed, a rare opportunity, especially for someone who by the calendar is well into retirement years.

While for a physics researcher, high T_c itself was the *opportunity of a lifetime*, and revisiting the old milestones within is a genuine nostalgia trip, it is also important to reflect on the purpose of such a book as this. Now that the great majority of experimental high-T_c studies are finished, who could be interested in such a book? To give some kind of answer to this, let me pinpoint the 2006 M^2S Conference in Dresden.[1] Not only was it at this conference that the idea for this book was hatched, but the conference itself evinced an astonishingly high level of interest in high-T_c issues, especially the theoretical aspects. Since it is now a number of years since a review of high-T_c studies with NMR has appeared, there may be a need on the part of theorists and other interested parties for an organized summary of results and of the ideas which have been advanced to bind them together.

The foregoing offers a partial answer to the "Why now?" question, but let me note that excellent answers to that question also emerge from the subject, itself. Although many issues were settled as of ten years ago, new ideas and even new phenomena have appeared since that time. Let me offer two examples. The confirmation by N. Curro and coworkers of a dynamical exponent crossover in the high temperature phase diagram opens up a new arena of potential experimental activity. Pines and coworkers had predicted such an effect; other theorists, including Varma, had predicted a phase boundary in that region. Another interpretive breakthrough is to be credited to Uldry and Meier, who in 2005 created a new, general method for

---

[1]This was the 8th International Conference on Materials and Mechanisms of Superconductivity and High Temperature Superconductors, Dresden, July 9–14, 2006.

analyzing relaxation (T_1) data in terms of temperature-dependent spin-spin correlation coefficients for close-neighbor copper sites. Discussed at some length in this volume, this method rests on a very firm theoretical foundation. It has the great virtue that it separates these correlation effects from the temperature dependence of the quasiparticle dynamics, itself, which theoretical modelers have been trying to get at. If this extraordinarily simple method had been "available" fifteen years ago, I suggest that our understanding of these issues would be greatly advanced from where it is.

As for regrets, they come mainly under the heading of omissions. There are many excellent pieces of work which are not included, simply for want of time and space. The book was conceived of as a monograph and is in no sense an encyclopedia, not even within the narrow purview of high-T_c NMR. As an example, I greatly regret omitting the large and fascinating subject of impurity doping in cuprates, from which we have nonetheless learned a lot.

Let me conclude by mentioning just a few of the people who have contributed, directly and indirectly, to the initiation and creation of this book. First let me thank Prof. Hiroshi Yasuoka, for bringing me to Tokai-mura in Japan for a five-plus year extension of my career in solid-state NMR, and through which I also became involved in the European physics scene. Let me thank Prof. Drs. Frank Steglich and Jürgen Haase, for their support during my extensive stay in Dresden in 2006 (and later). Invaluable support came from the Department of Physics at the University of Michigan in providing work space and the library facilities required for a project such as this. I want especially to thank Prof. Chandra Varma of the University of California at Riverside, for shepherding this book into existence through his role as editor for Springer Verlag, and for his reading of much of the manuscript, giving penetrating, enlightening, and invaluable comments on the presentation. I also thank Jinan Yang for help with computer issues and with preparation of the figures.

Ann Arbor, USA Russell E. Walstedt
June 2007

Contents

Chapter 1
Introduction

The discovery of high-temperature superconductors among hole-doped CuO_2 structures was one of those rare events in the scientific sphere which launches a whole new world of activity for hundreds of theorists and experimentalists alike. It is now more than 20 years since the New York City APS meeting dubbed the "Woodstock of Physics" witnessed the announcement of the original high-T_c results [112], and the level of activity on this topic remains extraordinarily high.

For the nuclear magnetic resonance (NMR) probe, the high-T_c materials have offered a panorama of new phenomenology to be analyzed and categorized in ways which had not previously come into use. It is the purpose of this volume to give a systematic account of many of the phenomena which have come to light in the process. The style of presentation is intended to be partly historical, partly tutorial for the non-specialist, and partly a review. There are by now a very large number of well-established ideas and effects descriptive of the cuprates to be addressed. However, there is also the feeling, even 20 years on, that this presentation is only a snapshot of a vigorous and ongoing enterprise.

We begin by presenting in this introductory chapter a broad perspective on the phenomenology and materials to be addressed, as well as on the NMR probe itself in the context of the cuprates. This is followed by an organizational synopsis of the remaining chapters.

Note on the Second Edition: Now 10 years on, the condensed matter research picture has changed markedly. Certain issues addressed as unresolved in the First Edition have been substantially clarified, and sections of the presentation have been updated as a consequence. Comments are inserted in the following subsections where appropriate. Further, while research on high-T_c materials continues, other areas have become attractive for study, in particular other examples of strongly interacting electron systems, including many systems involving $4f$ and $5f$ electrons. Accordingly, this edition will include two additional chapters, one on actinide oxides and one on

© Springer-Verlag GmbH Germany 2018

R.E. Walstedt, *The NMR Probe of High-T_c Materials and Correlated Electron Systems*, Springer Tracts in Modern Physics 276,
https://doi.org/10.1007/978-3-662-55582-8_1

the Kondo effect and its consequence in the bulk sphere, namely Heavy Fermion systems. These subjects have been particularly interesting *vis–a–vis* the NMR probe. There are many other potential topics to consider for future editions. The two added topics for this edition are reviewed briefly in turn in the following section.

1.1 The Basic Phenomenology of High-T_c Materials

The concept of a half-filled "conduction band" which cannot conduct on account of on-site Coulomb repulsion forces, was inaugurated in 1949 by N. F. Mott—the "Mott insulator" [122]. Mott, who discussed the case of NiO, commented, "Experiments on the conductivity of cupric salts would be of great interest" [122]. Nearly 40 years later a "cupric salt", with a modest amount of hole-doping in the conduction plane, emerged as a superconductor, and "high-T_c" was born. Just how such a system conducts—and superconducts—then became the investigative preoccupation of hundreds, possibly thousands of researchers around the world for many years to follow.

The basic process whereby the entire system of localized, $S = \frac{1}{2}$, Cu^{2+} moments becomes a set of itinerant quasiparticles with the addition of a far smaller number of mobile hole "carriers" is one of great fascination, upon which even now there seems to be no clear theoretical consensus. As a basic mechanism, however, the Zhang-Rice singlet picture [125] provides at least a conceptual framework. Since the quasiparticles which result are known (through NMR studies) to pair off at low temperatures into a ground state of magnetically inert Cooper-pair singlets, the resulting scenario of fully itinerant character for the solitary holes on the Cu^{2+} sites seems at this point incontrovertible.

Furthermore, these quasiparticles are known to possess a Fermi surface, with respect to which angle-resolved photoemission studies (ARPES) can be used to measure the superconducting energy gap. However, in many cases, gapping is also found to occur in the normal state as well [211–213], in what is known as the "pseudogap" [187]. Many of the high-T_c systems also exhibit poor, but metallic-like conduction in the normal state, characterized by a non–Fermi–liquid temperature dependence $\rho \propto T$ instead of the usual T^2.

While the cuprates exhibit certain metallic attributes, they also clearly possess some features characteristic of the undoped insulating host, namely a powerful antiferromagnetic (AFM) exchange interaction between nearest-neighbor Cu^{2+} moments. The effects of these AFM couplings have emerged from virtually all inelastic neutron scattering (INS) studies of the cuprates, and are a major feature of theoretical modeling and experimental fits to nuclear relaxation data as well (see Sect. 1.3).

1.2 Carrier Doping and the Master Phase Diagram

There are by now hundreds of host structures known which contain CuO_2 quadratic layers that superconduct under suitable doping conditions. A partial list is given in the review by Hazen [214]. In this volume, in order to limit the extent of material to be covered, we focus on the three principal systems $La_{2-x}Sr_xCuO_4$ (LSCO:x), $YBa_2Cu_3O_{7-x}$ (YBCO7-x), and the related stoichiometric compound $YBa_2Cu_4O_8$ (Y248). These three were chosen, because they emerged early on as systems easy to synthesize and characterize. As a consequence, there is an extensive literature built up around them both in terms of experimental data and, naturally, of theoretical modeling as well. Even with our purview narrowed to these three systems, there are unfortunately many interesting studies which had to be omitted.

One of the most important characteristics of the YBCO7-x and LSCO:x are their phase diagrams, plotted with coordinates T versus x. An early version of the LSCO:x phase diagram is shown in Fig. 1.1, where the superconducting phase lies under a dome extending from $x = 0.05$ up to $x \sim 0.35$ [3]. At lower x values there is an AFM phase which as $x \to 0$ occurs at $T_N \simeq 325$ K. The line extending above and to the left of the superconducting dome gives the crystallographic boundary between the high-temperature tetragonal phase and orthorhombic phase below. Other phase lines have been drawn in a similar location for other putative transitions and crossovers (such as the pseudogap temperature—see Sect. 1.4).

There is, in principle, a phase diagram similar to Fig. 1.1 for every cuprate structure. It happens for LSCO:x that the hole doping level can be varied across the entire superconducting range by varying x. In YBCO7-x, on the other hand, the doping is controlled via the oxygen content. At $x = 0$ (YBCO7) the structure is stoichiometric and the hole-doping level is ~ 0.33 per Cu ion, which is just slightly above optimal doping [2, 4]. As x is increased, the doping diminishes all the way into the AFM insulating phase. However, the overdoped phase which would correspond to negative values of x simply does not exist.

The stoichiometric compound Y248 is simply YBCO7 with an extra chain layer. It has a very stable oxygen content up to relatively high temperatures, but its doping

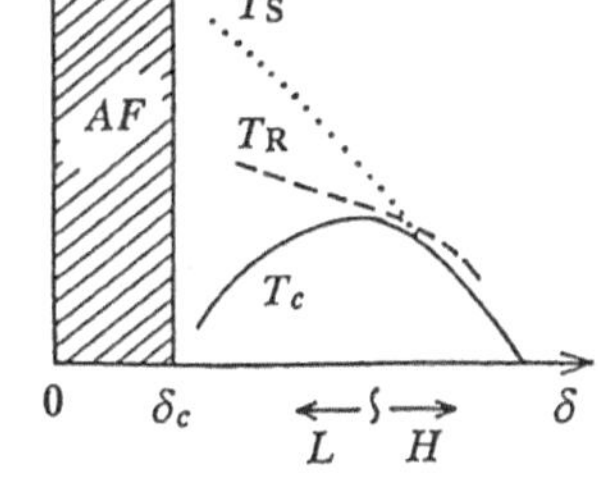

Fig. 1.1 The temperature versus x phase diagram of $La_{2-x}Sr_xCuO_4$, where x is the number of Sr^{2+} dopant ions per formula unit. A roughly parabolic region of superconducting behavior is seen to be centered on $x \simeq 0.15$. See text for additional discussion

level giving $T_c \simeq 84$ K cannot be altered without decomposition. It has been an attractive system for study and will be discussed in Chaps. 4, 5 and 6.

Characterization of samples for experimental investigation is a matter of great importance, since a large body of reliable data for mainstream structures is on hand. Sample characterization includes morphology of the sample material, x-ray crystallography for phase purity, superconducting diamagnetism (Meissner effect) including the value and sharpness of T_c, and magnetic susceptibility for excessive local moment content. The NMR and NQR studies themselves include sample characterization tests, to be mentioned in Sect. 1.3 to follow.

For articles discussing structures and a variety of bulk and other measurements, see the reviews edited by Ginsburg [1].

1.3 The NMR Probe

The NMR probe is a local, real–space probe, where the behavior of nuclear spins can be monitored on a site–by–site basis. The hyperfine (HF) interaction of a nucleus couples it to magnetic entities in its immediate environment. This interaction rarely goes beyond nearest-neighbors, though dipolar interactions may have some effects beyond that range.

One of the remarkable HF effects in cuprates is the extraordinarily large coupling with neighboring (Cu^{2+}) electron spin moments, which is comparable with on–site interactions. There is, however, an intervening O^{2-} ligand, so this is a second-neighbor interaction. Such an effect was unprecedented. The effect of HF couplings is also strongly affected by site symmetry, where strong, but equal couplings to antiferromagnetically correlated neighbor spins may be rendered ineffective. Such effects are well known for insulating antiferromagnets (e.g. ^{19}F in $KMnF_3$).

The measurements performed and type of data received are also very much conditioned by the properties of the naturally occurring nuclear species in the system investigated. Thus, the 100% abundant, $I = \frac{1}{2}$, ^{89}Y in YBCO7, with its nearly isotropic NMR shift, total absence of quadrupole effects, and weak, but adequate relaxation times was the subject of the first NMR study on cuprates to be published [127]. The ^{63,65}Cu isotopes with $I = \frac{3}{2}$, (nearly) axial site symmetry, and very substantial quadrupolar energies which allow both NMR and NQR to be performed, were quick to follow [128, 129, 131, 132].

Studies of ^{17}O, which only appeared after an interval of many months, were hampered by the fact that ^{17}O is present in only trace quantities in natural oxygen. Further, it is too expensive to incorporate into the sample synthesis and has to be diffusively exchanged into sample materials at an elevated temperature. Such a process is critically dependent on diffusion rates in any given material. In LSCO:x, for example, diffusion is fast and single crystals of moderate size are easily doped with ^{17}O. For YBCO7-x, however, this is only feasible with powders and crystals of the smallest size. With Y248 it is also very difficult, requiring lengthy high-temperature anneals [218]. Nonetheless, ^{17}O doping eventually became a routine element of technique.

A Sampling of NMR Measurements and their Interpretation

We give a brief survey of types of NMR observations made and their significance for the behavior of cuprate materials. First, the NMR spectrum is fundamental and is a major element of sample characterization. The width and distribution of NMR/NQR resonance lines tells a great deal about the phase purity and microscopic quality of sample material. If there are crystallographically inequivalent sites, they usually have different spectra. Quadrupolar frequency shifts and broadening and their field/frequency dependence can usually be analyzed to yield values of coefficients from the HF Hamiltonian. These matters are reviewed in some detail in Chap. 2.

The NMR shift, which is measured as a frequency shift proportional to the applied field, is fundamentally a measure of the various terms in the magnetic polarization of the sample, conveyed to the nuclear spins by HF coupling fields. The NMR shift is an important source of information on intrinsic susceptibilities. For example, the decay of the spin susceptibility in the superconducting state can usually only be measured via the NMR shift, where direct magnetic measurements would be swamped by the much larger diamagnetic response.

There are various contributions to NMR shifts in metals. The s-contact shift, for which we reserve the term "Knight shift" [32], is a mainstay effect in sp-band and d-band metals. In cuprates this term is oddly absent. In its place are the core–polarization shift, orbital shift, dipolar shift, and even "spin-orbit shift", a second-order effect which arises in the spin Hamiltonian picture. There are also very important transferred HF fields giving rise to NMR shifts, originating in the Mila-Rice [118] and Shastry [146] treatments of cuprate quantum chemistry. NMR shift measurements are a very important tool for determining the HF parameters, since any such determination is independent of model assumptions required for the dynamics.

NMR dynamics, as represented by the spin-lattice relaxation time T_1, are a key diagnostic of quasiparticle dynamics in metals and in cuprates in particular. T_1 has been linked for many years to $\chi''(\boldsymbol{q}, \omega)$ via the fluctuation dissipation theorem [163]. In cuprates, T_1 measurements have been interpreted (for the first time!) through theoretical modeling of $\chi''(\boldsymbol{q}, \omega)$. What makes this process particularly interesting is the interplay between the strong AFM fluctuation effects which typically produce an AFM peak near $\boldsymbol{q} = (\pi, \pi)$, and the transferred HF couplings at ^{17}O(2,3) and ^{89}Y in, e.g. YBCO7-x, which cause their relaxation processes to avoid the AFM peak. Meanwhile, the ^{63}Cu(2) are quite sensitive to the AFM peak. In effect one has, then, a "$\boldsymbol{q}$–space spectrometer" for quasiparticle fluctuations. These effects occupy a substantial fraction of the discussion in this volume. T_1 measurements are also very sensitive to the difference between s–wave and d–wave gap characteristics in the superconducting state. Calculations of d–wave T_1 characteristics below T_c were among the first strong evidences for d–wave singlet character of quasiparticle pairing in cuprates [160, 252].

A second relaxation time, the spin-echo decay time T_{2g}, has taken on a major significance in the effort to characterize the dynamic susceptibility of cuprate compounds. This decay time is actually used to measure a *static interaction* which occurs in cuprates, namely a c-axis indirect spin-spin coupling of the form $\mathcal{H}_{zz} = \sum_{ij\neq} \alpha_{ij} I_{zi} I_{zj}$, which was first identified by Pennington et al. [139]. How this comes about is discussed in Appendix A.1 and in Chap. 6 of the book. Interpretation of $\mathcal{H}_{zz}$ in terms of $\chi'(\boldsymbol{q}, 0)$ for $\boldsymbol{q}$ in the vicinity of (π, π) was established by the Pennington–Slichter model calculation [31]. Through measurements of T_{2g} the indirect coupling has become a major element in characterizing the strength of AFM fluctuations in cuprates. In addition, the interpretation of T_{2g} in the superconducting state by Bulut and Scalapino [252] again provided dramatic evidence in favor of d–wave pairing several years before it was widely accepted by the research community.

The foregoing is just a sampling of interesting effects which have come to light through NMR studies of cuprate superconductors. We refer the interested reader to sources on the fundamental principles of NMR [5, 6], on NMR studies of conventional metals and superconductors [7–9], as well as on NMR of high–T_c superconductors [13–17]. We now give a brief summary of these remaining chapters.

1.4 Outlines of the Remaining Chapters

This section gives a brief summary of the content of Chaps. 2–8.

1.4.1 Summary of Chapter 2—NMR Review

In Chap. 2 we give a review of the elements of NMR technique and interpretation, where basic definitions of key NMR parameters are given and some of the pitfalls and complexities associated with powder samples are discussed. The use of spin echoes—the standard tool for conducting NMR measurements—is described, with a detailed derivation including spin echo decay mechanisms presented in Appendix A.1. This leads into a summary of basic relaxation mechanisms for simple metals, including the Moriya theory of electron–electron interaction effects [58] and the BCS-based interpretation of the spin-paramagnetic shift and T_1 behaviors in the superconducting state. The Anderson–Farrell resolution of spin-orbit coupling effects for $T < T_c$ is covered in some detail. These effects were struggled over for many years, but almost all problems were finally resolved.

Discussion of the foregoing is extended to d–band metals, including the discovery and establishment of orbital shifts and relaxation effects. The interpretation of NMR spectra and relaxation effects in the mixed state of type-II superconductors is discussed at some length, with data from the literature. Many of the associated problems have cropped up again with the cuprates. The discussion of, e.g. the V_3X's is offered as a foundation for understanding cuprate mixed-state NMR.

1.4.2 Summary of Chapter 3—Preliminary Cuprate NMR

Chapter 3 gives an account of the early stages of high-T_c NMR research. In the first 2 years, almost all studies were done with either random powder samples or microcrystals with very low signal amplitude. There was an intense focus on YBCO7, because of its high T_c ($\sim$ 90 K) and because of its relative ease of preparation. In random powders, ^{63}Cu(2) NMR spectra (just the $\frac{1}{2} \leftrightarrow \frac{1}{2}$ transition) were ~5 kg wide in an 8 T laboratory NMR magnet, taking several days of signal averaging to perform a single scan. T_1 measurements on the ^{63,65}Cu were done mainly with NQR, once the frequencies had been extracted from NMR spectra.

^{63}Cu(2) NMR shifts persisting into the superconducting state were identified as Van Vleck orbital shifts. The anisotropy and magnitude of these shifts fit well with a simple *spin Hamiltonian* model of the copper ionic state, which appears to resemble that of a Cu^{2+} ion in an insulating host for the purpose of analyzing HF effects. This picture has held up even until now. Moreover, the $d_{x^2-y^2}$ ground state orbital dictates that the normally dominant orbital T_1 mechanism effectively vanishes in the cuprates. In Sect. 5.6 a proposal for its revival via the *marginal Fermi liquid theory* is presented [222].

The apparent vanishing of the c–axis spin HF field led to the Mila-Rice proposal: A massive transferred HF coupling from *nn* Cu^{2+} ions nearly 4Å away. This unprecedented HF term made sense of the T_1 and NMR shift data for the ^{63}Cu(2). Their watershed paper, including calculated quantum chemistry, provided a framework for interpretation of cuprate NMR data which has held up ever since [118].

One of the main elements of Chap. 3 is the general derivation of spin–lattice relaxation rates based on $\chi''(\boldsymbol{q}, \omega)$ and the Fluctuation–Dissipation theorem. This is preparatory to doing T_1 calculations in Chap. 4 based on INS data for $\chi''(\boldsymbol{q}, \omega)$. It naturally incorporates temperature–dependent spin–spin correlations into the calculation in keeping with the formulation of Mila and Rice [118].

The chapter closes with an account of the first interpretation of T_1 data in the superconducting state using d–wave pairing with BCS theory, by Monien and Pines [160]. This landmark result gave a clear indication of the correctness of d–wave and the inadmissibility of s–wave, even though it would be five more years before the correctness of this picture was widely accepted.

1.4.3 Summary of Chapter 4—The Pseudogap

Even before the Mila-Rice resolution of anomalous NMR shift behavior appeared, workers had begun to explore the highly unusual behavior in the underdoped region of the phase diagram in Fig. 1.1. There it was found that low-frequency fluctuations which drive the T_1 process, as well as the uniform susceptibility, were beginning to collapse at temperatures far above T_c. It was as though the superconducting energy gap were beginning to develop at some T^* well above T_c in anticipation of the actual transition to superconductivity. In the spirit of "spin–charge separation", an idea that dominated early thinking on cuprate quasiparticle behavior [52], this effect was at first dubbed the "spin gap". There was a tacit assumption that charge states were unaffected. Nonetheless, the spin pseudogap was soon recognized as a universal feature of the underdoped phases. Moreover, in a most astonishing bit of prescience, the appearance of such a precursive energy gap in the cuprates had actually been predicted by Friedel [187], who coined the name "pseudogap".

Pseudogap characteristics via NMR shift and T_1 data are documented in detail for several compounds in Chap. 4. Several years after the spin fluctuation properties of pseudogap systems were well established, it came to light through both electronic specific heat and ARPES measurements that charge properties were also profoundly altered in pseudogap systems. In this chapter we also review these effects briefly. With ARPES one finds that certain regions of the Fermi Surface, e.g. the quadratic axes, are gapped while while the diagonal regions are not. Since the dynamic susceptibility is a summation over the Fermi surface, this means that the gap reflected by $\chi''(\boldsymbol{q}, \omega)$ may vary significantly with $\boldsymbol{q}$. It follows that pseudogap T_1 behavior can be quite different at the ^{63,65}Cu site from what it is at the ^{17}O sites. These differences are clearly observed in a number of cases discussed.

The pseudogap behavior of LSCO:x is particularly interesting. In what was actually the first pseudogap behavior to be reported [189], the remarkable scaling of susceptibilies for different x values in this system was at first interpreted with a model suited to 2D insulating antiferromagnets. This is possible, since the latter systems exhibit a susceptibility temperature maximum which resembles pseudogap behavior. The disparate behavior of $1/T_1T$ for the ^{63,65}Cu and the ^{17}O clearly tags this system as having a pseudogap. The low-T $\chi_0(T) \propto T^{1/2}$ behavior of LSCO:x is in accord with the Varma theory of pseudogap behavior [222], as is that of Y248 [219].

Substantial insight into the pseudogap is provided by comparison of dynamic susceptibility data for optimally doped LSCO and underdoped YBCO with T_1 measurements on both planar ^{63}Cu and ^{17}O in these systems. Quantitative INS data for $\chi''(\boldsymbol{q}, \omega)$ has become available for both of these systems, along with extensive bodies of T_1 data for both the planar ^{63}Cu and ^{17}O for both of the compounds mentioned. Using the fluctuation–dissipation theorem, Moriya formalized a quantitative relation between $\chi''(\boldsymbol{q}, \omega)$ and T_1 which can now be realized *with no adjustable parameters* [58]. Calculations testing this relationship are now spelled out in considerable detail in Chap. 4 for both optimally doped LSCO and underdoped YBCO ($T_c \sim 60$ K). In both cases excellent agreement is found near or just above T_c. At higher temperatures, however, there is clearly a large piece of $\chi''(\boldsymbol{q}, \omega)$ that is simply missing from the INS data for both systems. Possible remedies are discussed. One possibility is that there is a broad, flat intensity covering the BZ. Such an effect would be, of course, very difficult to resolve in an INS study.

1.4.4 Summary of Chapter 5—T_1 Models

From the earliest days of high-T_c NMR one of the main challenges has been to understand the disparate behaviors of T_1 at various lattice sites. In addressing this question, there are several important branching points in the flow diagram of deduction so to speak. Magnetic dipole processes are assumed to be predominant. First, then, one must decide whether spin fluctuations alone will suffice to explain everything, or whether orbital relaxation, severely inhibited by the $d_{x^2-y^2}$ ground–state orbitals, will play a role as well. Many workers have opted for the spin fluctuation picture, while the existence of a possible orbital mechanism has been pointed out by Varma [222]. A summary of the latter calculation is given in Sect. 5.6.

In the purely spin-dynamical picture, one must next decide whether a single dynamical variable per Cu^{2+} site, namely the itinerant $S = \frac{1}{2}$ hole, is sufficient, or do the doped–in hole carriers form a second dynamical entity? The general approach to this question has been to see whether a complete account of the data can be given with one variable per site. Several T_1 modeling efforts based on the latter assumption are described in Chap. 5, where it is presumed with Moriya [152] that any successful scenario can be expressed in terms of $\chi''(\boldsymbol{q}, \omega)$. Three of these employ the random–phase approximation in order to incorporate the (mandatory) electron–electron interactions, and one is simply based on the shape and width of the the AFM peak in $\chi''(\boldsymbol{q}, \omega)$ determined by INS studies.

A novel, but related approach has been pioneered by Uldry and Meier [195], in which they attempt to systematize T_1 data on several cuprates using electronic spin–spin correlations as parameters which vary with T and in which there is a universal spin fluctuation frequency (correlation time), also a function of T, which applies to all T_1 processes. The other central assumption is that the HF Hamiltonian has the Mila–Rice form

$$\mathcal{H}_{HF} = \sum_{\alpha}\sum_{i} S_{i\alpha}\left[A_{\alpha} + B\sum_{j(nn)} I_{j\alpha}\right] \tag{1.4.1}$$

for the planar $^{63,65}Cu$ nuclei, with similar forms for the ligands. With these ingredients all of the data are fitted in a self-consistent way. The form (1.4.1) and the admissibility of the spin–spin correlations and correlation time as temperature–dependent parameters are the main assumptions here. A theoretical basis for this scheme is derived using $\chi''(\boldsymbol{q}, \omega)$ in Chap. 3. It yields an internally consistent picture, in which the dynamically–derived HF constants basically agree with those derived from the shift. This scheme is useful for evaluating theoretical models as well as for analyzing experimental data.

1.4.5 Summary of Chapter 6—The Dynamical Susceptibility

In this chapter three topics related to modeling and characterization of $\chi(\boldsymbol{q}, \omega)$ are presented and discussed. First, large indirect spin-spin couplings in cuprates were noted early on and modeled quantitatively by Pennington and Slichter [31]. Further, the strength of indirect spin–spin couplings $\mathcal{H}_{zz}$ noted earlier can be measured by means of spin-echo decay if the entire nuclear spin transition observed can be inverted by the spin echo refocussing pulse ($\theta_2 \simeq \pi$). Measurements of this type have been carried out on a number of systems, yielding a quantitative measure of $\chi'(\boldsymbol{q}, 0)$ $\boldsymbol{q} = (\pi, \pi)$. This has become an important source of input data for the dynamical susceptibility.

Another major data source for $\chi''(\boldsymbol{q}, \omega)$ is INS. The compounds LSCO:x with $x \simeq 0.15$ and $YBCO_{6.6}$ became important test cases for $\chi''(\boldsymbol{q}, \omega)$ in Chap. 4. These systems bear the distinction of having been measured with INS *in absolute units* over most of the first Brillouin zone (all known structure). Further, LSCO:x was measured over energies ranging from 1 to 200 meV, and at temperatures ranging from below T_c to nearly 300 K [278]. The dynamical susceptibility determined in this way was found to be somewhat incommensurate. These data were used to calculate T_1 for $^{63,65}Cu$ and for ^{17}O, with interestingly mixed results. See Chap. 4 for a detailed discussion.

The final section of Chap. 6 offers a brief review of the behavior of T_1 and T_{2g} in the normal state region above and principally on the left side of the superconducting dome in Fig. 1.1. Sokol, Barzykin, and Pines have proposed a crossover in the value of the dynamical exponent from z = 1 (quantum critical phase) to z = 2 (non-scaling region) [240, 273]. The line dividing these two behaviors is suggested to take place where the correlation length $\xi(T) = 2a$ (a = lattice constant). This idea has been tested up to 700 K with Y248 by Curro et al. [242]. A clear transition from z = 1 to 2 was found at T = 500 K, giving support to this scaling theory.

1.4.6 Summary of Chapter 7—NMR Studies of Actinide Oxides

This chapter gives a synopsis of actinide NMR research carried out over the last two decades or so. The discussion is limited here to oxides, which illustrate several important advances in NMR technique and phenomenology that emerged from the actinide studies. Actinide hyperfine couplings are much larger than those of spd–electron compounds. As a consequence, typical spin–lattice relaxation times (T_1) are far shorter than could possibly be resolved with conventional pulsed NMR technique. Interestingly, this is why ^{235}U, which has the smallest nuclear gyromagnetic ratio γ ($\gamma/2\pi \simeq 0.784$ MHz/T), was the first actinide NMR to be observed directly [297]. What was observed was actually AFNMR in the antiferromagnetic ground state of UO_2. Because of the immense static hyperfine field of over 400 T at ^{235}U in UO_2, the AFNMR is at $\nu \sim 200$ MHz and T_1 was short enough below T = 14 K for direct observation of the nuclear spin echo. This, combined with studies of the ^{17}O NMR in both the paramagnetic and AFM states, enabled a very extensive study of this system to be conducted. A detailed description is given of the "3k" AFM ordering of the U^{4+} moments in this system, yielding a substantial internal field at the ^{17}O sites. Measurements of the latter field and the NMR shift in the AFM state are described in detail.

The next actinide beyond uranium is neptunium, which in its oxide form has been the subject of a fascinating study of one of the first magnetic octupolar ground states to be identified. Both the paramagnetic and ordered states of NpO_2 give fascinating results for different reasons. The ordered state is probed using the ^{17}O NMR, which behaves almost normally, but which on careful analysis exhibits field–induced octupolar effects that match relevant theory very nicely. Thus, the octupolar ordering is confirmed rather indirectly, since there is no known direct probe that reflects the presence of ordered octupoles.

T_1 for ^{17}O in the paramagnetic state of NpO_2 reveals the presence of a cross–relaxation (CR) effect. This takes the form of a field–dependent T_{1O}, which gets shorter *as the field is reduced*. The upshot is a T_1 process transferred from fluctuating ^{237}Np nuclear spins. Analyzing this process (T_{1CR}) yields some interesting results. First, plotting T_{1CR} versus ω_{17}^2 gives a straight line, verifying that this process is driven by ^{237}Np nuclear spin (T_1) fluctuations. Second, evaluating the strength of the CR process reveals that there are anomalously large ^{237}Np–^{17}O nuclear spin–spin couplings, which in second moment form are as much as three orders of magnitude larger than the classical dipolar contribution. The origin of this is clearly an anomalous and novel *indirect spin-spin coupling*, the nature of which is easy to guess (see Chap. 7).

And finally, the quantitative analysis of T_{1CR} yields a value $T_{1Np} \simeq 40$ nsec that is essentially constant throughout the paramagnetic state. Since the fluctuating dipolar moments that drive these processes collapse abruptly at $T_0 \sim 30$ K, T_{1CR} also disappears at that point. Interestingly, T_{1CR} starts up again at low temperatures, reflecting what may be a process driven by the ^{237}Np T_2 fluctuations. This process is weaker, but is *constant with T* at many values of applied field at He temperatures. The mystery of this effect has yet to be deciphered.

Next, we recount the first observation of ^{239}Pu NMR *in a solid host compound* in PuO_2. The ground state of Pu^{4+} in PuO_2 is nonmagnetic, so that the immense 5f hyperfine couplings are disabled from rendering this NMR signal unobservable. On the contrary, T_1 is extremely long and is one of the principal obstacles to resolving the ^{239}Np NMR signal. A gyromagnetic ratio for ^{239}Pu is obtained after many years of waiting. However, a difficult–to–determine orbital NMR shift present in this host may be making a substantial contribution to γ_{239}. At least, this elusive prize was finally achieved in 2012 [299].

Finally, we recount briefly an effort to study the NMR of ^{17}O in AmO_2. The half–life of ^{243}Am is so short ($\sim$ 4, 000 years) that a synthesized crystal turns into an amorphous structure in just a few days. NMR spectroscopy reflects this change in a dramatic way [316].

1.4.7 *Summary of Chapter 8—The Kondo Effect and Heavy–Fermion Behavior*

A couple of decades before the surge of interest caused by the high–T_c discovery there was a similar surge, though smaller, of course, caused by the Kondo effect. The discovery by J. Kondo in 1959 that conduction electron exchange scattering from localized moments diverges logarithmically with declining temperature set the stage for the Kondo "condensation" of localized moments in metallic hosts. While there are many examples of this effect in dilute alloys with "magnetic" impurities, in contrast with the high–T_c surge, it took more than a decade for the Kondo condensation to become a hot research topic. Even in the early seventies, however, there was still no reliable theory for just how a localized moment "condensed" at low temperatures into merely an enhanced susceptibility. There was, in fact, great controversy over the physics of how this happened.

In the first subsection of Chap. 8 we discuss the dispute over whether local moments disappear by acquiring a quasiparticle to cancel the moment or whether such an induced moment and the spin polarization in its vicinity scale uniformly as $1/(T + T^*)$. That the latter process actually occurs was demonstrated brilliantly by the Boyce–Slichter study of ^{63}Cu NMR satellites from lattice sites surrounding Fe impurities in the classic Kondo system [323]. These results and discussion are reviewed as one of the most important NMR studies of the Kondo era.

Following the Kondo effect of isolated impurities came the "Kondo lattice" with a high concentration of Kondo scatterers. The Kondo effect was soon realized to be the basis for "heavy fermion" behavior in many f–electron intermetallic compounds.[1] After some decades of heavy fermion studies, it was realized that no first–principles theory was forthcoming. In place of that, the phenomenological two-fluid model by Nakatsuji, Pines and Fisk has provided a framework for the understanding of many basic properties of these systems [330]. We review this model in some detail in Chap. 8. Further, it formed the basis for understanding a very general NMR shift scaling effect for heavy fermion systems, which is also reviewed [331].

Two other topics close out Chap. 8. First, we illustrate the extent to which the dynamic mean–field theory of Shim et al. [333] accounts for the development of heavy–fermion susceptibility in, e.g., $CeIrIn_5$ [334]. Second, the dynamical formalism of Chap. 3 is employed to interpret the anomalous behavior of c–axis fluctuations that give highly disparate results for T_1 at different ^{115}In sites in the latter compound. The key ingredient here is dynamical correlations among the $4f$ moments in the heavy fermion state of this system. Unfortunately, there don't seem to be any neutron data on SRO for $CeIrIn_5$ to confirm the dynamical correlation modification of T_1.

[1] As a general reference and for an excellent historical sketch see the book by Hewson [327].

1.4.8 *Summary of the Appendix: The Properties of Spin Echoes*

The Appendix presents an extended discussion of the properties of spin echoes. First, the echo signal is calculated in detail using a simple classical model. Behavior of the echo signal as a function of the relative phase of the two excitation pulses is addressed, as is the excitation of echo signals under various experimental circumstances. The direct relation of the integrated area under the echo to the NMR lineshape function is derived. A summary of the density matrix calculation of spin echo excitation is presented. The comprehensive spin echo decay calculation given in 1967 by Alloul and Froidevaux [294] is introduced, and the relation to spin echo oscillations is highlighted. A special case of the indirect spin-spin coupling derived by Pennington and Slichter [31] is evaluated in detail. Results for the relaxation of spin echoes by spin–lattice processes are presented. Finally, a useful scheme to use phase toggling of echo refocusing pulses along with digital add/subtract of echo signals to eliminate spurious background effects is described in detail and illustrated with an actual case study.

Chapter 2
Introduction to NMR Studies of Metals, Metallic Compounds, and Superconductors

NMR studies of high-T_c materials and other correlated-electron systems developed in a context of many years of NMR studies of conventional superconductors as well as of d-band transition metals, alloys, and intermetallic compounds. It is therefore appropriate to begin this volume with a brief review not only of the NMR probe, itself, but of the general physical phenomena probed by NMR in metals, alloys and intermetallic compounds, in both the normal and superconducting states. The published literature on this subject matter is, of course, vast, and the references cited here will necessarily be rather limited. We begin by citing the standard works on nuclear magnetism by Abragam [5] and Slichter [6], where many basic NMR phenomena are derived and discussed. These sources also present an extensive guide to the early literature, which the interested reader will find highly informative. For NMR of metallic systems, there are thorough reviews by Narath [7] and Winter [8] and for early NMR studies on superconductivity, there is a detailed and informative review by MacLaughlin [9]. We shall also have recourse to the treatments of superconductivity by Tinkham [18] and Schrieffer [19]. Relevant reviews of high-T_c NMR studies will be cited at the beginning of Chap. 3.

2.1 The Fundamental Elements of NMR

The foundation of nuclear magnetic resonance (NMR) rests on the fact that many stable isotopes in the periodic table possess a nonzero spin quantum number I and, thus, angular momentum $\hbar \boldsymbol{I}$ and a magnetic moment $\boldsymbol{\mu} = \gamma \hbar \boldsymbol{I}$, where γ is the nuclear

© Springer-Verlag GmbH Germany 2018

R.E. Walstedt, *The NMR Probe of High-T_c Materials and Correlated Electron Systems*, Springer Tracts in Modern Physics 276,
https://doi.org/10.1007/978-3-662-55582-8_2

gyromagnetic ratio. In an applied magnetic field $\boldsymbol{H}$, a nuclear moment $\boldsymbol{\mu}$ has an energy $-\boldsymbol{\mu}\cdot\boldsymbol{H}$ and in classical terms undergoes a torque $\boldsymbol{\tau}=\boldsymbol{\mu}\times\boldsymbol{H}$.[1] Equating this torque to $\hbar d\boldsymbol{I}/dt$, one finds an equation leading to Larmor Precession of the component of $\boldsymbol{\mu}$ transverse to the field $\boldsymbol{H}$. The latter may be written $\boldsymbol{\mu}-\hat{\boldsymbol{u}}(\boldsymbol{\mu}\cdot\hat{\boldsymbol{u}})$, where $\hat{\boldsymbol{u}}=\boldsymbol{H}/|\boldsymbol{H}|$ is a unit vector parallel to $\boldsymbol{H}$. In this way, the Larmor Precession Frequency emerges as $\omega_0=\gamma$H. This is the basic nuclear magnetic resonance (NMR) frequency. In the contrasting case of nuclear quadrupole resonance (NQR), there is no Larmor precession effect, because the nuclear energy splitting is provided by quadrupolar coupling with a local electric field gradient (EFG). This effect may, however, still be called a "magnetic resonance", because the resonance is excited with an rf field that couples with the magnetic moment $\boldsymbol{\mu}$, and the signal is received by magnetic induction.

If we add to the foregoing torque equation of motion the terms $-\hat{\boldsymbol{u}}(\boldsymbol{\mu}\cdot\hat{\boldsymbol{u}}-\mu_0)/T_1$ and $-[\boldsymbol{\mu}-\hat{\boldsymbol{u}}(\boldsymbol{\mu}\cdot\hat{\boldsymbol{u}})]/T_2$ to embody the longitudinal (T_1) and transverse (T_2) relaxation processes, one arrives at the celebrated Bloch equation model for NMR spectroscopy [10]. Here, $\mu_0=\gamma^2\hbar^2 I(I+1)H/3k_BT$ is the thermal equilibrium nuclear moment in a field H. The Bloch equations lead to nuclear resonance spectra of Lorentzian form with a width parameter $\langle\Delta\omega\rangle=(1/T_1+1/T_2)$ and are well suited to the study of NMR lines in liquids [11]. For NMR of nuclei embedded in condensed matter, however, there arise many complexities that go beyond the Bloch equations. A basic discussion of these effects is given below in this section of Chap. 2.

The essence of pulsed NMR is to excite nuclear induction signals through the application of radiofrequency (rf) pulses of oscillating magnetic field $2H_1\,cos(\omega t)$ along an axis perpendicular to **H**. We illustrate here very briefly the action of such pulses [10] with the simple case of a narrow NMR line excited at its precise NMR frequency $\omega=\omega_0(1+K)$, where K is the NMR shift, here taken to be isotropic for simplicity. The action of the pulse must take place in a time t_w short compared with $T_{1,2}$, after which the nuclear magnetization will take the value $M_0[\hat{\mathbf{u}}\,cos(\gamma H_1 t_w)+(\hat{\mathbf{u}}\times\mathbf{H}_1)\,sin(\gamma H_1 t_w)/H_1]$ in a reference frame rotating in synchrony with the rotating field H_1. The pulse width t_w may be adjusted so that $\gamma H_1 t_w=\pi/2$ to obtain the maximum transverse component followed by a free-induction signal, or it may be adjusted so that $\gamma H_1 t_w=\pi$, after which there is no transverse component, but a maximum inverted magnetization $-\hat{\mathbf{u}}M_0$, as the first step in a T_1 measurement. The first of these is known as a "$\pi/2$–pulse", and the second is known as a "π–pulse" or an inversion pulse. In practice, circumstances frequently don't permit such precise conditions to be achieved, because of linewidths comparable to or greater than the value of H_1 used, or non-uniform values of H_1 over the sample. In spite of small problems such as these, the practitioner frequently speaks of "$\pi/2$–pulses" and "π–pulses" as an idealization of what actually occurs. The action of (rf) pulses under more general circumstances is discussed in Appendix A, where spin echoes are also described in some detail.

---

[1] Correctly, the field in the magnetic torque expression is the magnetic induction. We use the symbol H in keeping with the traditional literature.

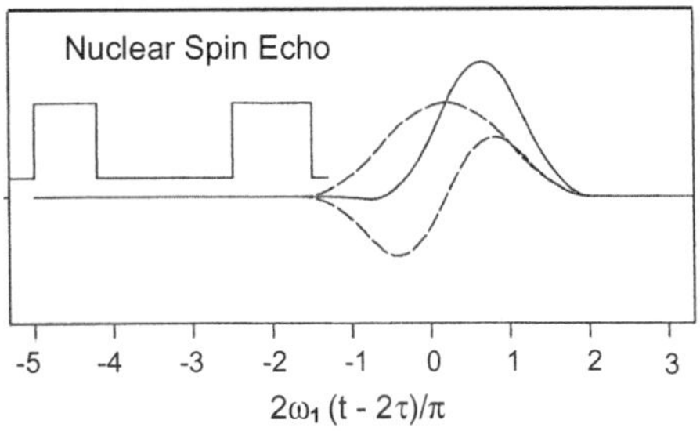

Fig. 2.1 A simulated spin echo signal is shown (solid curve), corresponding to the $(0.8\pi/2, \pi/2)$ pulse pair (shown) applied to a broad, inhomogeneous distribution of NMR frequencies. The echo is the sum of two contributions which are symmetric and antisymmetric, respectively, (dashed lines) with respect to the center point of the echo. The latter point is at time $2\tau + t_{w2} - t_{w1}$ from the leading edge of the first pulse, where τ is the time interval between the leading edges of the pulses and $t_{w1,w2}$ are the respective pulse widths [24]

The basic effects one typically measures and interprets in NMR studies of solids consist of the shape and structure of the spectrum, the frequency shift tensor $\mathbf{K}_{\alpha\beta}$, and the longitudinal (T_1) and transverse (T_2) relaxation processes. The acquisition and interpretation of these basic data will be the preoccupation of most of this volume, since each of these elements can provide important information about the solid state environment surrounding nuclear sites. Here we wish simply to introduce them, and in this section (2.1) the basic definition, measurement and interpretation of these parameters will be reviewed. A detailed description of spin echo excitation and behavior are given in Appendix A.

2.1.1 *Observation of NMR/NQR Signals*

In the early days of solid-state NMR, it was customary to employ continuous wave (CW) methods to observe NMR signals. Techniques such as the marginal oscillator and the popular crossed-coil *rf* bridge spectrometer manufactured by Varian Associates, as well as home-built *rf* bridges, were used to obtain a wealth of useful NMR data. Compared with pulsed NMR this method has, however, a number of inherent drawbacks, such as (i) it is not useful for severely broadened lines, (ii) it cannot distinguish between static and dynamic line broadening, so that (iii) it cannot measure the dynamic transverse relaxation time T_2, nor short values of T_1, to name just a few. Thus, it gradually gave way to pulse methods for solid-state NMR studies [22]. While we do not discuss CW NMR methods in any detail here, it must be noted that important NMR data on solids have been reported that were taken using CW techniques. A prominent example is a study of the Kondo effect by Boyce and Slichter [323, 324] discussed in Chap. 8.

Free Induction and the Spin Echo

Following a single *rf* pulse as described above in the introduction, one can often observe a *free-induction decay* signal with the form

$$F(t) = F_0 cos[(\omega_c - \omega_{rf})t] \int_{-\infty}^{+\infty} G(\omega_c + \omega) cos(\omega t) d\omega \,, \tag{2.1.1}$$

where $G(\omega_c + \omega)$ is a symmetric lineshape function centered on ω_c, and ω_{rf} is the frequency of the *rf* pulse and of the phase detector used to display the signal. In (2.1.1), the *rf* excitation pulse field is assumed to be large compared with the linewidth, so the pulse angle is uniform across the spectral width. In this simplified case, it is seen that $F(t)$ is simply the cosine transformation of the lineshape $G(\omega)$ and vice versa. The difficulty of determining $G(\omega)$ this way is that the early part of $F(t)$ is masked by the dead time of the receiver. Not infrequently, the decay time T_2^* is so short that very little of $F(t)$ can actually be recorded. For these reasons, the preferred method for extracting $G(\omega)$ with pulsed NMR is via the *spin echo*.

Nowadays, the *spin echo* [23] is used almost universally for measurements and observation of NMR/NQR signals in solids. Here we describe the basic features of the spin echo and illustrate its use in a simple case. A much more detailed discussion of echo formation and of the echo decay processes addressed below is given in Appendix A. Following an initial ($t = 0$) pulse such as that described above, a second pulse called the refocusing pulse is applied at the same frequency at $t = \tau$. For simplicity, we also assume the latter pulse has the same phase as the first pulse.[2] We also assume uniform excitation across the whole spectrum as in (2.1.1).

Taking the leading edge of the first pulse as time zero, then the second pulse at time τ is followed by a spin-echo signal at time 2τ. The mechanism of this effect is very simple. The second pulse effectively inverts the dephasing that has occurred during time τ, so that after the second pulse, the phase deviations of the precessing nuclei unwind and return to zero at time $2\tau + t_{w2} - t_{w1}$. In this way the macroscopic transverse magnetization that generates an observable NMR signal is restored.

We give here the expression for a spin echo for the same simple case of a uniformly excited, symmetric lineshape function $G(\omega)$ that was assumed for the free-induction signal in Ex. (2.1.1). Thus, we again assume that the *rf* field H_1 is much greater than the width of $G(\omega)$, *i.e.* $\langle \Delta\omega \rangle / \gamma$ (in field units). In that limit, the precessing nuclear magnetization that makes up the principal component of the echo signal is given by

$$E(t) = -\frac{1}{2} sin(\omega_1 t_{w1})[1 - cos(\omega_1 t_{w2})] cos[(\omega_c - \omega_{rf})(t - t_E)] \int G(\omega_c + \omega) cos[\omega(t - t_E)] d\omega \,, \tag{2.1.2}$$

where t_{w1} and t_{w2} are the widths of the two *rf* pulses, respectively, and $t_E = 2\tau + t_{w1} - t_{w2}$ is the center point of the spin echo, measured from the beginning of the first pulse. The maximum amplitude of the echo is obtained with $\omega_1 t_{w1} = \pi/2$ and

[2]Phase manipulation on the refocusing pulse to achieve special effects with spin echo studies is discussed in detail in Appendix A.

$\omega_1 t_{w2} = \pi$, giving the same expression as for the FID in (2.1.1), except with its sign inverted. Furthermore, (2.1.2) allows both positive and negative values of $t - t_E$, so that the shape of the echo signal in time is very much like two free-induction signals back to back. In this way, one is able to move the signal away from the excitation pulses so that it can be measured and recorded without interference from pulse transients.

We note two additional requirements for the observability of spin echoes in any given case. The echo signal represented by (2.1.2) will always decay as time τ increases. This decay time is denoted T_2 in the literature and is known as the *dynamic decay time*. See the next subsection for a discussion of this parameter. Thus, (2.1.2) only holds as written for $T_2 \gg 2\tau$.

A second requirement for observability of spin echoes is that the nuclear spins concerned must have a source of *static broadening* that is greater that any relevant dynamic decay process. A detailed discussion of such matters is given in the next subsection. Thus, the NMR line under study must have a "static width" greater than its "dynamic width" for spin echoes to be feasible. If the *rf* field H_1 is greater than the static width of the line, then the *rf* pulse will excite an echo signal from all of the nuclei. Otherwise, it will excite just a subset of nuclei centered on the carrier frequency of the pulse. In many cases the static width of the spectrum is much greater than H_1. A remarkable property of the spin echo is that it can be employed to very accurately record the detailed shape of such a static spectrum. The basis for such a capability is discussed next.

We touch here briefly on an important point regarding the frequency resolution of spin-echo spectroscopy, which is discussed in detail in Appendix A. Since the NMR signals are excited using pulse fields H_1 of typically 100 Gauss amplitude, one might suppose that these large fields would limit the available frequency resolution to $\sim \gamma H_1$. However, this is not the case. It is straightforward to demonstrate that the total area under the spin echo waveform is proportional to the line intensity at the frequency of excitation to within a very narrow bandwidth. The resolution obtained in this way is formally independent of the magnitude of the excitation field H_1, but is actually determined by the time frame over which the echo can be integrated. See Appendix A for more details. An example of a (simulated) spin echo generated somewhere in the center of a broad, inhomogeneous NMR line is displayed in Fig. 2.1. On the other hand, in the opposite limit that H_1 is actually much greater than the static linewidth, an excellent way to record the NMR spectrum is simply to Fourier transform the echo waveform as suggested by (2.1.2).

The question of observability of spin-echo signals is an important one for the NMR spectroscopist. We note here that solid-state NMR signals are typically of the order of microvolts, with expected raw signal-to-noise ratios which are typically of the order of unity or less. Thus, one cannot overemphasize the importance of digital signal averaging methods to the success of modern pulsed NMR technique. The great majority of solid-state NMR studies reported nowadays have used digital signal averaging to improve the signal-to-noise ratio. It is also possible to estimate the available NMR signal strength given reasonable estimates of the experimental parameters.

2.1.2 *Definition and Interpretation of NMR Parameters: T_2 Processes*

T_2 processes are the decay of NMR signals in time, *i.e.*, of coherent, transverse nuclear magnetization. There are two T_2 parameters in common use, namely the spin echo decay time constant T_2 and that of the free induction decay T_2^*. T_2 decay is strictly a dynamic effect, while T_2^* combines static and dynamic components. When, as is often the case, one has $T_2 \gg T_2^*$, then T_2^* will be dominated by purely static broadening processes. It is important to note that for (2.1.2) to represent an actual spin echo, one has to have $T_2 \gg T_2^*$ for that to be possible. And while there are many very simple compounds where that condition is not fulfilled, it is rare for solid state systems with interesting and even exotic properties to have the foregoing condition fail.

Let us examine the types of hyperfine interactions that contribute to T_2 and T_2^* decay effects. We note at the outset that a useful formal treatment of spin echo decay has been given by Alloul and Froidevaux [294]. Here, we give a brief discussion of dynamic processes that contribute to T_2 decay. There are several general classes of hyperfine fluctuation that can be distinguished, namely spin-spin interactions, spin-lattice interactions, and spin-flip interactions that are imposed by applied *rf* fields. Finally, since spin echoes are relaxed by any change in the local magnetic environment at the relevant nuclear spin sites, one must include physical diffusion of ions that carry the nuclear spins. We shall not, however, discuss this mechanism except simply to mention it.

Let us first consider nuclear spin-spin couplings. These consist mainly of the dipolar,[3] pseudo-dipolar, [28] and RKKY [27] exchange interactions. There are several points about these interactions that need to be addressed in the context of spin-echo decay. First, which interactions are 'allowed' in such a discussion? We take the z axis as the 'quantization axis' in the problem, i.e. the Zeeman term $\mathcal{H}_Z = -\hbar\omega_0 \sum_i I_{zi}$ is the principal nuclear spin energy term. Then, spin-spin coupling terms are only 'secular' (i.e., effective) if they commute with $\mathcal{H}_Z$. Thus, terms of the following form are generally admissible.

$$\mathcal{H}_{ss} = \sum_{i,j} A_{ij} I_{zi} I_{zj} + \sum_{i,j} B_{ij}(I_{+i}I_{-j} + I_{-i}I_{+j}) + \sum_{i,j} D_{ij} I_{zi} I'_{zj} + \cdots, \quad (2.1.3)$$

where the prime indicates a different, non-resonant nuclear species in the same crystal structure, and the dots indicate that other terms involving non-resonant ("unlike") spins are omitted. However, what effect would, e.g., the D_{ij} terms have on the decay process of (a) coherently precessing nuclear magnetization represented by $I_x = \sum_i I_{xi}$ or (b), on a spin echo generated via the unprimed, resonant nuclei? The answers for these two cases can be markedly different.

---

[3]Dipolar couplings are discussed in detail by Abragam [5], Chap. IV, and by Slichter [6], Chap. 3.

With typical dipolar couplings the FID and spin-echo decays are similar, but not identical. What can make a great deal of difference is inhomogeneous broadening that is non-uniform on a local level. The Hamiltonian term would have the form $-\sum_i \Delta\omega_i I_{zi}$, where $\Delta\omega_i$ is a static quantity that varies by an amount equal to or greater than dynamic, interactive broadening terms at the local level. Such terms will contribute substantially to T_2^* decay, but will tend to inhibit spin-echo decay, making T_2 longer. Terms like this arise from local crystalline disorder, imposing modulation on the NMR shift mechanism. They prevent terms such as the B_{ij} terms in (2.1.3) from driving local spin fluctuation processes, because variations in the local resonant frequencies (i.e., the $\Delta\omega_i$ terms) prevent flip-flop fluctuations from conserving energy. The resulting static interaction environment does not contribute to the spin-echo decay process.

A second type of local coupling that can relax the FID, but may not affect the spin-echo is the "zz" unlike spin term D_{ij} of the previous paragraph. If the primed spin system has local fluctuations driven by $I'_{\pm i} I'_{\mp j}$ fluctuations, then the D_{ij} terms shown there will contribute to both the T_2 and T_2^* relaxation processes. However, it is not unusual for such unlike spins to be essentially static, whereupon they will make no contribution to the T_2 (spin-echo) decay. Of course, long-range inhomogeneity also contributes to T_2^* decay, but not to T_2. When such broadening becomes greater than γH_1, then $1/T_2^* \sim \gamma H_1$.

A second contributor to spin-echo decay is reversal of neighboring nuclear-spin orientations by the refocusing pulse. This mechanism makes use of the A_{ij} terms in $\mathcal{H}_{ss}$. If such a set of $I_{zi} I_{zj}$ terms are rendered static by their environment, such as happens in certain high-T_c systems,[4] the only way such terms can be made to contribute to T_2 decay is via the refocusing pulse. The magnitude of this effect depends on the pulse angle used for the measurement, with maximum effect taking place for a π pulse. This variation of the $I_{zi} I_{zj}$ broadening can also be used to separate this effect from other T_2 processes that do not vary with pulse angle.

Another interesting possibility is to employ terms of the D_{ij} form in a *spin-echo double resonance* scheme by applying *rf* pulses at the NMR frequency of an unlike, *i.e.*, primed spin species while monitoring the echo signal from the unprimed spins. In the event that an appreciable fraction of the primed spins are reversed by the second pulse, a measurable degradation of the unprimed-spin signal will be recorded. By scanning the frequency of the primed-spin pulse, one can map out the primed-spin NMR line without ever observing that NMR signal directly. This has been used in studies of nuclear species that are difficult to observe directly [270]. This technique is known as spin–echo double resonance (SEDOR).

A third mechanism of T_2 decay occurs through T_1 processes. Only in a rather small subset of cases is T_1 a significant factor in the T_2 process, simply because in many instances $T_1 \gg T_2$. As suggested by the Bloch equations mentioned above, $1/T_1$ is a contribution to the linewidth in general, but the situation is a bit more nuanced than that. The T_1 process contributes, of course, to both $1/T_2$ and $1/T_2^*$. This is not a particularly useful effect, but mostly just something one has to watch out for. There is one small corner of parameter space where the T_1 effect on spin-echo decay is

[4] See the discussion of the Pennington-Slichter indirect coupling and its measurement in Chap. 6.

a bit more interesting and useful. This is in the case of odd half-integral nuclear-spin quantum numbers I, where all but the central $\pm 1/2 \leftrightarrow \mp 1/2$ transition are heavily broadened by electric-field gradient noise (i.e., "quadrupolar broadened"), so that only the central transition is resolved. This circumstance arises in a surprising number of cases in practice. The upshot of it is that one finds a contribution to the echo decay that is enhanced by a factor $(I + 1/2)^2$ [284]. Thus,

$$1/T_2 = (1/T_2)_{spin-spin} + (I + 1/2)^2/T_1, \tag{2.1.4}$$

where the first term is the sum total of the other decay processes discussed above. In metallic host materials the T_1 term can often be identified, because it is Korringa-like, i.e. $1/T_1 \propto T$, while the spin-spin term is typically independent of T. There are corresponding enhancements for other $m \leftrightarrow m + 1$ transitions if their spin-echo signals can be observed in isolation, but it is only when all transitions of a given nuclear spin can be excited at once that the echo decay enhancement is simply $1/T_1$. The latter holds whenever $I = 1/2$.

Phenomena related to observation and relaxation of spin echoes are also discussed in more formal detail in Appendix A [294].

2.1.3 *The Basic Structure of NMR Spectra in Solids*

While many NMR phenomena can be understood in classical terms (e.g. the Bloch equations), nuclear spins are fundamentally quantum in nature, and their physics must be discussed in quantum mechanical terms to obtain a complete picture. As such, they can interact with their microscopic surroundings in highly complex ways, which is what allows the study of NMR properties to provide important microscopic information on the physical properties of host solids. Nuclei with spin quantum numbers $I > \frac{1}{2}$ also possess electric quadrupole moments, denoted Q, which then couple to static and fluctuating electric field gradients (EFG) generated by electric charge distributions in their environment. This adds to the richness and complexity of NMR spectra.

The basic nuclear spin Hamiltonian for spectra is given by (2.1.5), including both the Zeeman term representing the effect of an applied field $\boldsymbol{H}$ and a quadrupolar term representing the interaction of the nuclear quadrupole moment with the electrical field gradient (EFG) tensor $V_{uv} = \frac{\partial^2 V}{\partial u \partial v}$ at the nuclear site in question (see [5], Chap. VI). Thus,

$$\mathcal{H}_{spec} = -\gamma\hbar \boldsymbol{I} \cdot (\hat{\mathbf{1}} + \hat{\mathbf{K}}) \cdot \boldsymbol{H} + \frac{e^2 q Q}{4I(2I-1)}[3I_z^2 - I(I+1) + \frac{\eta}{2}(I_+^2 + I_-^2)], \tag{2.1.5}$$

where $\hat{\mathbf{K}}$ is the NMR shift tensor and $\hat{\mathbf{1}}$ is the unit tensor. In this equation we use conventional notation, where $q = |\frac{\partial^2 V}{\partial Z^2}| = V_{ZZ}$ is the principal component of the

EFG tensor, and the EFG asymmetry parameter is defined as $\eta = (V_{XX} - V_{YY})/V_{ZZ}$. The (X, Y, Z) are the principal axes of the tensor, where we define $|V_{XX}| \leq |V_{YY}| \leq |V_{ZZ}|$ and note that LaPlace's equation gives $V_{XX} + V_{YY} + V_{ZZ} = 0$. It follows that $0 \leq \eta \leq 1$. $\mathcal{H}_{spec}$ will generate the vast majority of NMR spectra found in practice.

Although (2.1.5) is frequently quoted, it has the disadvantage that it assumes that the applied field is along the principal axis of the EFG tensor. Since one generally works with the field along *some* principal axis of the EFG, it is useful to generalize (2.1.5) for an arbitrarily chosen principal axis. In principal axis coordinates, the quadrupolar Hamiltonian becomes [5]

$$\mathcal{H}_Q = \frac{eQ}{2I(I+1)} \sum_{\alpha} V_{\alpha\alpha} I_{\alpha}^2. \tag{2.1.6}$$

In terms of the $V_{\alpha\alpha}$, (2.1.5) becomes in traceless form [6]

$$\mathcal{H}_{spec} = -\gamma\hbar \boldsymbol{I} \cdot (\hat{\mathbf{1}} + \cdot\hat{\mathbf{K}}) \cdot \boldsymbol{H} + \frac{eQ}{4I(2I-1)}[V_{\alpha\alpha}(3I_{\alpha}^2 - I(I+1)) + \frac{1}{2}(V_{\beta\beta} - V_{\gamma\gamma})(I_{+}^2 + I_{-}^2)], \tag{2.1.7}$$

where α is the quantization axis and (α, β, γ) are an arbitrary permutation of (X,Y,Z). In this form it is clear that first-order splittings will be proportional to $\mid V_{\alpha\alpha} \mid$, and the second-order splittings will be proportional to $(V_{\beta\beta} - V_{\gamma\gamma})^2$.

While (2.1.5) and (2.1.7) are useful for generating spectra, it should be noted that a number of other terms are necessary for a complete description of NMR phenomena. Some of the omissions are (a) terms to represent fixed internal hyperfine (HF) fields which may occur in ordered magnetic systems, (b) spin and orbital fluctuation terms which drive T_1 processes, (c) dipolar and indirect spin-spin coupling terms which generate T_2 processes, and (d) radiofrequency (*rf*) excitation terms. Such terms will be invoked when the need arises.

We make a number of general remarks about the use of (2.1.5) and (2.1.7) to generate and analyze spectra.

- (i) Experiments are usually carried out in one of two limiting cases, namely, $H = 0$, so that quadrupolar energies alone determine the spectrum (NQR), and secondly, the *high field limit* $\gamma H \gg \frac{e^2qQ}{4\hbar I(2I-1)}$, wherein the quadrupolar terms may be treated with perturbation theory.
- (ii) For $I = \frac{1}{2}$ the nuclear quadrupole moment Q vanishes identically, according to the Wigner-Eckart theorem (e.g. see [6], Chap. 10).
- (iii) For nuclear spin sites with cubic symmetry, one has $V_{XX} = V_{YY} = V_{ZZ} = 0$, greatly simplifying the spectrum as with $I = \frac{1}{2}$. For quadrupolar nuclei, the effects of stray EFG's in nominally cubic structures may still play a role in the spectrum, however. Crystalline disorder caused by impurities, dislocations, etc., often cause a distribution of stray EFG's to occur at the nuclei of interest. Thus, one often finds *cubic* NMR lines with *complete first-order broadening*, so that only the $(-\frac{1}{2} \leftrightarrow \frac{1}{2})$ transition is easily resolved [20]. Such cases will be discussed as they arise.

- (iv) In the almost universally encountered high-temperature limit defined by $\hbar\gamma H \ll k_B T$, Boltzmann factors are always small compared with unity. The sign of qQ in (2.1.5) is immaterial to the spectra and is, in fact, indeterminate. For convenience, we shall always take qQ to be positive. Note also that the term "$-I(I+1)$" in (2.1.5) has been inserted only to render the quadrupole tensor traceless. It plays no role in the spectral transition frequencies and may be omitted or replaced by some other quantity for calculational convenience.
- (v) In the most frequently occurring cases of non-cubic site symmetry, i.e. tetragonal or trigonal, the EFG symmetry is termed "axial", giving $V_{XX} = V_{YY}$. Thus, $\eta = 0$ and the quadrupolar Hamiltonian is greatly simplified. We shall focus primarily on such cases and comment on how the results are affected if $\eta > 0$.
- (vi) In the high-field limit with $\eta = 0$, we comment briefly on how the EFG term affects the spectrum. If we let the angle between $\boldsymbol{H}$ and the EFG principal axis Z be θ, then the simplest case is when $\theta = 0$. The eigen-energies then have the exact form $E_m = -\hbar\gamma H(1+K)m + h\nu_Q m^2/2$, where $\nu_Q = \frac{3e^2qQ}{2hI(2I-1)}$, m is the $\langle I_z \rangle$ quantum number, and we have assumed a scalar shift K. In such a case, the energy levels are shifted as shown in Fig. 2.2. In the spectrum shown, the $(-\frac{1}{2} \leftrightarrow \frac{1}{2})$ transition is not affected by the ν_Q term, and the $(\pm m \leftrightarrow \pm(m+1))$ transitions are displaced by $\pm\nu_Q, \pm 2\nu_Q, \ldots$, for $m = \frac{1}{2}, \frac{3}{2}, \ldots$, respectively, as shown in the figure. The lines displaced by $\pm n\nu_Q$ are known as *first-order satellites*.
- (vii) For $\theta \neq 0$ the eigenfrequencies can often be estimated satisfactorily with second-order perturbation theory (see [5], Chap. VI). The first-order resonance frequencies become

$$\nu_L + \nu_{\pm n}^{(1)} = \nu_L \pm n\nu_Q(3cos^2\theta - 1)/2, \tag{2.1.8}$$

$n = 1, 2, \ldots$. Here, $n = 1$ corresponds to the $(\pm\frac{1}{2} \leftrightarrow \pm\frac{3}{2})$ transitions and $n > 1$ corresponds to transitions with progressively higher values of m. In the case of powder samples with a uniform distribution of θ over the unit sphere, the first-order quadrupole shift stated above gives rise to a *powder pattern* intensity distribution

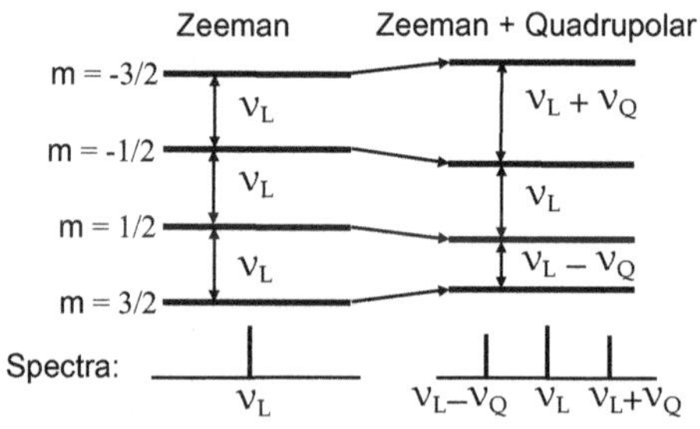

Fig. 2.2 On the left, the nuclear Zeeman term gives uniformly spaced energy levels and a single NMR line. On the right, first-order quadrupole splittings produce "first-order satellites" from the $(\pm 1/2 \leftrightarrow \pm 3/2)$ transitions. The $(-1/2 \leftrightarrow 1/2)$ transition is not affected to first order

over frequencies ranging from $\nu_L - n\nu_Q/2$ to $\nu_L + n\nu_Q$, where the $\theta = 0$ satellite displacement is $n\nu_Q$. The normalized distribution function is

$$\rho_q(\nu) = [6\nu_Q(\nu - \nu_L + \nu_Q/2)]^{-\frac{1}{2}}, \tag{2.1.9}$$

so that the $\theta = 0$ line at frequency $n\nu_Q$ becomes a powder pattern with a square-root singularity at $\nu = \nu_L - n\nu_Q/2$. Here, the Larmor frequency $\nu_L = \gamma H(1 + K)$. We note that m is no longer strictly a good quantum number, so there are second-order frequency shifts as well. Interestingly, the second-order frequency shifts of the satellites at $\pm n\nu_Q(3cos^2\theta - 1)$ are equal, so that the second-order correction to their pairwise separation $2n\nu_Q(3cos^2\theta - 1)$ vanishes. If $\eta > 0$, however, the singularities all disappear and the peak at that position becomes progressively broader as η increases.

- (viii) For the often-used $(-\frac{1}{2} \leftrightarrow \frac{1}{2})$ transition, the frequency to second order is given by

$$\nu^{(2)}_{\pm\frac{1}{2}} = \nu_L - \frac{\nu_Q^2}{16\nu_L}\left[I(I+1) - \frac{3}{4}\right] sin^2\theta\,(9\,cos^2\theta - 1). \tag{2.1.10}$$

In the case of a random powder sample, (2.1.8) and (2.1.10) give rise to characteristic *powder patterns* which are easy to recognize. Such a pattern is shown in Fig. 2.3, where a small amount of Gaussian broadening has been added. In cases where $\eta > 0$, these powder patterns are altered significantly. Also, if there is shift anisotropy $K(\theta)$ with a range comparable to the second-order quadrupole shift of the $(-\frac{1}{2} \leftrightarrow \frac{1}{2})$ transition, its powder-pattern shape will also be altered significantly, whereas the first-order peaks will all reflect the shift value $K(\pi/2)$. Further discussion of these effects is given below.

- (ix) When $H = 0$ we have the case of *nuclear quadrupole resonance* (NQR). The nuclear spin energies are determined entirely by Q, I, and the local EFG's. If ν_Q

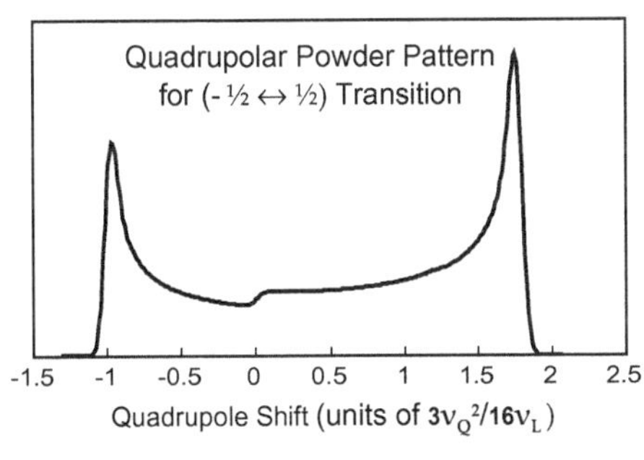

Fig. 2.3 Quadrupolar powder pattern for the $(-1/2 \leftrightarrow 1/2)$ transition based on the second-order quadrupolar energy given in 2.1.10 (axial case). Minimal Gaussian broadening has been applied. The coefficient given is for $I = 3/2$. For arbitrary half-integral I the coefficient becomes $(4I(I+1) - 3)\nu_Q^2/64\nu_L$

(see item (vi) above) is large enough, i.e. in the MHz range, then NQR spectroscopy is feasible. Considering the case of $\eta = 0$, (2.1.5) then shows that $E_{|m|+1} - E_{|m|} = (2|m| + 1)\nu_Q$. Thus, there is a series of $I - \frac{1}{2}$ NQR lines at frequencies ν_Q, $2\nu_Q, \ldots$. The m-levels are doubly degenerate, since E_m is independent of the sign of m. The two parallel transitions produce identical signals under the usual excitation conditions [21]. If $\eta > 0$, the levels are still doubly degenerate, but the transition frequencies are now a function of η. If there are two or more NQR frequencies, their ratio can be used to determine η.

2.1.4 Definition and Interpretation of NMR Parameters: The Shift Tensor $\mathbf{K}_{\alpha\beta}$

NMR frequency shifts are useful for characterizing a resonance line as well as analyzing different components in the magnetic susceptibility of a metallic compound. NMR shifts are useful probes, because they always reflect one or more components of magnetic polarization in the system, such as spin or Van Vleck paramagnetism. There can, of course, be magnetic polarization without a palpable NMR frequency shift, but there is no magnetic NMR shift without a source of polarization, i.e. susceptibility, to drive it. There are complications, of course, when there are fixed sources of internal magnetic polarization, such as in the *mixed state* of a type-II superconductor. Such cases are considered in Sect. 2.3. In general, we have omitted consideration of cases which involve magnetic ordering. Shifts, then, are defined to be linear in the applied magnetic field H. Analysis of shifts can be quite complex with random powder samples when the shift is anisotropic, and even more so if they are combined with inherently anisotropic quadrupolar splittings. Such cases occur frequently, and it is important to know how to analyze and characterize them.

Defining the NMR Shift

In general terms, the NMR shift is a measure of how the magnetic induction B differs at a nuclear site in a metallic specimen from its value in free space when the sample is removed. In practice, shifts are referred to bare values of the nuclear gyromagnetic ratio γ, where γ is the ratio

$$\gamma = \frac{\mu}{\hbar I} \tag{2.1.11}$$

of the nuclear magnetic moment μ to its angular momentum $\hbar I$. "Bare" values of γ are measured in compounds or other environments which are as magnetically neutral as possible. For nuclei of elements which are highly active magnetically, defining γ can be difficult and somewhat arbitrary. On the other hand, even compounds lacking a form of paramagnetism almost always have so-called chemical shifts, which are typically of the order of tens or hundreds of *ppm*. Such shifts are orbital in origin and have been discussed in detail in [6], Chap. 4.

NMR shifts in metals are typically two orders of magnitude larger than chemical shifts, so that the latter are often ignored. In high-T_c compounds, however, we shall see that, *e.g.* for ^{17}O and ^{89}Y, chemical shifts may play a role in the discussion. In d-band metals, however, the NMR shift is dominated by the spin-paramagnetic and Van Vleck terms mentioned earlier. In Sect. 2.2 we discuss a detailed formulation of these effects as well as methods for analyzing NMR shift data. Here we focus on defining and measuring the shift tensor.

Given a suitable reference γ, then, the NMR shift K is defined with respect to a change in the resonant frequency $\nu_{res} = \gamma H(1 + K)$, or

$$K = \frac{\nu_{res}}{\gamma H} - 1. \tag{2.1.12}$$

Since a lot of NMR spectroscopy is done by scanning the magnetic field, it is important to note that the fractional shift in the resonant field, which is $-K/(1 + K)$, not only has the opposite sign, but a palpably different magnitude than K if the shift is large. The unmistakable hallmark of an NMR frequency shift or of a spectrum of shifts for a powder is that they scale precisely with the applied field. If this is not the case, then one is dealing with quadrupole effects or other complications.

The NMR Shift: Tensor or Scalar?

In practice, most shifts are characterized as scalar parameters. However, formally the NMR frequency shift tensor $\mathbf{K}_{\alpha\beta}$ is defined to be the α component of the internal field at the nuclear spin divided by the β component of the applied field ([5], Chap. VI). In general, then, the induced, internal shift field is not parallel to the external field. This can be a complication in the interpretation of large shift effects. However, in cases where the shift anisotropy is 1% or less, it is a good approximation to replace the shift tensor with a scalar shift given by $K(\theta, \phi) = K_X sin^2\theta \cos^2\phi + K_Y sin^2\theta sin^2\phi + K_Z cos^2\theta$, where the $K_{X,Y,Z}$ are the principal values of the shift tensor and (θ, ϕ) give the orientation of the field $\boldsymbol{H}$ in the principal axis coordinates. Another approach, which involves no approximation, is to confine shift measurements on oriented powder or single crystal samples to field orientations along the principal axes. The (X, Y, Z) axes will generally coincide with crystalline axes of symmetry.

In metals with cubic crystal structure or with single crystals, then, shift measurements are usually straightforward. If microscopic disorder is reasonably low, then line-broadening effects will be 0.01% or less and a 1% shift can be reliably measured to good accuracy. If the nuclear site is non-cubic, however, there is the possibility of an anisotropic shift as well as the virtual certainty of quadrupolar shifts and broadening for $I > \frac{1}{2}$. For such cases it is very helpful to have either single crystals or oriented powders for non-cubic systems [40]. Otherwise, data will have to be obtained by analyzing powder patterns, i.e. broadened spectra which result from a sample made up of randomly oriented particles. The functional form of the axial shift powder pattern is the same as for the first-order quadrupolar satellite pattern (2.1.9), namely,

$$\rho_K(K) = [6K_{ax}(K - K_{iso} + K_{ax}/2)]^{-\frac{1}{2}}\,, \tag{2.1.13}$$

where K is the shift variable, which is distributed over the range $K_{iso} - K_{ax}/2 \leq K \leq K_{iso} + K_{ax}$. For a crystal with applied field at an angle θ with respect to the axis of symmetry, the shift may be written $K_{iso} + K_{ax}(3cos^2\theta - 1)/2$, where K_{iso} and K_{ax} are the isotropic and axial components of shift, respectively. If the three principal shift values are all different, then the symmetry is lower than axial and the singularity disappears. The distribution will have *edges* at the maximum and minimum shift values, with a peak at the intermediate principal shift value. It is possible, although it may be somewhat laborious, to extract anisotropic shifts as well as quadrupole frequency parameters from powder patterns. For the $(-\frac{1}{2} \leftrightarrow \frac{1}{2})$ transition the change of these two effects with field are opposite, so that field dependence can be employed to determine their separate values. An example of this is shown in Fig. 2.4.

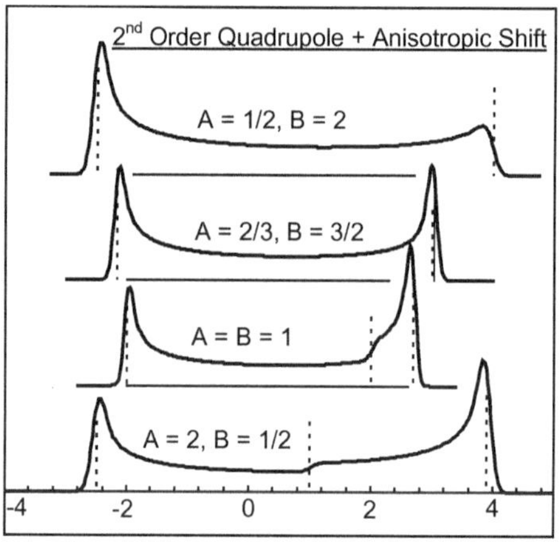

Fig. 2.4 Powder patterns for a $-\frac{1}{2} \leftrightarrow \frac{1}{2}$ transition, showing several cases of combined effects of an anisotropic shift and second-order quadrupolar broadening. The total broadening perturbation is $A\,sin^2\theta\,(9\,cos^2\theta - 1) + B\,(3cos^2\theta - 1)$. In applied field H, $A \propto 1/H$ and $B \propto H$. In reduced units we let $B = H$, and the frequency displacement is in units of $3\nu_Q^2/16\nu_L$ (for $I = 3/2$). Plotted spectra then show the progression of powder patterns as H is varied from 1/2 up to 2. The dashed vertical lines show the positions of the three edges in the absence of broadening. The prominent upper singularity (see Fig. 2.4) converges with the $\theta = 0$ edge with increasing field and disappears above $B/A = 8/3$. However, there is still a visible maximum for the top example, where $B/A = 4$. Note that these spectra are actually oriented the way they would appear in a field sweep increasing to the right, and, therefore, the anisotropic shift coefficient is *negative*. With $B/A < 0$, the patterns would be quite different. Studies such as are shown here allow the practitioner to extract values of A and B from experimental data

2.1.5 Definition and Interpretation of NMR Parameters: T_1

T_1 is the time constant which describes the exchange of nuclear spin energy with the surrounding lattice, hence it is known as the *spin-lattice relaxation* time. Referring to the energy levels in Fig. 2.2, T_1 determines the rates at which nuclei make transitions between these levels. Magnetic T_1 processes are generated by fluctuating HF fields. Each magnetic HF interaction[5] has a fluctuation spectrum with an amplitude and an autocorrelation time. The corresponding T_1 process is driven by the fourier component of the latter spectrum at the NMR/NQR resonance frequency. The magnetic dipole T_1 processes are divided broadly into those driven by spin and orbital HF fields.

The formulation of the various relaxation processes for metallic systems including superconductors has been a major work in progress since NMR was first discovered. Below we review the interpretation of T_1 effects for simple metals and Type I superconductors (Sect. 2.2) and for d-band metals and Type-II superconductors (Sect. 2.3). The interpretation of T_1 processes for the cuprate superconductors will be discussed extensively in Chaps. 3–6. Particularly important is the role played by electron-electron interactions in T_1 phenomena. Accordingly, we review Moriya's theory of correlated electron effects in the T_1 processes of simple metals in Sect. 2.2.

Rather more rare in occurrence are quadrupolar T_1 processes generated by fluctuating components of EFG. A general theory of such effects caused by the *phonon Raman scattering process* was put forward by Van Kranendonk in the early days of NMR [30]. Such processes are extraordinarily weak, however, and have rarely been identified in metals. Nonetheless quadrupolar relaxation processes for the ^{63,65}Cu isotopes have occasionally been claimed to exist in solids, using the isotopic ratio of T_1 values as a diagnostic.[6] When quadrupolar T_1 effects occur in metals, it is our conjecture that they are much more likely to be caused by ionic motion (e.g. diffusion) than by phonons.

The Magnetic Dipole Relaxation Process

Nuclear spin transitions generated by the HF interactions in (2.2.1) are known as *magnetic dipole transitions* (as opposed to electric quadrupolar transitions). Here, we analyze the magnetic dipole T_1 process in some detail. We review methods to measure it as well as certain pitfalls to be avoided in conducting such measurements. Whether measuring T_1 with high-field NMR or in zero field with NQR, the nuclear spin energy levels E_m will be labeled with the I_z quantum number m. To the extent that m is a good quantum number, the transition rate between states m and m' may be written

$$|\langle m'|I_+|m\rangle|^2 W = |\langle m|I_-|m'\rangle|^2 W = (I-m)(I+m+1)W\delta_{m',m+1} \qquad (2.1.14)$$

---

[5]See Sect. 2.2 for a comprehensive definition of the magnetic HF interactions.

[6]The ratio $^{63}T_1/^{65}T_1$ is a good diagnostic for this purpose, since it is $(\gamma_{65}/\gamma_{63})^2 = 1.148$ for magnetic relaxation and is $(Q_{65}/Q_{63})^2 = 0.854$ for electric quadrupolar relaxation.

If $I = \frac{1}{2}$, then there is only a single rate W involved, where $T_1 = (2W)^{-1}$. For $I > \frac{1}{2}$ there are $2I$ rate constants and relaxation eigenmodes. The formula (2.1.14) will be correct for the cases (a) when $I = \frac{1}{2}$, (b) for arbitrary values of I when $\eta = 0$ and the field is applied along the principal axis of the EFG tensor, and (c) for NQR when $\eta = 0$. It is approximately correct for arbitrary η values when $\nu_Q \ll \gamma H/2\pi$ and for NQR when η is small. In other cases m-state mixing effects become important and it may be necessary to solve the spin Hamiltonian (2.1.5) numerically to obtain accurate matrix elements of $I_\pm$. In this discussion of T_1 analysis, we shall assume (2.1.14) to be valid, and we shall also neglect nuclear quadrupole relaxation effects. They are, as noted above, quite rare in metallic systems.

Detailed Balance in T_1 Processes

T_1 in metals is simplified by the fact that fluctuation spectra for itinerant fermions are constant over the range of typical NMR frequencies. T_1 values are therefore found to be independent of the Zeeman splitting and are the same for NMR and NQR measurements alike.[7] We shall assume, further, that the lattice constitutes an infinite heat reservoir at equilibrium at temperature T. To illustrate the incorporation of detailed balance into the rate equations, we analyze a two-level system with energies $E_2 > E_1$, defining $\Delta E = E_2 - E_1$. The populations of these levels are designated $p_{1,2}$ and will be treated as continuous variables. The rates at which up and down transitions occur are defined to be $\mathcal{W}_{21}$ (up) and $\mathcal{W}_{12}$ (down), respectively. At thermal equilibrium the principle of detailed balance gives $\mathcal{W}_{21} = \mathrm{e}^{-\beta\Delta E}\mathcal{W}_{12}$ ([6], Chap. 5) ($\beta = 1/kT$). The rate equations for populations are then

$$\frac{\mathrm{d}p_{1,2}}{\mathrm{d}t} = \pm[p_2\mathcal{W}_{12} - p_1\mathcal{W}_{21}], \tag{2.1.15}$$

where $p_1 + p_2 = N$ is the total number of nuclei in the ensemble. When the $p_{1,2}$ are at equilibrium we have $\mathrm{d}p_{1,2}/\mathrm{d}t = 0$, giving $p_2/p_1 = \mathcal{W}_{21}/\mathcal{W}_{12} = \mathrm{e}^{-\beta\Delta E}$. Thus, the Boltzmann distribution of spin populations follows from detailed balance.

Writing $p_{1,2} = \pm\frac{1}{2}(p_2 - p_1) + N/2$, the (2.1.15) may be combined to give

$$\mathrm{d}(p_2 - p_1)/\mathrm{d}t = (p_2 - p_1)(\mathcal{W}_{12} + \mathcal{W}_{21}) - N(\mathcal{W}_{12} - \mathcal{W}_{21}).$$

From this relation we see that the relaxation rate of $(p_2 - p_1)$ is $(\mathcal{W}_{12} + \mathcal{W}_{21})$. Since this quantity must be independent of ΔE, we set $(\mathcal{W}_{12} + \mathcal{W}_{21}) = 2$ W, where W is a rate constant which depends only on T. Detailed balance then leads immediately to

$$\mathcal{W}_{12} = \frac{2W}{(1 + \mathrm{e}^{-\beta\Delta E})}; \quad \mathcal{W}_{21} = \frac{2W\mathrm{e}^{-\beta\Delta E}}{(1 + \mathrm{e}^{-\beta\Delta E})}. \tag{2.1.16}$$

Using (2.1.16), (2.1.15) then becomes

[7] Note, however, that NMR and NQR reflect different relaxation modes, as will be seen below. For example, if $I = \frac{3}{2}$, for NMR $1/T_1 = 2$ W, whereas the single rate measured with NQR is 6 W, or one-third the length of the conventionally defined T_1.

$$\frac{\mathrm{d}p_{1,2}}{\mathrm{d}t} = -2W\left[p_{1,2} - \frac{N\mathrm{e}^{-\beta E_{1,2}}}{\sum_i \mathrm{e}^{-\beta E_i}}\right]. \tag{2.1.17}$$

Even though p_1 and p_2 appear to act independently, they are not, of course, independent. The general solution to (2.1.17) is seen to be

$$p_{1,2}(t) = (p_{1,2}(0) - p_{01,02})\mathrm{e}^{-2Wt} + p_{01,02}\,, \tag{2.1.18}$$

where $p_{01,02} = N\exp(-\beta E_{1,2})/\sum_i \exp(-\beta E_i)$ are the equilibrium population values. This result also satisfies $p_1(t) + p_2(t) = N$ if $p_1(0) + p_2(0) = N$, which must be the case.

Relaxation Equations for I > $\frac{1}{2}$

Generalization of the foregoing result to multi-level nuclear relaxation equations is straightforward. They can always be written in terms of the *difference* variables

$$d_m = p_m - p_{0m}, \quad \text{with } p_{0m} = N\exp(-\beta E_m)/\sum_m \exp(-\beta E_m), \tag{2.1.19}$$

where m is the quantum number for the energy eigenvalues. We note that the value of p_{0m} does not depend on the choice of the zero-point of energy. In general, then, one can measure T_1 by creating a non-equilibrium population distribution, then observing the decay toward equilibrium of the population difference of some suitable pair of levels using spin-echo excitation. The latter pair must be chosen to have $|\Delta m| = 1$ for magnetic excitation, and they must be chosen to reflect a non-equilibrium population difference during the equilibration process.

If the initial non-equilibrium condition is created by applying an *rf* pulse to some pair of levels, note that the recovery can often be observed using spin-echo signals from a *different* pair of levels.[8] This degree of experimental freedom is not often used in NMR. It could be used, for example, to avoid multiple echo effects in a T_1 measurement when T_1 is very short.

Relaxation Eigenmodes for I > $\frac{1}{2}$

For multi-level nuclear spin Hamiltonians ($I > \frac{1}{2}$) [42] there is an eigenmode structure to the relaxation equations with multiple relaxation eigenrates. We illustrate this situation with the simple example of $I = \frac{3}{2}$, for which there are four energy levels as shown in Fig. 2.2. The dynamical equations for the differences d_m are then

[8]This requires the ability to be able to apply *rf* pulses to the sample at two separate frequencies, which can be a very useful adjunct to the usual spectrometer setup.

$$\frac{dd_{3/2}}{dt} = -3Wd_{3/2} + 3Wd_{1/2} \tag{2.1.20a}$$

$$\frac{dd_{1/2}}{dt} = 3Wd_{3/2} - 7Wd_{1/2} + 4Wd_{-1/2} \tag{2.1.20b}$$

$$\frac{dd_{-1/2}}{dt} = 4Wd_{1/2} - 7Wd_{-1/2} + 3Wd_{-3/2} \tag{2.1.20c}$$

$$\frac{dd_{-3/2}}{dt} = 3Wd_{-1/2} - 3Wd_{-3/2} \tag{2.1.20d}$$

We seek solutions of the form $d_m = c_m \exp(-\lambda W t)$. Substitution into (2.1.20) then yields the matrix equation

$$\begin{bmatrix} \lambda-3 & 3 & 0 & 0 \\ 3 & \lambda-7 & 4 & 0 \\ 0 & 4 & \lambda-7 & 3 \\ 0 & 0 & 3 & \lambda-3 \end{bmatrix} \begin{Bmatrix} c_{3/2} \\ c_{1/2} \\ c_{-1/2} \\ c_{-3/2} \end{Bmatrix} = \begin{Bmatrix} 0 \\ 0 \\ 0 \\ 0 \end{Bmatrix}$$

This matrix is solved straightforwardly. The characteristic equation is

$$\lambda^4 - 20\lambda^3 + 108\lambda^2 - 144\lambda = 0\,, \tag{2.1.21}$$

for which the solutions are $\lambda = (0, 2, 6, 12)$. The corresponding decay rates are $\lambda_0 = 0$, $\lambda_2 = 2W = T_1^{-1}$, $\lambda_6 = 6W = 3T_1^{-1}$, and $\lambda_{12} = 12W = 6T_1^{-1}$. Ignoring the null mode, the corresponding eigenvectors for T_1, $T_1/3$ and $T_1/6$ may be written {3, 1, −1, −3}, {−1, 1, 1, −1}, and {−1/3, 1, −1, 1/3}, respectively.

The solutions to (2.1.20) may be applied to different techniques for measuring T_1. We illustrate their use in the simple case of a cubic environment where $E_m = -\hbar\gamma_n(1+K)Hm$. Since $|E_m| \ll k_B T$, we may write (2.1.19) $p_{0m} \approx \frac{N}{2I+1}(1-\delta m)$, where $\delta = \hbar\gamma_n H(1+K)\beta$. Then if we apply an *rf* pulse with pulse-angle θ to the spins, the resulting initial condition is

$$p_m(0) \approx \frac{N}{2I+1}(1 - \delta \cos\theta\, m).$$

As a result, the initial values are $d_m(0) \approx \frac{N}{2I+1}(1 - cos\theta)\, m$. The $d_m(0)$ have the form {3, 1, −1, −3} of the "T_1" mode, so that the solution can be written

$$d_m(t) \approx \frac{Nm\delta}{2I+1}(1 - cos\,\theta)\, e^{-t/T_1}.$$

If $\theta = \pi$, then one has a perfect *inversion*, which gives the largest excursion in signal amplitude. The time variation of the spin echo signal is then proportional to

$$p_m(t) - p_{m+1}(t) = \frac{N\delta}{2I+1}[1 - (1 - \cos\theta)\,e^{-t/T_1}]. \tag{2.1.22}$$

If $p_m(0) = p_{m+1}(0)$ for all applicable m, then we have a condition known as *saturation*, which also yields a single-exponential recovery $\propto (1 - \exp(-t/T_1))$.

Recovery Mode When Only the $(-\frac{1}{2} \leftrightarrow \frac{1}{2})$ Transition is Excited

If there are quadrupolar splittings, then a single pulse applied to the $(-\frac{1}{2} \leftrightarrow \frac{1}{2})$ transition results in a $\{d_m(0)\}$ vector with the form $a_0\{0, 1, -1, 0\}$. Writing the decay mode as a superposition of the foregoing eigenmodes, we have

$$\{d_m(t)\} = \frac{1}{10}a_0[\{3, 1, -1, -3\}\exp^{-t/T_1} + 9\{-1/3, 1, -1, 1/3\}\exp^{-6t/T_1}].$$

The spin echo signal for the $(-\frac{1}{2} \leftrightarrow \frac{1}{2})$ transition will therefore exhibit the well-known decay function

$$E(t) = E_0\,(0.9\exp^{-6t/T_1} + 0.1\exp^{-t/T_1})\,. \tag{2.1.23}$$

Such a precise mathematical form occurs only for excitation of a single pair of levels over a time interval which is short compared with $T_1/6$. If relaxation begins to take place during the initial excitation phase, then the simple form of the initial condition is lost and the decay will be rather more complicated. In the foregoing example, $d_{3/2}(t)$ would be proportional to $(\exp -t/T_1 - \exp -6t/T_1)$, so that it would deviate very quickly from its initial value of zero. As we noted earlier, the relaxation measurement could be conducted in this fashion, i.e. by monitoring a different transition frequency from the one used for the initial excitation.

The T_1 analysis described here for $I = \frac{3}{2}$ is straightforwardly extended to larger (half-integer) values of I. There will be $2I + 1$ equations such as the (2.1.20) above, so the solutions will have added complexity. But solving the determinants for $I > \frac{3}{2}$ is simplified by the fact that as increments of 1 are added to I, the eigenvalues for the previous case are carried forward.

The NQR Case

Continuing our discussion of an $I = \frac{3}{2}$ nucleus, we consider finally the NQR case, where the $\pm\frac{1}{2}$ levels and the $\pm\frac{3}{2}$ levels are degenerate pairs separated by an energy difference of $h\nu_{NQR} = h\nu_Q(1 + \eta^2/3)^{1/2}$. A single excitation pulse at the NQR frequency then creates an initial condition given by $\{d_m\} = a_0\{-1, 1, 1, -1\}$. This effectively 2-level system relaxes entirely via the *symmetric mode* with a single decay rate $3/T_1$. One must be careful to take account of the relaxation mode effects when comparing T_1 data obtained with different methods.

We make a few additional remarks about the decay mode effects described above for the $I = \frac{3}{2}$ system. First, there are many nominally cubic systems where the observed signal is dominated by the $(-1/2 \leftrightarrow 1/2)$ transition, because transitions to higher m-values are broadened by stray first-order quadrupolar splittings. Such splittings are the consequence of impurities and other forms of crystalline disorder. If such broadening is severe, magnetization recovery may take place via (2.1.5) rather than with the expected single-exponential T_1 decay. Since there is a dramatic difference between these two cases, it is very important to be aware of such broadening effects.

A second point we wish to emphasize is that the analysis of T_1 relaxation curves given here for $I = \frac{3}{2}$ is entirely independent of the nature of the NMR spectrum. The functional behavior depends only on which pair of levels is used to excite a non-equilibrium initial condition, and on the fact that the quantum numbers $m = \langle I_z \rangle$ are relatively pure in characterizing the energy eigenstates. If the latter condition is not fulfilled, then numerically evaluated eigenfunctions may be needed to generate sufficiently accurate relaxation rates.

Measuring T_1 in the Presence of Spectral Diffusion

We consider briefly some well-known problems encountered in T_1 measurements. Two of the most troublesome are the related problems of inhomogeneous broadening and spectral diffusion. These problems occur frequently in d-band metals and intermetallic compounds. The key to accurate T_1 measurements is the experimenter's ability to create a strongly non-equilibrium population difference between a pair of nuclear energy levels as an initial condition. The T_1 recovery will then be accompanied by a spin-echo signal which undergoes a substantial change in amplitude. In cases of severe inhomogeneous broadening such as that shown in Fig. 2.5, a difficulty arises from the fact that a single pulse of rf field will only affect a region of width γH_1, where H_1 is the amplitude of the *rf* pulse field. The narrow region saturated by a single pulse is shown as a dotted line in Fig. 2.5a. If *the broadening is entirely static*, it may be possible to focus the measurement on the polarization behavior at precisely the irradiation frequency, i.e. the center frequency of the population disturbance shown in the figure. As noted in Sect. 1.2.2, this can be done by carefully integrating the entire area under the spin echo, so that the measured recovery curve will accurately reflect what is taking place at $\omega = \omega_{rf}$.

However, such broadening may actually have a partially *dynamical* character owing to the presence of spin-spin couplings of the form

$$\mathcal{H}_{ff} = B I_{i\pm} I_{j\mp} \tag{2.1.24}$$

between neighboring nuclei. Under circumstances which occur frequently, the $\mathcal{H}_{ff}$ interactions lead to *flip-flop* dynamic processes between neighboring spins, which then leads to a rapid smearing of the band of excited spins over a much wider range of frequencies than those excited directly by the *rf* pulse(s). This process, known as *spectral diffusion*, is illustrated with dashed lines in Fig. 2.5b. There results what appears to be a rapid recovery of the polarization at the excitation frequency. However, such a recovery has nothing to do with the T_1 process, but is easily mistaken for it. Often it is much faster than T_1, and results in the experimenter's inability to achieve inversion of the polarization [43]. This highly undesirable effect has been known in adverse cases to render T_1 measurements unfeasible.

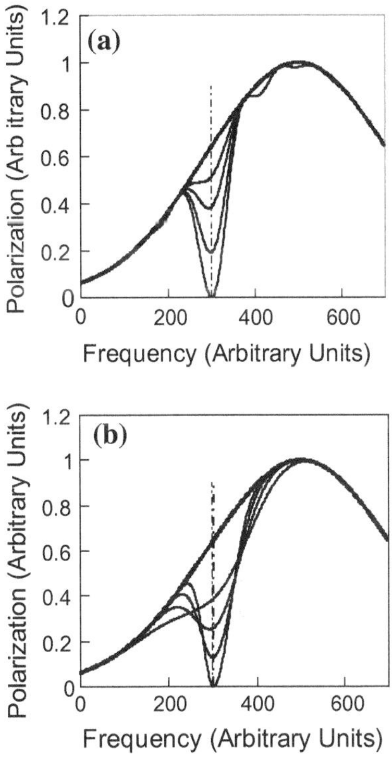

Fig. 2.5 **a** The saturation effect of a $\pi/2$ pulse applied at frequency ω_0 is shown for an effective pulse field amplitude of 20 in units of the frequency scale. The polarization curve is also shown at several later times as T_1 recovery takes place *without spectral diffusion*. **b** Development with time of the polarization curve is shown at several stages following the same initial saturation pulse as in (**a**). The curves are a Gaussian model in which there is no T_1 recovery effect, thus the total polarization, plus its first moment $\int(\omega - \omega_0)P(\omega)d\omega$, are conserved over the time interval shown. Note the spurious "recovery" of the polarization at frequency ω_0 as a function of time

We consider here the circumstances which give rise to spectral diffusion effects and offer a few suggestions on how to deal with this problem. The flip-flop process $(m, m') \rightarrow (m \pm 1, m' \mp 1)$ between (primarily) nearest-neighbor (nn) spins has been recognized since the early days of NMR as essential for spin diffusion, as well as for maintaining thermal equilibrium within the nuclear spin system. Spectral diffusion is, in fact, a process whereby a disturbance in a narrow frequency range is able to diffuse over the entire NMR line so as to create a *uniform spin temperature*. However, since an energy-conserving flip-flop transition can only take place between neighbors with nominally the same Larmor frequency, i.e. the same shift value, it is interesting to consider how polarization can migrate over large frequency intervals such as depicted in Fig. 2.5b. With coupling energy of the form of (2.1.24), it is plausible that a flip-flop transition could take place between neighboring spins with a Zeeman frequency mismatch of the order of $\Delta\omega \sim B$ and still conserve energy, the difference in Zeeman energies being taken up by the transverse terms and eventually dissipated by other flip-flop processes. Over a large number of such fluctuations, the narrow polarization disturbance created by an *rf* pulse could broaden symmetrically over frequencies well beyond the confines of the initial excitation profile. To have this happen, of course, the inhomogeneous broadening must be local in character and be otherwise configured to facilitate the process.

One method for measuring T_1 under these circumstances is to endeavor to saturate the entire line with a string of pulses known as a *comb* instead of a single pulse. Were this possible, then all polarization variables would relax back to equilibrium with the single time constant T_1. If pulses at a single frequency do not produce the desired saturation effect, then applying frequency modulation to cover the entire resonance line is known to be effective [44]. At low temperatures, however, long strings of pulses may produce undesirable heating effects. If T_1 is long enough, another more exotic (but widely used) method is to use *field switching* to create a non-equilibrium initial condition for the T_1 measurement. The field at the sample can be switched mechanically or electronically.

The most obdurate cases occur when T_1 is too short, even at low temperatures, for field switching, and pulse combs are also not feasible. One's only option may to be hope that spectral diffusion will run its course in a time less than T_1, and that the final stage of the polarization recovery process can be analyzed to extract the T_1 rate. Needless to say, such an approach also has limited applicability.

2.2 NMR Probe of sp-Band Metals and Type I Superconductors

In order to put high-T_c materials into context, this section gives as background material a brief summary of the basic static and dynamic magnetism of "simple" metals as sensed by the NMR probe. These systems give a historical perspective of the subject and provide foundation for the treatment of the more complex d-electron metals. In typical *good* metals we think about electrons as itinerant fermions which occupy Bloch wavefunctions $u_k(\boldsymbol{r})e^{i\boldsymbol{k}\cdot\boldsymbol{r}}$. Such orbitals are often far removed in character from the atomic states they are actually composed of. Calculations of NMR shift, relaxation, and indirect spin-spin couplings in metals were originally formulated in this band-theoretical language. The s-character of band electrons is always present and often plays a major role. In contrast, high-T_c systems are *bad metals*, often with just enough mobile carriers to render the d-holes on the Cu^{2+} sites itinerant. Some properties remain atomic-like in character and can be thought of usefully in terms of localized wavefunctions. Perhaps by accident, there is very little s-character in the wavefunctions of most high-T_c ions, allowing the less familiar p and d orbitals to dominate.

In this section we review the energy-band, k-space picture used to describe the static and dynamic magnetism sensed by the NMR probe in both the *normal* (non-superconducting) and *superconducting* (SC) states. We begin the discussion with the idealized Korringa treatment of NMR shifts and relaxation, then discuss important modifications of this picture for actual sp-band metals.

Historically, the first NMR studies of superconductivity were done on this class of materials. We review problems encountered in reconciling both the shift and relaxation behavior with the BCS theory [39] for these systems. It was with elemental metals such as Hg, Sn, Pb, and Al that questions such as that of the role played by spin-orbit coupling in the behavior of the spin susceptibility and corresponding NMR

(Knight) shift first came to light. They also gave us our first clear understanding of the BCS coherence peak in the T_1 rate [64].

Since even before the BCS theory of superconductivity [39] was announced, NMR shifts have been employed to investigate the nature of the superconducting state. In the early days the only known metallic NMR shift was the *s-contact Knight* shift. Moreover, the original BCS ground state was an eigenfunction of S_z, leading to the naive expectation that $K(T) \propto \chi_{spin}(T)$. The actual reality was rather more complex. This interesting piece of NMR history will be touched upon in Sect. 1.2. Measurement of such effects has become a standard application for NMR studies. Observation of shift behavior is also very important in the study of high-T_c systems and for any of the latter-day unconventional superconductors.

2.2.1 NMR Shifts and Relaxation in Simple Pauli Paramagnets

Following on the discovery of the s-contact NMR shift by Knight [32], one of the first theoretical milestones for the NMR study of metals was the Korringa relation between the latter shift and the product T_1T, which was also shown to be independent of temperature [41]. These results were expected to apply to cubic sp-band metals, where it is noteworthy that the p-electrons with their far smaller HF couplings have a nearly negligible effect on both the NMR shift and T_1. The surprise of substantial deviations from the Korringa relation for simple systems such as the alkali metals were therefore a stimulus for Moriya's treatment of electron-electron interactions in these systems. The result was important modifications to the Korringa relation which were verified experimentally. This was a very important step in the treatment and understanding of electron interactions in metals. We review these developments in the following subsections.

The Comprehensive HF Coupling Hamiltonian in Solids

As a preliminary step we introduce here the comprehensive HF Hamiltonian

$$\mathcal{H}_{HF} = 2\gamma\hbar\mu_B \boldsymbol{I} \cdot \left[\frac{\boldsymbol{l}}{r^3} - \frac{\boldsymbol{s}}{r^3} + \frac{3\boldsymbol{r}(\boldsymbol{s}\cdot\boldsymbol{r})}{r^5} + \frac{8}{3}\pi \boldsymbol{s}\delta(\boldsymbol{r}) \right], \tag{2.2.1}$$

where $\gamma\hbar\boldsymbol{I}$ is the nuclear moment located at $\boldsymbol{r} = 0$ and $\boldsymbol{l}$ and $\boldsymbol{s}$ are the orbital and spin angular momentum operators, respectively, of an itinerant fermion located at $\boldsymbol{r}$ (see [5], Chap. VI). Equation (2.2.1) can be written $-\gamma\hbar\boldsymbol{I} \cdot \boldsymbol{H}_e$, where $\boldsymbol{H}_e$ is an effective HF-field operator which is the ultimate source of all magnetic shift and relaxation effects in the metallic environment. There are three distinct types of interactions, namely the orbital field, the dipolar field, and the s-contact interaction. Although it took a number of years for them to emerge, calculations for the metallic shifts listed above were all ultimately based on (2.2.1). Evaluating the thermally averaged effects of $\boldsymbol{H_e}$ in metallic systems is a matter of great subtlety and complexity. In a very real sense, this is the essence of *the high-T_c problem* for the NMR probe.

The Knight Shift: s-contact HF Coupling in an sp-Band Metal

The so-called s-contact HF field and its corresponding NMR shift in copper metal were discovered by Knight [32]. This shift is simply given by the expectation value of the final term of (2.2.1) when spin polarization from an external field H is present. There results a temperature-independent NMR shift given by

$$K_s = \alpha_s \chi_{sp}, \tag{2.2.2}$$

where $\alpha_s = \frac{8\pi}{3}\langle|u_{\boldsymbol{k}}(0)|^2\rangle$, $\langle\rangle$ indicating an average over the Fermi surface. $\chi_{sp} = \frac{1}{2}(g\mu_B)^2 n(E_F)$ is the spin susceptibility of the *sp* band, where $n(E_F)$ is the Fermi surface density of states per atom for one direction of spin [75]. We also note that the dipolar HF coupling, i.e. the second and third terms of (2.2.1), also yields a palpable NMR shift in many cases. It vanishes at sites having cubic symmetry, and is typically an order of magnitude or so smaller than the contact shift. In high-T_c materials, where contact shifts are often absent, the dipolar terms take on a much greater significance than for cubic metals.

Interestingly for the case of copper metal, it later developed that some 40% of the orbital character of the Fermi surface is d-like [33]. As a result, a significant fraction of the observed Knight shift for copper is attributable to the orbital Van Vleck term discussed in the next subsection [34].

The Korringa Calculation of T_1

The idealized viewpoint regarding a simple metal is epitomized by the seminal calculation of NMR shift and relaxation time by Korringa [41]. In this case, noninteracting electrons were assumed, with a density of states at the Fermi surface describable by band theory. HF interactions with the relevant nuclear spins were limited, not unreasonably, to the s-contact interaction (2.2.1), the p-electron dipolar coupling being negligibly weak in comparison. Cubic symmetry allowed quadrupolar effects and p-electron contributions to the NMR shift to be ignored, and any gradient to the density of states deemed too small to alter low-temperature behavior. The shift expression is given above in (2.2.2).

Next, we need to develop an expression for T_1 at the same level of approximation. We use the basic *golden rule* expression from time-dependent perturbation theory, where the perturbing interaction is the s-contact term from (2.2.1), $\mathcal{H}_{sc} = -\frac{8\pi}{3}\hbar^2\gamma_n\gamma_e\delta(\boldsymbol{r})\boldsymbol{I}\cdot\boldsymbol{s}$. If we consider the elemental transition from state $|\boldsymbol{k}\uparrow m\rangle$ to state $|\boldsymbol{k}'\downarrow m+1\rangle$, where m is the initial nuclear spin quantum number, the rate for such a transition may be written [6]

$$\mathcal{W}_{\boldsymbol{k}'\downarrow m+1,\boldsymbol{k}\uparrow m} = \frac{2\pi}{\hbar}|\langle\boldsymbol{k}'\uparrow m+1|\mathcal{H}_{sc}|\boldsymbol{k}\downarrow m\rangle|^2\delta(E_k - E_{k'}). \tag{2.2.3}$$

To complete the calculation this transition rate must be summed over initial electron states $|\boldsymbol{k}\uparrow\rangle$ which are occupied and final states $|\boldsymbol{k}'\downarrow\rangle$ which are unoccupied.

The operative term in $\mathcal{H}_{sc}$ is the one $\propto I_+ S_-$, whereupon the squared matrix element becomes

$$|\langle \boldsymbol{k} \uparrow m|\mathcal{H}_{sc}|\boldsymbol{k}' \downarrow m+1\rangle|^2 = \left[\frac{4\pi}{3}\gamma_n\gamma_e\hbar^2|u_{\boldsymbol{k}}(0)u_{\boldsymbol{k}'}(0)|\right]^2 (I-m)(I+m+1). \tag{2.2.4}$$

The principal approximation is to assume that all states in the vicinity of the Fermi surface have approximately the same probability density at the nucleus. Thus, we may replace the quantity $|u_{\boldsymbol{k}}(0)u_{\boldsymbol{k}'}(0)|^2$ with the squared Fermi surface average $\langle |u_{\boldsymbol{k}}(0)|^2\rangle^2$. Introducing the restricted sums on $\boldsymbol{k}$ and $\boldsymbol{k}'$ and noting further that $1/T_1$ is twice the coefficient of $(I-m)(I+m+1)$ (see Eq. (2.1.14)), this gives [63]

$$1/T_1 = \frac{4\pi}{\hbar}\left[\frac{4\pi}{3}\gamma_n\gamma_e\hbar^2\langle|u_{\boldsymbol{k}}(0)|^2\rangle\right]^2 \sum_{\boldsymbol{k},\boldsymbol{k}'}^{(r)} \delta(E_{\boldsymbol{k}} - E_{\boldsymbol{k}'}). \tag{2.2.5}$$

The restricted sums $\sum_{\boldsymbol{k},\boldsymbol{k}'}^{(r)}$ can be replaced with energy integrals where, for example, the sum on $\boldsymbol{k}$ of states with energy E becomes $\int n(E)dE$, where $n(E)$ is the density of states, and the occupation restriction is enforced with the Fermi occupation function $f(E) = [\exp\beta(E-E_F)+1]^{-1}$.

The summation in (2.2.5) then becomes

$$\int dE \int dE' n(E)n(E')f(E)(1-f(E'))\delta(E-E') = \int n(E)^2 f(E)(1-f(E))\,dE. \tag{2.2.6}$$

The final step is to note that the function $f(E)(1-f(E))$ is very sharply peaked at E_F at the usual experimental temperatures, so that $n(E)^2$ can be replaced with $n(E_F)^2$. Then we note that the area under the peak in $f(E)(1-f(E))$ is easily shown to be $k_B T$. The relaxation rate becomes

$$1/T_1 = \frac{4\pi k_B T}{\hbar}\left[\frac{4\pi}{3}\gamma_n\gamma_e\hbar^2\langle|u_{\boldsymbol{k}}(0)|^2\rangle\right]^2 n(E_F)^2. \tag{2.2.7}$$

This expression may be combined with the result in (2.2.2) to give the *Korringa product*, which we write as

$$\frac{1}{T_1 T K_s^2} = \frac{4\pi k_B \gamma_n^2}{\hbar\gamma_e^2}. \tag{2.2.8}$$

Equation (2.2.8) suggests that $1/T_1T$ divided by K_s^2 for these kinds of metals should equal an expression made up of physical constants which is independent of the other properties of the metal. As we shall see below, the result of Moriya's theory is to multiply the right-hand side of (2.2.8) by a constant factor <1. Setting that aside for the moment, however, let us see how well the Korringa relation is obeyed in simple metals. In Table 2.1 shift and relaxation data from the literature are listed for the alkali metals and the noble metals Cu, Ag, and Au [7].

These elemental metals all have cubic crystal structures, so that dipolar shift effects do not come into play. Using the measured shift values, values of T_1T extracted from (2.2.8) are compared with experimental values. It is seen that the latter values are generally smaller than those obtained from (2.2.8). Including p-electron dipolar hyperfine fields in the relaxation calculation would only make this discrepancy larger. Thus, the most serious defect in the Korringa treatment appears to be the neglect of electron-electron exchange enhancement effects, which we shall discuss next.

2.2.2 *The Moriya Theory of Exchange Enhancement in Simple Metals*

There are various corrections one could make to the Korringa treatment of relaxation. The electron-electron interaction potential has the effect of enhancing the spin susceptibility, and thus the NMR shift [57], in close analogy with the enhancement effect of ferromagnetic exchange among localized moments. However, it was pointed out by Moriya [58] that exchange fluctuations would also enhance the T_1 process. Moriya's calculation of this effect in the *random phase approximation* showed this enhancement effect to be substantial, but not as great as that of the shift and susceptibility. This is consistent with the values shown in Table 2.1, where the experimental Korringa product (Relative Korr. Prod.) is consistently smaller than the result in (2.2.8). Thus, the measured Korringa products can be used with the Moriya theory to estimate the magnitude of these enhancement effects (see below).

We give an outline of the random-phase approximation calculation here, and refer the interested reader to Refs. [7, 58] for a more detailed account. As noted

Table 2.1 NMR shift and relaxation data are tabulated for alkali metals and for the noble metals, along with values of T_1T derived from (2.2.8) from the compilation by Narath [7]. To test the validity of the Korringa relation, a relative Korringa product defined as $(T_1TK_s^2)^{-1}$ divided by $4\pi k_B\gamma_n^2/\hbar\gamma_e^2$ is listed in column five

Metal and isotope	$\mathbf{K_s}(\%)$	T_1T (expt.) (s K)	T_1T (Korringa) (s K)	Relative Korr. Prod.
^{7}Li	0.0263	45.0 ± 2	26.0	0.58
^{23}Na	0.112	4.8 ± 0.1	3.1	0.65
^{85}Rb	0.650	0.81 ± 0.08	0.63	0.78
^{133}Cs	1.47	0.13 ± 0.01	0.069	0.53
^{63}Cu	0.232	0.9 ± 0.2	0.7	0.78
^{109}Ag	0.522	9.0 ± 1.0	4.5	0.50
^{197}Au	1.4	4.0 ± 0.5	4.5	1.13

Table 2.2 A tabulation of enhancement parameters α derived for a selection of s-band metals from experimental data using (2.2.9). For comparison, calculated values given by Pines [57] are listed

Elemental metal	α from $\mathcal{K}(\alpha)$	α calc. (Pines)
^{7}Li	0.36	0.37
^{23}Na	0.30	0.24
^{85}Rb	0.19	0.15
^{109}Ag	0.43	0.32

above, the shift has a simple exchange-enhancement factor, $K_s = K_{s0}/(1-\alpha) = \alpha_s\chi_{s0}/(1-\alpha)$, where $\alpha = \rho_0\chi_{s0}$, and $\rho_0 = 2\pi\zeta/\hbar^2\gamma_e^2$. Finally, ζ is the q-independent (by assumption) electron-electron interaction potential. The single parameter ζ drives the enhancement of both static and dynamic spin susceptibilities in the random phase approximation picture. For T_1 the fluctuation enhancement is q-dependent. In Sect. 3.5 these magnetic properties will be discussed in terms of q and ω-dependent susceptibilities. Here, we simply state the result for a spherical Fermi surface:

$$\frac{1}{T_1 T K_s^2} = \frac{4\pi k_B \gamma_n^2}{\hbar\gamma_e^2}\mathcal{K}(\alpha), \tag{2.2.9}$$

where $\mathcal{K}(\alpha) = 2(1-\alpha)^2 \int_0^1 dx\, x/[1-\alpha G(x)]^2$. Here $G(x)$ is the Lindhard function for free electrons given by $\frac{1}{2}\{1 + [(1-x^2)/2x]ln|(1+x)/(1-x)|\}$, with $x = q/2k_F$. $G(x)$ is integrated over the range of $\boldsymbol{q}$ values which span the Fermi surface. We see in (2.2.9) that the Korringa product of (2.2.8) is simply multiplied by the quantity $\mathcal{K}(\alpha)$. The enhancement factor for K_s is that for $\boldsymbol{q} = 0$ $(G(0) = 1)$, while that for $1/T_1$ is for $|\boldsymbol{q}| > 0$ $(G(|\boldsymbol{q}| > 0) < 1)$. Therefore, $\mathcal{K}(\alpha)$ declines from 1 to 0 as α varies from 0 to 1 [61].

If other effects which modify the Korringa product are largely absent [62], the experimental value of $\mathcal{K}(\alpha)$ can be used to estimate the electron-electron exchange (i.e. the value of α) in sp-band metals. A few examples are given in Table 2.2, where α values in the range of $0.2 \leq \alpha \leq 0.4$ are seen to be typical. Rather surprisingly, susceptibility enhancement effects are present to the degree of 25% to more than 50% in these *nearly free-electron metals*. The Korringa product is often used as a diagnostic tool to examine the nature of the T_1 process.

2.2.3 *NMR in the Superconducting State of Simple Metals*

We present here a brief account of the superconducting properties of sp-band metals. Historically, several fundamental issues were resolved with these systems following initial uncertainties regarding both theory and experiment. A clear and comprehensive review of early work was given by MacLaughlin [9]. Some elemental metals addressed in the early days were Hg(4.15K), βSn(3.74K), Pb(7.22K), and Al(1.18K), where their respective values of T_c are given in parentheses. They are all Type I superconductors, which means that their coherence lengths $\xi(0)$ are long compared with their London penetration depths $\lambda(0)$ [60].

With the announcement of BCS theory [39], there were a number of interesting issues to investigate. The principal ones for the NMR probe were (a) the expected collapse of the spin susceptibility at $T < T_c$ owing to the SC energy gap for quasiparticle excitations and (b) the existence of a singularity, which becomes the *coherence peak*, just below T_c in a plot of $1/T_1$ versus T. Because of the large superconducting diamagnetism which occurs below T_c, virtually the only means of measuring the relatively tiny Pauli susceptibility term is through the associated Knight shift K_s (Eq. 2.2.2). We shall discuss the superconducting shift measurements first.

The Knight Shift for T < T$_c$: Verifying the Yosida Function

Performing NMR measurements in the SC state of Type I superconductors was a particular experimental challenge, because of the limited magnetic field penetration depth $\lambda \sim 500$ Å for these systems. In order to get a reasonably uniform internal field for NMR studies, experiments were performed on fine particles or thin films with at least one dimension which was small compared with λ. The resulting samples generally exhibited superconductivity with T_c near the bulk value and NMR lines with relatively little diamagnetic shift or inhomogeneous broadening. Particles with dimension $d \ll \lambda$ have greatly reduced diamagnetism in the SC state. Other than that, it was generally believed that these samples exhibited *bulk NMR properties*.

However, the peculiar results obtained at first for $K(T)$ versus T for $T < T_c$ demonstrated that the foregoing statements were not entirely correct. To see this clearly, we must first describe the results that were expected from the shift measurements, then what was actually found, and finally, how to account for the observed data. The BCS theory prescribed an energy gap $\Delta(T)$ above the superconducting ground state for quasi-particle excitations when $T < T_c$. As $\Delta(T)$ increases below T_c the spin susceptibility $\chi_{sp}(T)$ and $K_s(T) = \alpha_s \chi_{sp}(T)$ were expected to decline and asymptotically approach 0 as $T \to 0$. The theoretical functional form of $\chi_{sp}(T)$ for $T < T_c$ is known as the Yosida function [49].

Experimentally, however, large residual $T \to 0$ shift values were reported. For the ^{199}Hg NMR in SC Hg, for example, $K_s(T \to 0)/K(T > T_c) \simeq 2/3$ was found [48], with similar results for ^{119}Sn in βSn [59], ^{207}Pb in Pb [45], and ^{27}Al in Al [47]. These results caused great consternation and confusion, and it was a number of years before a satisfactory understanding was reached [50].

Spin-Orbit Scattering: The Ferrell–Anderson Picture

In the end, it was realized that the theoretical preparation was incomplete, and further developments with the *Al* metal samples showed that the earliest results were incorrect. We shall return to the *Al* story below. The problem with the Yosida/BCS calculation was that it had neglected spin-orbit coupling and its role in generating spin-flip scattering processes. The situation was eventually addressed and for the most part resolved in short papers by Ferrell [51] and Anderson [52], as well as in a longer paper by Abrikosov and Gorkov [53].

We consider briefly the physical picture presented in Anderson's treatment [52]. In the presence of spin-orbit coupling one must bear in mind that $\langle S_z \rangle$ for the electrons is no longer a good quantum number, and that the spin-up and spin-down states are mixed into eigenstates which no longer have any particular average value of spin. However, a very subtle point is this: In *clean, bulk regions of the pure metal* there is no spin-flip scattering even in the presence of strong spin-orbit coupling. Thus, the spin susceptibility will still vanish as $T \to 0$ in the SC state. However, what modifies this ideal result is a spin-flip scattering lifetime τ_{so} which is short enough that the quantity $\hbar/\Delta(0)\tau_{so}$ is no longer small. In other words the smearing of the eigenstate energies from τ_{so} becomes comparable with the SC energy gap. Spin-flip scattering occurs only where ordinary potential scattering is strong, e.g. near surfaces and near impurities in the bulk. Ironically, then, this effect will be important near surfaces where *rf* fields penetrate and NMR signals are excited, but not necessarily in bulk regions which cannot be probed.

Anderson treated the spin-orbit scattering problem using his theory of *dirty superconductors* [54]. His result for the $T \to 0$ susceptibility ratio is an expression which can be integrated numerically, from which the following limiting cases are given [52]:

$$\frac{\chi_{so}(T \ll T_c)}{\chi_{so}(T > T_c)} \approx \frac{1}{6}\frac{\hbar}{\Delta(0)\tau_{so}}, \quad \Delta(0) \gg \hbar/\tau_{so} \tag{2.2.10a}$$

$$\frac{\chi_{so}(T \ll T_c)}{\chi_{so}(T > T_c)} \approx 1 - 2\frac{\Delta(0)\tau_{so}}{\hbar}, \quad \Delta(0) \ll \hbar/\tau_{so}. \tag{2.2.10b}$$

Equations (2.2.10) show that when τ_{so} is long there is only a small residual value of $\chi_{so}(T \ll T_c)$, but when it is short, i.e. strong spin-orbit scattering near surfaces, $\chi_{so}(T \ll T_c)$ is only slightly perturbed from its normal state value. The latter case is often found for heavy metals such as Hg [48] and Pb [45], and was also found for βSn [46, 59]. These authors performed various doping experiments, which were also consistent with (2.2.10). Ferrell obtained a result similar to (2.2.10b) by means of a physical argument [51].

In a metal as light as elemental *Al* one might expect spin-orbit effects to be very small. Since colloids of *Al* are difficult to prepare, shift studies were performed on thin film samples. The first results gave, surprisingly, residual shift values of ~75% [47]. It appears that orientation of their multi-film samples was quite critical, as subsequent work [55] yielded a very small residual shift. Later measurements by Fine, et al., on a single thin-film sample [56] gave results in good agreement with the Yosida function [49]. These data are displayed in Fig. 2.6, finally giving verification to the BCS theory-based prediction.

T_1 Behavior Below T_c: The BCS Coherence Factor

The behavior of T_1 for an sp-band (Type I) superconductor according to the BCS theory was first investigated, theoretically and experimentally, by Hebel and Slichter [64]. We sketch the calculation of T_1 in the SC phase following the derivation in [9]. First, recall from Sect. 1.2.1 that the T_1 process consists of scattering fermions

Fig. 2.6 Data for the ^{27}Al Knight shift in Al metal from [56] are shown, plotted in units of the normal state shift as a function of T/T_c. The solid curve is the Yosida function, i.e. behavior expected on the BCS theory

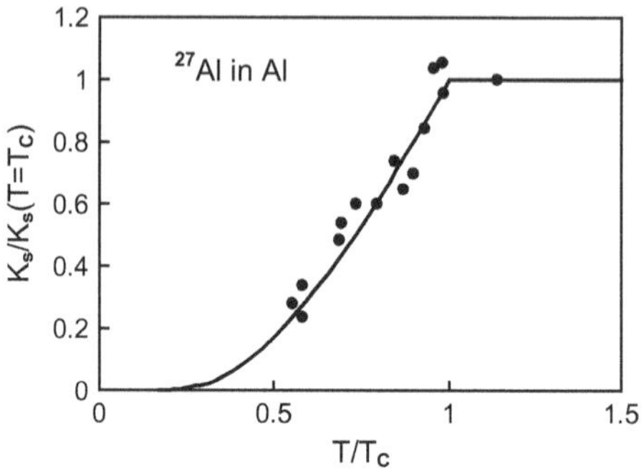

from an initial state $|\boldsymbol{k}\sigma\rangle$ to a final state $|\boldsymbol{k}' - \sigma\rangle$, while the nuclear spin goes from m to $m \pm 1$ so that angular momentum is conserved. This outline is maintained in the SC state, but other circumstances change drastically. For $T < T_c$ there is some fraction of fermions which are paired ($|\boldsymbol{k}\sigma\rangle$ and $|-\boldsymbol{k} - \sigma\rangle$) in the ground state and do not scatter. The quasiparticles not condensed into the ground state are thermally excited above the energy gap $\Delta(T)$ and are available to scatter as sketched above. The main differences are that, first, there appears in the summation in (2.2.5) a factor $C_{\pm}(\boldsymbol{k}, \boldsymbol{k}') = \frac{1}{2}[1 \pm (\Delta^2/E_{\boldsymbol{k}}E_{\boldsymbol{k}'})]$ known as the *coherence factor*, and second, the normal state density of states factors in the integrals leading to (2.2.7) are replaced by the BCS density of states $n_{BCS}(E) = n(E_F)E/(E^2 - \Delta^2)^{1/2}$, where $\Delta \leq E \leq \infty$. The coherence factor arises because of the pairing of electrons in the condensed phase, and is characteristic of many dynamic properties treated with BCS theory [19]. The choice of sign for $C_{\pm}(\boldsymbol{k}, \boldsymbol{k}')$ depends on whether the scattering perturbation is invariant under time reversal applied to the fermion bath. If it is, as in the case of phonons (e.g. ultrasonic attenuation) then $C_{-}(\boldsymbol{k}, \boldsymbol{k}')$ applies. If it is not invariant, such as for electromagnetic absorption and magnetic dipolar nuclear relaxation, then $C_{+}(\boldsymbol{k}, \boldsymbol{k}')$ applies.

For the superconducting nuclear relaxation rate $1/T_{1s}$ one finds, in units of the normal state rate $1/T_{1n}$,

$$\frac{T_{1n}}{T_{1s}} = \frac{2}{k_B T}\int_{\Delta}^{\infty} \frac{dE\,(EE' + \Delta^2)\, f(E)\,(1 - f(E'))}{[(E^2 - \Delta^2)(E'^2 - \Delta^2)]^{1/2}}, \tag{2.2.11}$$

where $E' = E \pm \hbar\gamma_n H$. Below T_c the excitation probability is the Fermi function $f(E_g) = [\exp(\beta E_g) + 1]^{-1}$ of the gapped energy $E_g = (E^2 + \Delta^2)^{1/2}$. Since the nuclear Zeeman energy $\hbar\gamma_n H$ is minuscule on the scale of the quasiparticle energy, there is in fact a logarithmic singularity in (2.2.11) for the case of the coherence factor $C_{+}(E, E')$. The most satisfactory explanation for how this singularity is "tamed" is by means of anisotropy of the gap energy $\Delta(T)$ [9, 65, 66]. The nuclear Zeeman energy will be neglected henceforward.

The behavior of T_{1n}/T_{1s} in (2.2.11) may be divided into two regions, $k_B T \ll \Delta(T)$ and $k_B T > \Delta(T)$. For $k_B T \ll \Delta(T)$ the integral is dominated by the peak in

$n_{BCS}(E)$ near the gap, which becomes very nearly constant $\Delta(T) \approx \Delta(0)$. In this limit the factor $f(E)(1 - f(E)) \approx \exp(-\beta\Delta(0))$, so that $1/T_{1s}$ decays exponentially at low T. In the other limit $k_B T > \Delta(T)$ (T near T_c), the factor $f(E)(1 - f(E))$ will have nearly its normal-state value, and the singularity in (2.2.11) has the potential to elevate T_{1n}/T_{1s} to values considerably in excess of 1. This is the source of the well-known BCS *coherence peak* in $1/T_1(T)$ at $T < T_c$. This peak was first measured on ^{27}Al in Al metal and identified by Hebel and Slichter [64], establishing a major milestone in experimental confirmation of the BCS theory.

It is important to note that for measurements such as ultrasonic attenuation the appropriate coherence factor to use is $C_-(\boldsymbol{k}, \boldsymbol{k}')$, whereupon the factor $(EE' + \Delta^2)$ in the numerator of (2.2.11) is replaced by $(EE' - \Delta^2)$. This effectively cancels the singularity at $E \simeq \Delta$. As a result, the ultrasonic attentuation coefficient was expected to fall abruptly below T_c, as had been confirmed experimentally [67]. This striking contrast between results obtained with probes having different symmetry properties was a very important confirmation of the correctness of the BCS picture.

Following the Hebel-Slichter results, Masuda and Redfield [65] reported further T_1 measurements on ^{27}Al in Al metal taken at temperatures ranging from above to well below T_c. Their results (Fig. 2.7) offered a full panorama of the BCS-theoretical behavior, with a coherence peak (in $1/T_1$) rising above the normal state Korringa curve by a factor ~ 2, then descending into the superconducting state in a clearly exponential fashion. These data were obtained with a field-switching technique similar to that used in [64], the main difference being that pumped ^3He cooling allowed them to extend the measurements to 0.35 K, i.e. far below $T_c = 1.18$ K for Al metal.

Analyzing T_1 Data in the Superconducting State

Since the formulation of (2.2.11) has been widely employed to analyze SC-state data on high-T_c materials as well as all manner of unconventional superconductors, we recount here the analytic procedure used to fit the Al data in some detail [9, 65]. These results are referred to in other sections of the book as well.

In order to control the singularity in (2.2.11), Hebel and Slichter [64] postulated a broadening of the electronic energy levels. While such a device enabled them to obtain a reasonable result for the coherence peak, it lacked a clear physical basis. Masuda and Redfield, on the other hand, suggested that the observed peak was conditioned by a mild anisotropy of the gap parameter $\Delta(T)$. Not only does this give a better fit to the peak data, but it also has a clear physical origin in the BCS equations. There, the gap parameter is formally a function of $\boldsymbol{k}$, and the $\boldsymbol{k}$-independent gap parameter is just a convenient approximation used to solve the equations [66]. Ample theoretical evidence has been reported for gap anisotropy effects from van Hove singularities and other causes [68].

To illustrate the T_1 analysis with (2.2.11) we summarize briefly the development in [9] and [65], and refer the reader to these sources for a more detailed discussion. The procedure is to incorportate the gap anisotropy into the BCS density of states factors in (2.2.11) at the point where sums on $\boldsymbol{k}$ and $\boldsymbol{k}'$ are converted to energy integrals. We therefore back up a step from (2.2.11) and rewrite it as

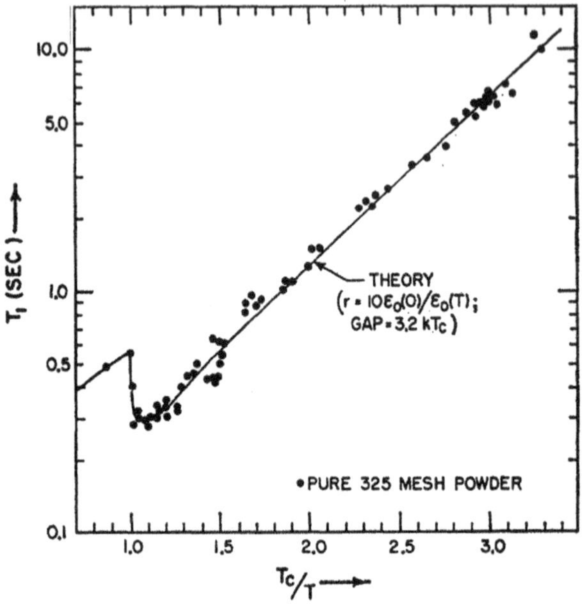

Fig. 2.7 A semilog plot of T_1 data for the superconducting state of Al metal versus T_c/T (dots) from [65]. Also shown is their fit of the data to the BCS theory (solid line), where the coherence peak (i.e. T_1 *minimum*) amplitude was adjusted using a distribution of energy gap values attributed to anisotropy. The low-temperature energy gap value is adjusted to the slope of the data, yielding $\Delta_0 = 3.2$ in reasonable accord the the BCS weak-coupling value of $3.52\,\mathrm{k_B T_c}$

$$\frac{T_{1n}}{T_{1s}} = \frac{2}{k_B T}\int_0^\infty [N_s(E)^2 + M_s(E)^2]\, f(E)\,(1 - f(E))dE, \tag{2.2.12}$$

where

$$N_s(E) = Re \int_{a_1}^{a_2} \frac{E\ P(a)da}{[E^2 - \Delta(T)(1+a)^2]^{1/2}}, \tag{2.2.13a}$$

$$\text{and } M_s(E) = Re \int_{a_1}^{a_2} \frac{\Delta(T)(1+a)\ P(a)da}{[E^2 - \Delta^2(1+a)^2]^{1/2}}. \tag{2.2.13b}$$

In (2.2.13) the variation of Δ_k over the Fermi surface is represented as a distribution $P(a)$ extending over $a_1 \le a \le a_2$, where $\Delta_k \Rightarrow \Delta(T)(1+a)$ and $\int a\ P(a)da = 0$ by definition. As noted in [9], because the averaging over a is followed by integration over E, the precise shape of $P(a)$ is functionally immaterial to the final results. In fitting the T_1 data of Fig. 2.7, $\langle a^2 \rangle$ determines the depth of the T_1 minimum and $\Delta(0)$ fixes the slope of the exponential tail. The fit shown corresponds to $2\Delta(0) = (3.2 \pm 0.2)\,\mathrm{k_B T_c}$ and $\langle a^2 \rangle^{1/2} \simeq 0.05$, in reasonable agreement with the BCS weak-coupling theoretical value $2\Delta(0) = 3.52\,\mathrm{k_B T_c}$. Comparable gap values are given by other probes [9].

There has also been extensive experimentation with doping effects on the SC T_1 behavior, yielding some interesting effects. For example, it was found that shortening the mean free path l of the quasi-particles led to even larger coherence peaks in the T_1 data for Al [69] and In-based [70] dilute alloys. This effect can be rationalized in terms of the theory of "dirty superconductors" [54], where potential scattering caused by impurities can become strong enough to cause effective mixing between Bloch functions on different parts of the Fermi surface. The result is a smoothing out of intrinsic gap anisotropy. Data for the Al-based systems was in good agreement with a theory of this effect [71], but for In-based alloys, there was only agreement for $l > \xi_0$, where ξ_0 is the coherence length. It was suggested that *gap inhomogeneity* might be suppressing the coherence peak in such a case. However, no further resolution of this point has apparently been reported.

As we shall see in the next subsection, there are a number of d-electron superconductors where T_1 has no coherence peak for reasons which are also not clear. Moreover, among high-T_c systems, suppression of the coherence peak in T_1 is universal, as is also the case for many *unconventional superconductors*. We defer further discussion of this point to subsections below, where fuller descriptions of the systems concerned are presented.

2.3 Static and Dynamic Magnetism in d-Band Metals

In this subsection we review the additional complexities which arise when a conductive solid has an open d-shell in addition to its sp-band effects. The d-electrons in solids are typically more tightly bound to their host ion, interacting more weakly with neighbors. As a result they have narrower, more highly structured energy bands with substantially higher densities of states at their Fermi surfaces. Moreover, they are subject to a *crystalline electric field* (CEF) which results from bonding and proximity to neighboring ions, which breaks the full rotational potential symmetry enjoyed by a free ion. As a result, the d-electron angular momentum "always" has a zero expectation value in the ground state. On the other hand, the greater densities of states of d-bands, when combined with the electron-electron exchange effects [57] which came quite clearly to light in the Moriya treatment of enhanced NMR shifts and relaxation [58], lead to strongly enhanced susceptibilities and even to permanently magnetized ground states in a number of pure metals (e.g. Fe, Co, and Ni in the $3d$ series). Moreover, ions with d-electrons as dilute impurities or dopants frequently possess a localized magnetic moment. As an example, the high-T_c materials do not exhibit localized magnetism *per se*, but the Mott insulators that they are derived from exhibit local-moment antiferromagnetism at low doping levels. When the d-holes become itinerant through doping, a localized d-electron spin Hamiltonian continues to be a reliable model for properties such as hyperfine (HF) effects, spin and Van Vleck paramagnetism, as well as hybridization effects witn p-electron neighbors.

Special Features of NMR in d-Band Metals and Superconductors

Keeping superconductivity in mind, we shall restrict our attention to d-electron systems which do not order magnetically. We shall discuss, as an example, the behavior of the V_3X series of compounds, which are superconductors as well. There is a wealth of interesting NMR phenomena which arise in such systems, including:

- Prominence of the *core-polarization* (CP) d-spin HF field in studies of d-band magnetism. Since d-electrons have no contact interaction, this isotropic, negative HF coupling, which was identified by Heine [35] and by Wood and Pratt [35, 77], takes its place. The core polarization plays a major role, since dipolar interactions are often smaller and are cancelled by cubic symmetry.
- The greatly enhanced importance of orbital effects, both the static (Van Vleck) shift and susceptibility and the robust orbital T_1 mechanism. The d-band orbital susceptibility was first formulated by Kubo and Obata [36]. Its contribution to the NMR shift was identified by Clogston et al. [37] in connection with the V_3X compounds, where it had a profound effect on the interpretation of superconducting shift data.
- The latter authors [38] also pioneered the use of the $K-\chi$ plot as a means of analyzing NMR shifts and their corresponding susceptibility components in d-band metals. This technique continues to be used very widely and has led to great insight into the magnetic infrastructure of metals and intermetallic compounds. We present their method here.
- When d-electron metals are superconductive, they are nearly always Type-II superconductors. Thus, NMR studies are very often conducted in the *mixed state*, i.e. in the presence of a large magnetic field and a vortex lattice. The field penetrates throughout the sample, dividing itself into as many flux bundles as possible, which occurs when each vortex contains just one fluxon. We shall review results on NMR shifts, lineshapes, and on T_1 studies as well. These effects are directly relevant to superconductive high-T_c NMR studies, since they are largely conducted under the same conditions.

2.3.1 The d-Electron HF Interactions

In general there are two types of d-electron HF interactions, namely spin and orbital, the latter arising from the first term in (2.2.1), and the former from the second two terms. In d-band metals there are also multiple bands with their own Fermi surface density of states and corresponding HF interaction terms. These terms will make additional contributions to the NMR shift and relaxation rates, rendering the task of analyzing experimental data for these parameters a bit more complex. By and large, however, d-electron bands are somewhat more "lively" than their rather prosaic s and p counterparts, and tend to dominate the nuclear spin properties. In high-T_c materials, depending on your point of view, there is scant evidence for actual sp-bands *per se*, so the d-electron HF effects are, by any measure, the dominant ones.

Anisotropic d-Spin Susceptibility and HF Tensor

Allowing for anisotropic d-spin HF coupling, we then take

$$\mathcal{H}_{ds} = \sum_{\beta=x,y,z} A_{d\beta} I_\beta S_\beta \text{ with } \chi_{d\beta} = \frac{N_A g_\beta \mu_B \langle S_\beta \rangle}{H} . \qquad (2.3.1)$$

The d spin shift coefficient is then defined as

$$\alpha_{d\beta} = \frac{K_{d\beta}}{\chi_{d\beta}} = \frac{A_{d\beta}}{N_A \gamma \hbar g_\beta \mu_B} . \qquad (2.3.2)$$

For ordinary 3d metals we shall take $g_\beta \sim 2$, whereupon $\chi_{d\beta}$ becomes isotropic. However, these definitions carry over into the discussion of cuprates in Chap. 3, where there is evidence for substantial g-shifts. We also note here that the HF coefficient can be quoted as simply $A_{d\beta}$ in energy units, as $A_{d\beta}/\hbar$ in units of s^{-1}, as $A_{d\beta}/\gamma\hbar$ in kG/spin, or as $A_{d\beta}/\gamma\hbar g_\beta$ in units of kG/μ_B. Although it may be somewhat confusing, one finds all of these units in use in the high-T_c literature. The d-spin HF field is also known as the core-polarization HF field (hence $A_d \to A_{cp}$), which we now discuss.

Core-Polarization Hyperfine Coupling: The Mechanism

Since d-electron wavefunctions vanish at the site of the nucleus, they possess no direct contact HF field from the final term in (2.2.1). However, they do activate a contact HF term indirectly in such a way as to generate an effectively isotropic HF coupling term with the nuclear spins. This mechanism came to light with the work of Heine [35] and of Wood and Pratt [77]. Detailed calculations have been given by Freeman and Watson [72]. The mechanism of the *core polarization HF field* is a simple consequence of the Pauli exclusion principle, and works as follows. In d-electron atoms there are two or more filled s-shells which are magnetically inert. Nonetheless, these filled s-shells undergo repulsive coulomb interactions with the Fermi-surface d-electrons which, generally speaking, lie outside of them spatially. Thus, coulomb repulsion drives the s-shell electron orbitals closer to the nucleus. This repulsive effect is, however, dependent on spin. Consider the majority d-spin moment to point up. The up-moment s-shell electrons are then known to avoid close proximity with the up-moment d-electrons in a Pauli-exclusion effect known as the *Fermi hole*. The down-moment s-electrons have no such effect. With the same wavefunction, then, the down-moment s-electrons will have a higher coulomb repulsion energy. For that reason, a down-moment s-shell orbital will "shrink" toward smaller radii relative to the up-moment s-shell orbital, creating a *net down moment of s-electron density at the nucleus*, i.e. a net negative contact HF interaction. The differential orbital distortion effect is small, but the core s-shell HF couplings are enormous, so that there is a substantial net HF coupling from this effect. For 3d-electrons the resulting Hf field is $\sim -125\,\text{kG}/\mu_B$, increasing gradually across the 3d series. For the 4d and 5d shells the effect is generally larger [72]. As an example, we note that the $5\,\mu_B$ S-state moment of Mn^{2+} in MnF_2 generates a HF field $\simeq -127\,\text{kOe}/\mu_B$ at the ^{55}Mn nucleus. This compound has purely 3d-spin magnetism.

2.3.2 Orbital Shift and Susceptibility

The orbital term in (2.2.1) comes into play for both the NMR shift and T_1. The orbital shift is driven by the Van Vleck orbital susceptibility, which was first formulated for band electrons in a simple model by Kubo and Obata [36]. The susceptibility was formulated in a tight-binding approximation, yielding the following expression for the β-axis susceptibility,

$$\chi_{orb} = N_A \Omega \mu_B^2 \int \frac{d\mathbf{k}}{(2\pi)^3} \sum_{nn'} \frac{f(E_{n\mathbf{k}}) - f(E_{n'\mathbf{k}})}{E_{n'\mathbf{k}} - E_{n\mathbf{k}}} \langle n\mathbf{k}|l_\beta|n'\mathbf{k}\rangle \langle n'\mathbf{k}|l_\beta|n\mathbf{k}\rangle\,, \quad (2.3.3)$$

where N_A is Avogadro's number, Ω is the atomic volume, and the summing indices n and n' represent the five d-orbital subbands as well as the spin quantum number. The quantity $f(E_{n\mathbf{k}}) = [\exp(\beta(E_{n\mathbf{k}} - E_F)) + 1]^{-1}$ is the Fermi occupation function. χ_{orb} is written here as a molar susceptibility.

Clogston et al. [76] have derived a companion expression to (2.3.3), which gives the induced orbital hyperfine field at the nucleus when the effects of the spin-orbit coupling and certain other terms in the general expression, deemed to be small, are neglected. The simplified expression is

$$H_{orb} = \Omega \mu_B^2 \int \frac{d\mathbf{k}}{(2\pi)^3} \sum_{nn'} \frac{f(E_{n\mathbf{k}}) - f(E_{n'\mathbf{k}})}{E_{n'\mathbf{k}} - E_{n\mathbf{k}}} \langle n\mathbf{k}|2l_\beta/r^3|n'\mathbf{k}\rangle \langle n'\mathbf{k}|H\, l_\beta|n\mathbf{k}\rangle, \quad (2.3.4)$$

where the field H is taken to be directed along the β axis. If we make the assumption that the radial functions of all the d-orbitals in the band are approximately the same, then we may extract $\langle 1/r^3\rangle$ from the latter expression, finding

$$K_{orb} = [2\langle 1/r^3\rangle / N_A]\, \chi_{orb}\,, \quad (2.3.5)$$

where the quantity in brackets is identified as the orbital shift coefficient α_{orb}. When using calculated ionic values for $\langle 1/r^3\rangle$ [78], the foregoing authors suggest that a correction factor $\xi = \langle 1/r^3\rangle_{metal}/\langle 1/r^3\rangle_{ion} \simeq 0.75$ be invoked for realistic estimates of the metallic shift coefficient α_{orb}.

It should be noted that the foregoing expressions for the orbital shift and susceptibility were derived with the applied field assumed to be lying along a particular spatial axis. However, these calculations were originally presented in application to cubic d-band metals, for which no appreciable spatial anisotropy was expected to occur. In applications to high-T_c materials with their nearly 2D structures, these expressions are easily adapted to reflect the expected anisotropy. In such applications, we note that the orbital shift coefficient in (2.3.2) will be generally assumed to be isotropic.

Analysis of NMR Shift and Susceptibility for d-Band Metals: The (K, χ) Plot With the foregoing additions to our list, we are now in a position to formulate a complete model of the shift and susceptibility of a d-band metal. The following model would apply to a pure metal from the 3d, 4d, or 5d transition series, but also is intended to represent a binary or ternary intermetallic compound as long as its Fermi surface can be modeled with two bands, an sp-band which belongs mostly to the transition metal atoms and a d-band which belongs exclusively to the transition metal atoms. The p-orbitals on the ligands are assumed to be strongly mixed into the d-band, while the s-orbitals are not, i.e. are presumably nearly filled [38]. We then write for the susceptibility,

$$\chi_{tot} = \chi_{dia} + \chi_s + \chi_d(T) + \chi_{orb} , \tag{2.3.6}$$

where χ_s and $\chi_d(T)$ are the spin-paramagnetic terms, χ_{orb} is the d-electron orbital term given by (2.3.3), and χ_{dia} is the core diamagnetic term. Among these, χ_s is quite small, though its associated Knight shift may not be, so that any Landau diamagnetism is very small, indeed. Neither the latter nor χ_{dia} has any appreciable NMR shift effect [76]. Molar susceptibilities are used throughout this volume [79].

The NMR shift of the transition ion, then, has only three significant terms [76],

$$K(T) = K_s + K_d(T) + K_{orb} = \alpha_s\chi_s + \alpha_d\chi_d(T) + \alpha_{orb}\chi_{orb} , \tag{2.3.7}$$

in conventional notation for the shift coefficients. Typically, the only temperature dependence originates in $\chi_d(T)$. Therefore an experimental plot of $K(T)$ versus $\chi_{tot}(T)$ with T as the implicit variable should yield a straight line of slope α_d. An experimental K versus χ plot is just the beginning of a full (K, χ) analysis of the various terms in (2.3.6) and (2.3.7). Such an analysis for the case of Pt metal was presented in [76], with the plotted result shown here in Fig. 2.8. We give a brief account of the analytical procedure.

First, any $K - \chi$ analysis must begin at the origin, and several parameters need to be estimated in order to proceed. We refer the reader to the original paper for full details. The first step is to plot $\chi_{dia} \simeq -28 \times 10^{-6}$ emu/mol at $K = 0$ (point 1), since there is no appreciable diamagnetic shift [76]. This value of χ_{dia} is taken from the literature [80]. From point 1 we construct a line of slope α_s to extend over the interval $\Delta\chi = \chi_s$. For this step, α_s is obtained from HF structure parameters derived from optical data, yielding $\alpha_s \simeq 2.12 \times 10^3$ (emu/mol)$^{-1}$. χ_s is estimated using a free-electron model, where the number $n_{s,d}$ of (s, d) electrons satisfies $n_s + n_d = 10$. With a parabolic model for the $5d$-band a value of $n_s = 10 - n_d = 0.2$ was obtained, giving $\chi_s \simeq 4.8 \times 10^{-6}$ (emu/mol). The s-band line in Fig. 2.8 then extends from point 1 to point 2. Next, we construct a line from point 2 with a slope equal to the estimated value of $\alpha_{orb} = 2\langle 1/r^3\rangle/N_A$. The ionic value of $\langle 1/r^3\rangle$ has been estimated to be 9.5 a.u. [79] from HF spectroscopic data, leading to the value $\alpha_{orb} \simeq 264$ (emu/mol)$^{-1}$, including a relativistic correction [76]. Drawing a line with slope α_{orb} from point 2, the intersection of that line with the line through the data (point 3) then partitions the remaining susceptibility into orbital and d-spin contributions as shown.

In the case of Pt metal, χ_{tot} is dominated by $\chi_d(T)$, and χ_{orb} is a relatively small fraction ($\simeq$ 15%) of χ_{tot}. In cases where χ_{orb} is a signficant fraction of the total, its temperature dependence may be a big enough effect so that a simple (K, χ) plot

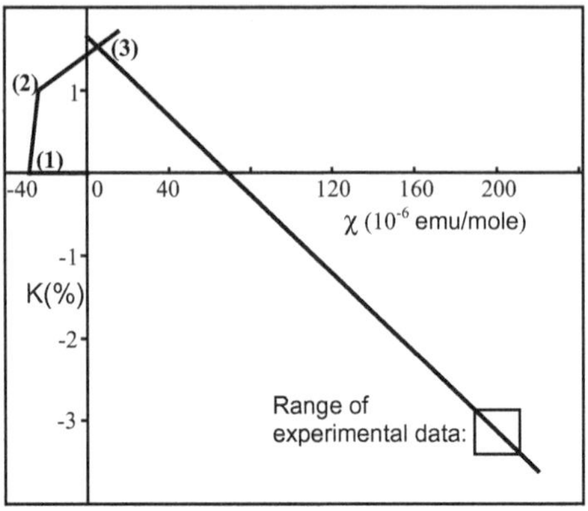

Fig. 2.8 A K-χ plot is shown for ^{195}Pt in Pt metal from [76]. Plotting successive contributions to (K,χ) carries one from the origin to point 1 (χ_{dia}), then from point 1 to point 2 (K_s, χ_s), and finally from point 2 with an estimated slope α_{orb}. The intersection at point 3 with the line through the experimental data completes the partition of the susceptibility. A detailed discussion of the analysis is given in the text

is no longer reliable [76]. Nonetheless, the (K, χ) method has been widely used to analyze NMR shift data with experimental susceptibilities. For example, it was used to determine that substantial χ_{orb} terms were present in the V_3X's and in V metal [37]. We reproduce in Table 2.3 the table of results from [37].

In this table, values of χ_{orb} from (K, χ) analyses are subtracted from experimental susceptibilities to estimate χ_d, which is given as $\chi_{expt} - \chi_{orb}$. The latter values are then to be compared with d-spin susceptibilities estimated from the specific heat coefficient γ, where $\chi_{sp.ht.} \simeq 3\mu_B^2\gamma/\pi^2 k_B^2$. A simple theory gives electron-phonon enhancement of γ by a factor $[1 + N(0)\mathcal{V}]$. $\mathcal{V}$ is the BCS electron-phonon interaction parameter. Since χ_d is not so enhanced [84], one may estimate $\mathcal{V}N(0) \simeq \chi_{sp.ht.}/\chi_d - 1$. In the last two columns of Table 2.3, values of $N(0)\mathcal{V}$ deduced in that way are compared with those derived from T_c using the BCS formula $N(0)\mathcal{V} = [ln(0.855\,\theta_D/T_c)]^{-1}$. Considering the difficulty of estimating the relevant

Table 2.3 Table of susceptibility components and BCS parameters for V_3X compounds and V metal. The estimates for χ_{orb} were determined using (K, χ) plot analyses. The values of $\chi_d \simeq \chi_{expt} - \chi_{orb}$ so deduced are checked using density-of-states estimates from specific heat data as corrected for electron-phonon enhancement by means of BCS theory. See text for detailed explanation. All susceptibilities are in units of 10^{-4} emu/mol

	$R \equiv \chi_d(0)/\chi_d(T_c)$	χ_{expt}	χ_{orb}	$\chi_{sp.ht.}$	$\chi_{expt} - \chi_{orb}$	$N(0)\mathcal{V}$ $(Sp.Ht.)$	$N(0)\mathcal{V}$ (BCS)
V_3Ga	$0 \leq R \leq 0.25$	17.9	5.2–7.5	14.0	12.7–10.4	0.13–0.39	0.35
V_3Si	$0 \leq R \leq 0.25$	15.7	5.2–7.5	10.7	10.5–8.2	0.09–0.35	0.35
V		3.08	2.11	1.22	0.971		0.244

parameters, this analysis gives a nicely consistent picture in which it is established that (a) χ_d at $T \ll T_c$ is a small fraction of its normal-state value, and (b), values of χ_d consistent with electron-phonon enhancement of γ are obtained. Note that $\chi_s \simeq 0.3$ (units of Table 2.3) has been ignored and is essentially negligible.

2.3.3 Spin-Lattice Relaxation Effects for d-Band Electrons

An open d-band brings added richness—and complications—to the T_1 process as well, contributing a total of three additional terms to the T_1 rate. These are the d-spin core polarization (CP), the d-spin dipolar, and the d-orbital terms. The d-orbital relaxation process was first discussed by Obata [81]; we shall spell out the results of that calculation below. The general treatment of the CP relaxation process, on the other hand, is an enterprise of great complexity. Under a set of straightforward restrictions, Yafet and Jaccarino have presented a very clear discussion of this problem [82], which has been useful under a variety of circumstances. Here we discuss briefly the approximations employed and give the results.

The stated approximations are as follows:

- Spin-orbit coupling is disregarded.
- The unperturbed states (i.e. without HF terms) are taken to be Hartree-Fock solutions of the simplified Hamiltonian.
- Exchange intergrals responsible for the CP effect are treated in first order.
- Exchange effects between conduction band orbitals are neglected.
- Conduction band states are self-consistent Bloch states, which are admixtures of s, p, and d orbitals.
- The conduction band d-states are treated in a tight-binding approximation.
- Finally, consideration is limited to the case of a cubic metal. As a result, the cross terms between sp and d-orbital matrix elements can be shown to vanish.

The resulting expressions for the shift and relaxation rate are:

$$K = \frac{4\pi}{3}(\gamma_e\hbar)^2[\langle\phi_s^2(0)\rangle(1 - F_d) - |\phi_{cp}(0)|^2 F_d]n(E_F) \tag{2.3.8}$$

and

$$\frac{1}{T_1T} = \frac{4\pi k_B}{\hbar}\left[\frac{4\pi}{3}\gamma_n\gamma_e\hbar^2\right]^2\left[\langle\phi_s^2(0)\rangle(1 - F_d)^2 + |\phi_{cp}(0)|^4 F_d^2(\frac{1}{3}f^2 + \frac{1}{2}(1 - f)^2)\right]n(E_F)^2, \tag{2.3.9}$$

where F_d is the fraction of d-character of all the states at the Fermi surface, and $n(E_F)$ is the total density of states (of all "bands") at the Fermi surface. Thus, even if there is appreciable s-d mixing in the Bloch functions the contributions to shift and relaxation will occur separately as shown. CP terms are only kept for the d-orbitals, even though they may occur in the calculation for the s and p orbitals. The reason for this is that the contact terms dominate the sp regions of the Fermi surface, and any CP terms are neglected for those states as small corrections.

The parameter f gives the relative weight of Γ_5 character for d-orbitals at the Fermi surface, while $1-f$ gives the proportion of Γ_3 character [83]. It is seen that the sp-contact and d-CP portions of the formulas for K and $1/T_1T$ obey separate Korringa-like relations (see (2.2.8)), where the Korringa relation for the CP terms has a reduction factor $q(f) = \frac{1}{3}f^2 + \frac{1}{2}(1-f)^2$, i.e.

$$\frac{1}{T_{1cp}T} = \frac{4\pi k_B}{\hbar}\left[\frac{4\pi}{3}\gamma_n\gamma_e\hbar^2|\phi_{cp}(0)|^2\right]^2 n_d(E_F)^2 q(f) = \frac{4\pi k_B\gamma_n^2}{\hbar\gamma_e^2}K_{cp}^2 q(f), \tag{2.3.10}$$

where the second equality shows the modified Korringa relation. $q(f)$ is essentially a degeneracy factor, having a value 1/2 for the two Γ_5 orbitals ($f=1$), 1/3 for the three Γ_3 orbitals ($f=0$), and 1/5 when all five d-orbitals are equally weighted ($f=0.6$). Thus the five d-orbitals contribute independently to $1/T_{1CP}$, very much like the fluctuations of separate magnetically-active neighbors contribute to the T_1 process of a ligand nucleus [62]. The difficulty in utilizing (2.3.10) is that the parameter f for d-band metals is generally unknown.

The orbital T_1 process was first discussed in detail by Obata [81], treating both orbital and dipolar relaxation rates in the same calculation. As with the CP calculation above, spin-orbit coupling is neglected, the d-orbitals are treated in a tight-binding approximation, and the system considered is a cubic metal. There are no cross terms between orbital and spin-scattering relaxation processes. It was noted in [82] that the dipolar process has interference with the contact and CP processes, but if the states at the Fermi surface can be separated into a d-band and an sp-band, then the interference terms vanish. For a cubic metal, both the orbital and dipolar relaxation rates are isotropic, but both depend on the symmetry parameter f, according to

$$\frac{1}{T_{1orb}T}, \left[\frac{1}{T_{1dip}T}\right] = \frac{4\pi k_B}{\hbar}[\gamma_n\hbar H_{orb}]^2 n_d(E_F)^2\, p(f), [d(f)], \tag{2.3.11}$$

where $H_{orb} = 2\mu_B\langle 1/r^3\rangle$, $p(f) = \frac{2}{3}f[2-\frac{5}{3}f]$, and $d(f) = \frac{5}{49}[\frac{1}{3}f^2 + \frac{2}{5}(1-f)]$.

Comparison of the dependences of $q(f)$, $p(f)$ and $d(f)$ on f is shown in Fig. 2.9. The orbital relaxation rate ($p(f)$) is seen to vary from zero, when the Γ_5 orbitals alone are occupied at the Fermi surface, to a maximum when all five orbitals are equally occupied, since the orbital matrix elements between the two symmetries are the largest ones. The dipolar relaxation rate ($d(f)$) varies only slightly across the full range of f, but it is only about 10% of the average value of $p(f)$. Thus, the dipolar relaxation is only significant when $p(f)$ is near zero.

The complete list of magnetic relaxation processes for d-band metals is then

$$1/T_1 = 1/T_{1s} + 1/T_{1cp} + 1/T_{1orb} + 1/T_{1dip}. \tag{2.3.12}$$

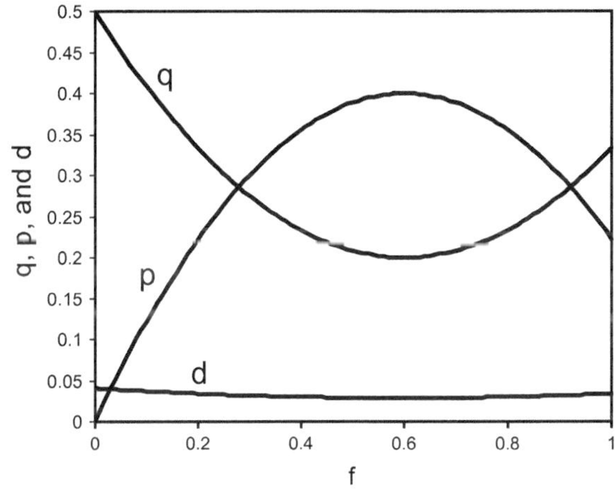

Fig. 2.9 The symmetry factors $q(f)$, $p(f)$, and $d(f)$ pertaining to CP, orbital, and dipolar relaxation rates discussed in the text, respectively, are plotted as a function of the fraction f of Γ_5 character in the relevant d-band orbitals at the Fermi surface. The point $f = 0.6$ represents equal weight of all five d-orbitals and is the maximum and minimum point of $p(f)$ and $q(f)$, respectively. In many cases the dipolar rate is negligible, as it is only $\sim$ 9% of the average value of the closely related orbital rate

We illustrate the analysis of experimental T_1 rates into these different contributions with calculations given by [82] for d-band metals V and Nb. In stating these results we define $H_s = (8/3)\pi\mu_B\langle\phi_s^2(0)\rangle$ and $H_{cp} = (8/3)\pi\mu_B|\phi_{cp}(0)|^2$, whereupon (2.3.9) becomes

$$\frac{1}{T_{1s,cp}T} = \frac{4\pi k_B}{\hbar}[\gamma_n\hbar H_{s,cp}]^2\, n_{s,d}^2\,[1, q(f)] \tag{2.3.13}$$

for the s-band and CP contributions. The parameters $H_{s,cp}$ are the HF fields per unit of spin from these sources. The T_1 contributions from the three main sources have been estimated [82] for V and Nb metals as listed in Table 2.4. For this purpose, H_s has been estimated using data from the literature [82], and H_{cp} and H_{orb} have

Table 2.4 Table of HF parameters and specific heat densities of states leading to estimates of $(T_1T)^{-1}$ for elemental V and Nb using the formulas given in the text. The total relaxation rates are stated for $f = 0.6$, but would not change significantly for other values of f. The theoretical estimate undoubtedly exceeds the experimental numbers because of the electron-phonon enhancement of the specific heat density of states, which does not affect nuclear relaxation rates. See text for data sources and additional discussion. HF fields are given in units of 10^6 Oe, densities of states in states/eV-atom, and relaxation figures in $(sK)^{-1}$

	H_s	H_{cp}	H_{orb}	$n_s(E_F)$	$n_d(E_F)$	R_s	R_{cp}	R_{1orb}	$(T_{1calc}kT)^{-1}$	$(T_{1exp}T)^{-1}$
V	1.12	−0.117	0.19	0.12	1.84	0.642	1.65	4.35	2.86	1.27
Nb	2.48	−0.21	0.285	0.14	1.66	3.76	3.74	6.9	7.55	2.8

been derived from Hartree-Fock calculations given by Freeman and Watson [78]. The total densities of states for these elements are taken from specific heat measurements, where the s-band components are estimated assuming for simplicity one s-electron per atom. The quantities R_s, R_{cp}, R_{orb} are defined to be the expressions given above for $1/T_1T$ from these sources *without the symmetry factors* $[1, q(f), p(f)]$. The experimental values of $(T_{1exp}T)^{-1}$ are data reported by Butterworth [85].

What this table shows is that the CP and orbital mechanisms of relaxation are important. For the lighter metals, where the s-band contribution is smaller, these d-electron terms may even be the dominant ones. On the other hand, we see that the specific heat density of states clearly overestimates the relaxation rates, pointing to a substantial electron-phonon enhancement effect for V and Nb metals (which are also superconductors). In [9] a quantitative discussion of the latter effect is given, finding reasonable agreement between the nuclear relaxation data and estimates from other measurements, somewhat like the comparisons in Table 2.3. As the details go beyond the intended scope of this discussion, the reader is referred to [9] and references therein for further discussion.

This completes our discussion of normal-state NMR shift and relaxation effects in transition metals. The mechanisms in (2.3.7) and (2.3.12) are the foundation of effects to be found in high-T_c materials, though the details remain to be worked out. We next consider the effects of Type II superconductivity in the d-band metals.

2.3.4 NMR Studies of 3d Metals as Type-II Superconductors

The type-II superconductivity of d-band metals and intermetallic compounds presents a rather different set of problems from those of the type-I systems discussed in Sect. 2.2.2. First, type-II systems exhibit great complexity of behavior in a magnetic field, showing three phases as a function of the magnitude of the applied field for $T < T_c$ and exhibiting two critical fields, H_{c1} and H_{c2}. Below H_{c1} there is the typical Meissner exclusion of all magnetic flux except for the usual penetration depth λ.

Between H_{c1} and H_{c2} there is a state of essentially complete, but inhomogeneous field penetration in the form of superconducting current vortices which allow a single fluxon of field to exist in a region of normal (i.e. nonsuperconducting) metal with a diameter $\sim \xi_0$, the Pippard coherence length. Of course, the magnetic field of a fluxon penetrates the surrounding material to a depth of $\sim \lambda$. Each fluxon comprises one flux quantum $\Phi_0 = hc/2e = 2.07 \times 10^{-7}\,\mathrm{G\,cm^2}$.

Because of the accommodation of type-II superconductivity to applied fields through the penetration of vortices, the superconducting state can persist up to fields much higher than the thermodynamic critical field H_c.[9] One can make a rough estimate of H_{c2} using a picture of vortices with normal cores of approximate radius ξ_0. In the mixed state with field H ($T \ll T_c$) the density of vortices is H/Φ_0. The fraction

[9] H_c is defined as the field giving the field energy density $H^2/8\pi$ equal to the condensation energy into the superconducting state [18].

of normal sample material is then $\sim \pi\xi_0^2 H/\Phi_0$. Setting this expression equal to 1 gives an estimate $H_{c2} \sim \Phi_0/\pi\xi_0^2$. This is only slightly larger than the expression deduced from Ginzburg-Landau (GL) theory,

$$H_{c2} = \frac{\Phi_0}{2\pi\xi_0^2}. \tag{2.3.14}$$

As was first demonstrated by Abrikosov [92], the normal-superconducting boundary wall has positive energy for $\lambda < \sqrt{2}\xi$. However, if *the GL parameter* $\kappa = \lambda/\xi > \sqrt{2}$, then the boundary wall has negative energy, so that the system will lower its free energy by making the boundary area as large as possible. Thus, the field arranges iself into individual fluxons, each surrounded by a normal-superconducting wall. In this way we see that the condition $\kappa > \sqrt{2}$ is the *definition of a type II superconductor*, κ being the GL parameter. Abrikosov predicted the existence of such a *mixed state*, in which the vortices arrange themselves (in reasonably clean material) into an ordered lattice if $\kappa = \lambda/\xi > \sqrt{2}$ [92]. The lowest energy state of such a lattice was shown to be minimized by a triangular arrangement of vortices by Kleiner et al. [93]. Below, we describe NMR spectra which are consistent with this theoretical conclusion.

It is interesting to look at simple expressions for ξ and λ to see why there is a general tendency for d-electron superconductors to exhibit type-II superconductivity. The expressions

$$\lambda_0 = \left[\frac{m\,c^2}{4\pi n_s\,e^2}\right] \quad \text{and} \quad \xi_0 = a\frac{\hbar v_F}{k_B T_c}, \tag{2.3.15}$$

where a is a constant of order unity, come from the London equations and from Pippard's uncertainty principle argument, respectively. Since d-bands tend to have large effective masses and high-T_c materials in particular have low carrier densities n_s, we expect λ_0 values to be larger than for sp-band superconductors. Furthermore, small values of E_F (and thus of v_F) as well as higher values of T_c will make for short ξ_0's. Thus, it is generally expected that $\lambda > \sqrt{2}\xi_0$. As an example, we mention V_3Si, a superconductor at $\sim$ 17 K which we shall discuss further below. Here, we only note that Greytak and Wernick [94] reported a measurement of $\lambda_0 = 1500$ Åand a related estimate of $\xi_0 = 25$ Å. Thus, $\kappa \simeq 60$ and from (2.3.14) we estimate $H_{c2} \simeq 530$ kG. A second estimate can be made from the thermodynamic critical field $H_c = 6370\,G$ [86] and the measured value of λ_0 using (4–62) from [18], $H_{c2} = 4\pi\lambda_0^2/\Phi_0$. This yields $H_{c2} = 550$ kG which is in surprisingly good accord with the previous number. In any case, V_3Si is a very "hard" superconductor, indeed.

Internal Field Distributions in the Mixed State from NMR Studies

NMR studies of type-II superconductors are typically done in the mixed state, i.e. for $H_{c1} < H < H_{c2}$. There is substantial NMR line broadening from spatially-dependent shielding supercurrents in the mixed state. We shall first review studies conducted to map the distribution of such currents, after which we will look at shift and relaxation studies. A number of internal field studies have been reported for V and Nb metals

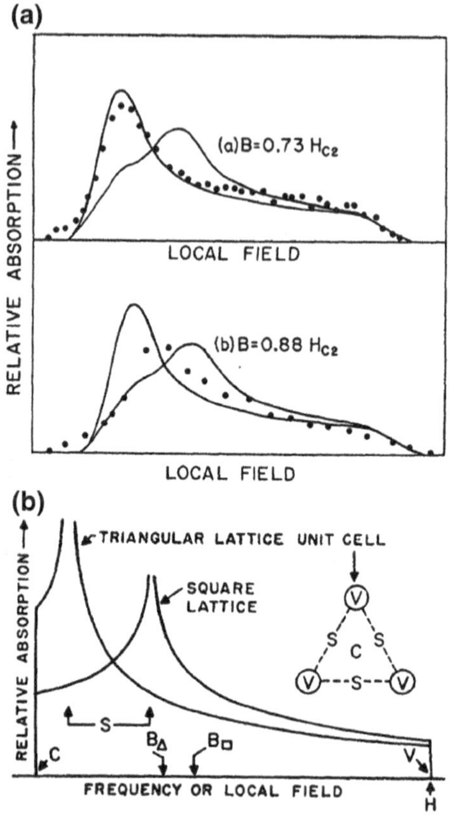

Fig. 2.10 **a** Distributions of diamagnetic shielding fields measured using *rf* saturation of the ^{51}V NMR in superconducting V metal, and **b** simulations of internal field distributions under similar circumstances for both square and triangular vortex lattices. In (**a**) results are given for $T \simeq 1.4\,\text{K}$ and for fields of 0.73 H_{c2} and 0.88 H_{c2} as shown, where $H_{c2} = 2720\,\text{G}$. For (**b**) the field simulations were done on the basis of solutions of the Ginzburg-Landau equations in the Abrikosov limit. The inset shows the triangular lattice unit cell, with points there and on the distribution labelled V, S, and C corresponding to a vortex (V), a saddle point (S) showing a divergence, and the minimum field (C)

[87–89]. We review briefly the results given by Redfield [87], who conducted the following field-switching NMR experiment on ^{51}V in V metal. The sample was polarized in $\simeq 10\,\text{kG}$, after which the field was switched to a value below $H_{c2} \sim 3,000\,\text{G}$ for a time of 0.1 s, during which an *rf* field was applied transverse to the steady field. The field was then switched back to 10 kG where the nuclear polarization was recorded. By scanning the frequency of the *rf*, the spectrum of internal fields was able to be mapped out. Results from this study are shown in Fig. 2.10a for two field values. These data have been fitted to calculated internal field distributions based on the GL equations in the Abrikosov limit, assuming both square and triangular lattices. The raw calculated distributions are shown in Fig. 2.10b. Fits are carried out by convoluting the latter distributions with a Gaussian of adjustable width, then adjusting the width and total area for best fit.

The triangular lattice unit cell is shown as an inset to Fig. 2.10(b), where the points V, C, and S are seen to correspond to the maximum field, minimum field, and the singularity in the distribution, respectively. The singularity appears much closer to C for triangular lattice than for the square lattice. Results for the lower field value ($\simeq 0.73H_{c2}$) are seen to be in good accord with the triangular lattice. The higher field value is rather closer to H_{c2}, where a kind of gapless superconducting state sets in and there are many corrections to the simple field-distribution calculation. Thus, it is not surprising that the simple triangular lattice calculation is no longer accurate [87]. Studies of effects near H_{c2} are reviewed in [9]. We shall, however, restrict our attention here to the region $H_{c1} \ll H \ll H_{c2}$.

It is interesting to compare the field distribution results from [87] with theoretical predictions. For example, the results from Kleiner, et al. [93] predict that $H_V - H_S = 1.35\,(H - B)$. From [87] we find for the conditions of Fig. 2.10 the experimental value $H_V - H_S = 285$ G, while under the same conditions $1.35\,(H - B) \simeq 165$ G, a considerable underestimate. On the other hand, according to the Abrikosov solution [9, 92] $1.35\,(H - B) = 1.35\,(H_{c2} - H)/\beta_a(2\kappa^2 - 1)$, which we find to be 520 G, which is a considerable overestimate. For this purpose we take $\beta_a = 1.16$ for the triangular lattice from [93] and obtain $\xi_0 = [\Phi_0/2\pi H_{c2}]^{1/2} = 348$ Å. The estimate $\lambda = 400$ Å[91] then gives $\kappa = 1.15$. Substantial corrections to these simple theoretical results have been given by Eilenberger [90].

The NMR Shift in Type-II Superconductors

In the foregoing study of elemental vanadium, the precision with which the vortex lattice could be characterized was not great enough to resolve changes in the d-spin shift $K_d(T)$ below T_c [87]. However, in a truly *hard* type-II system where $\kappa \gg 1$, it is possible to resolve small shift changes at field values such that $H_{c1} \ll H \ll H_{c2}$. The reason for this is that the NMR lineshape is then strongly dominated by the singularity from the saddle regions S (Fig. 2.10b), which is therefore shifted only slightly away from the applied field. Moreover, the NMR linewidth from diamagnetic shielding is typically only modestly greater than one might find in the normal state.

We illustrate these points with data and discussion of the closely related cases of V_3Si and V_3Ga, where V_3Si was touted earlier as a truly hard superconductor. Interestingly, somewhat before the nature of the the type-II diamagnetic shielding was investigated in detail in the 1960s, The NMR shifts in these compounds of both ^{51}V and ^{71}Ga were reported by Clogston et al. [37]. We reproduce their (K, χ) plot in Fig. 2.11. The interesting point here is that all NMR shift values here are consistent with the interpretation that, at $T = 1.8$ K, the corresponding d-band susceptibility is diminished to $\sim 25\%$ of its value at 20 K. Further, its low temperature value is smaller than that at room temerature. Since these NMR shifts were measured in the mixed state at a field of 14 kOe, where they found "negligible diamagnetic *rf* susceptibility" [37], one can only conclude that spurious frequency shift effects from the diamagnetic shielding are negligibly small in comparison. Note that the total range of ^{51}V shift values here is only $\sim 0.1\%$.

It is interesting to consider the nature of the internal field distribution in V_3Si given the values of its parameters ξ_0 and λ_0 (25 Å and 1500 Å, respectively) stated above. First, the second moment of this distribution has been calculated by Pincus et al. [86], yielding the expression

$$\langle \Delta H^2 \rangle_{av}^{1/2} \simeq \frac{\Phi_0}{4\pi^{3/2}\lambda^2} \tag{2.3.16}$$

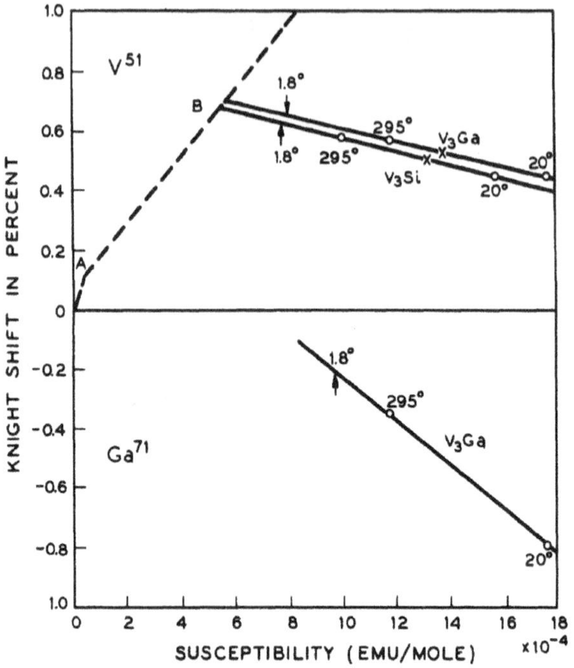

Fig. 2.11 Measured Knight shifts are plotted versus molar susceptibilities for ^{51}V in both V_3Si and V_3Ga and for ^{71}Ga in V_3Ga [37]. For the ^{51}V shifts a (K, χ) construction is shown, including the s-band contribution (point A) and the orbital contribution (points A–B), which intersects with lines through the data points. For the ^{71}Ga the results clearly show a transferred shift which is driven by the temperature-dependent d-band susceptibility. In both cases the location of the shift values measured at 1.8 K are shown, indicating that the negative CP shift is diminished to $\sim 75\%$ of its value at 20 K, and that the values of $|K_d(T)|$ in both cases are smaller than the values found at room temperature

under conditions where not only $H_{c1} \ll H \ll H_{c2}$ holds, but also $\lambda/d \gg 1$, where d is the separation of vortices in the mixed state. At $H = 14$kOe (i.e. for the data of Fig. 2.11), we find $d = [2\Phi_0/\sqrt{3}H]^{1/2} = 413$ Å for the triangular lattice, giving $\lambda/d = 3.6$. For these conditions (2.3.16) gives $\langle \Delta H^2 \rangle_{av}^{1/2} = 41$ Oe for a square lattice. This may be a slight overestimate, but is of the right order, since the measured peak-to-peak derivative linewidth is 39 ± 5 Oe [86]. It is interesting to note in this case, where $\kappa \gg 1$, the Abrikosov relation ([92]) yields $H_V - H_S = 86$ G, barely twice the second moment estimate.

We therefore find for the hard type-II superconductors that it is quite feasible to measure shift values with weak temperature dependences in the mixed state of superconductivity. This is a very important result for high-T_c NMR studies.

2.3.5 T_1 Phenomenology for Type-II Superconductors

In this final subsection of Chap. 2 we review T_1 behavior at $T < T_c$ for type-II d-band superconductors. Because of the complications which attend the measurement and interpretation of T_1 in the *mixed state* of these systems, our discussion of this topic will be limited to a few simple cases. A clear and thorough discussion of these effects, in terms of both theory and experiment, is given in [9]. The hallmarks of type-I system T_1 behavior are the coherence peak (in T_1^{-1}) just below T_c and the exponential decay $T_1^{-1} \propto \exp(-\Delta_0/k_B T)$ for $T \ll T_c$. For type-II systems these behaviors carry over in a modified way or not at all. The differences found between type-I and type-II behavior appear to be caused by the field penetration in the mixed state and the consequent spatial inhomogeneities.

The Coherence Peak Region, $T \le T_c(H)$

Caroli et al. have shown that in the vicinity of a vortex core the excited state energies are diminished to $\Delta E \sim \Delta_0^2/E_F \ll \Delta_0$ [107]. Thus, for points near the (H, T) phase boundary, i.e. for H near $H_{c2}(T)$, there is a gapless region which has received a detailed theoretical examination [95–97]. The fluctuation (i.e. T_1) properties in this region are different from a type-I system and appear to depend on whether the carriers have a long mean-free path $l \gg \xi_0$ ("clean") or whether it is short $l < \xi_0$ ("dirty"). In the simpler dirty limit case the electronic properties are local in character, and Cyrot [98] has derived the relation

$$\frac{T_{1n}}{T_{1s}} = 1 + \frac{ec}{\sigma k_B T_c}(H_{c2} - H)\, g(t), \tag{2.3.17}$$

where $t = T/T_c$, σ is the normal state conductivity and $H_{c2} = H_{c2}(T)$. This relation is only valid for $(H_{c2} - H) \ll H_{c2}$, so that it only gives the slope $T_{1n} d(T_{1s}^{-1})/dH$ near H_{c2}. The function $g(t) > 0$ for $t > 0.6$, while $g(t) < 0$ below that point. One must be very careful in using (2.3.17) to interpret actual data, since T_1 is usually measured by scanning temperature rather than field.

Masuda and Okubo have obtained T_1 data on ^{51}V in V_3Sn, a superconductor with $T_c \simeq 3.6$ K (their sample) and $H_{c2}(0) = 16$ kG [99, 100]. Their data are shown for three values of applied field in Fig. 2.12. These authors used (2.3.17) to show that

$$d(T_{1n}/T_{1s})/dT \propto g(t) \quad (T \le T_c(H)), \tag{2.3.18}$$

as well, and have evaluated the constant of proportionality. We note that for $H =$ (4470, 8940, 13410) G, $T_c(H) =$ (3.2, 2.6, 1.8) K and thus $t(H_{c2}) =$ (0.89, 0.72, 0.50) for the three data curves in Fig. 2.12. Thus, the slopes of the data plots at $T \le T_c$ are nicely consistent with these parameters, since $d(T_{1n}/T_{1s})/dT$ is positive for the first two, then turns negative for the third, where $t(H_{c2})$ falls below 0.6. It seems, then, that measurements at low (or zero) field on V_3Sn yield a conventional BCS coherence peak, but that it is gradually quenched out by fluxoids at higher field values.

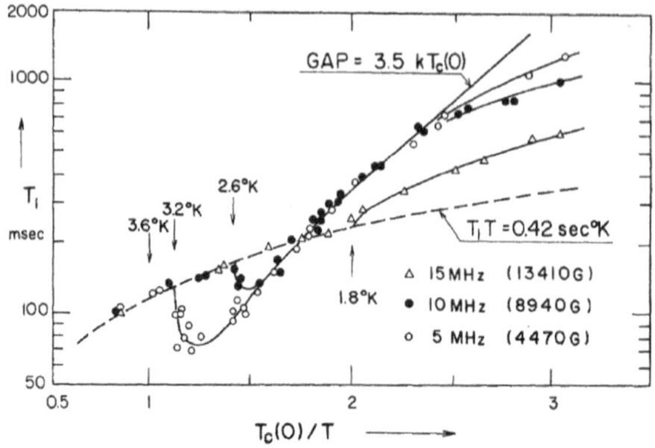

Fig. 2.12 A semilog plot of data for T_1 of ${}^{51}V$ in V_3Sn versus $T_c(0)/T$ in the superconducting state for three values of applied field as shown from [99]. The data shown exhibit coherence peaks for 1341 and 8940 G, but not for 13410 G. All three plots show deviations from the exponential BCS behavior at the lowest temperatures, an effect attributable to spin diffusion to the normal cores. See text for discussion

We mention three other cases of type-II T_1 studies at $T \leq T_c$. Each is somewhat anomalous in its own way. The first is the case of elemental V metal [101], where no convincing coherence peak was observed. The authors considered a number of possible explanations, finally suggesting that the T_1 peak may be suppressed by trapped flux. However, sometime after these results were published, the theory of T_1 in the clean limit emerged [95], and this interpretation was applied to the case of vanadium by Pesch [97]. The result was a good fit to the data for both samples [101], with no sign of a coherence peak. It seems, then, that coherence peaks will appear at low fields in the dirty limit, but not in clean samples.

No coherence peaks were observed in any of the V_3X compounds, where X = Ga, Si, Ge, and Pt, in the study conducted by Silbernagel et al. [102, 103]. The ostensible reason for this is disorder caused by the "Martensitic" structural transformation which was reported at temperatures just above T_c in these compounds [104, 105]. The authors offer no further discussion of the point [103].

The most anomalous case to be reported was that of niobium [106]. The relaxation of ${}^{93}Nb$ nuclear spins was studied in the SC state at field values varying from 2.40 kG up to 6.80 kG corresponding to T_c values which vary from 6.8 K down to 3.0 K. Using (2.3.14), H_{c2} corresponds to $\xi_0 \sim 200$ Å, so that the given mean-free path $l \simeq 140$ Å is on the borderline between the clean and dirty cases. In any case, the T_1 data were found to fit the gapless theory [96, 97] for the higher field values. For the lower two field values, however, the T_1 data were found to execute a deep minimum just below $T_c(H)$, contrary to the gapless theoretical fits. Such behavior is clearly reminiscent of the V_3Sn data in Fig. 2.12, and one might inquire as to whether the calculation of $d(T_{1n}/T_{1s})/dT$ given in [100] based on (2.3.17) could be applied in this case. The authors did such a calculation and found that the coefficient of $g(t)$ for Nb is smaller

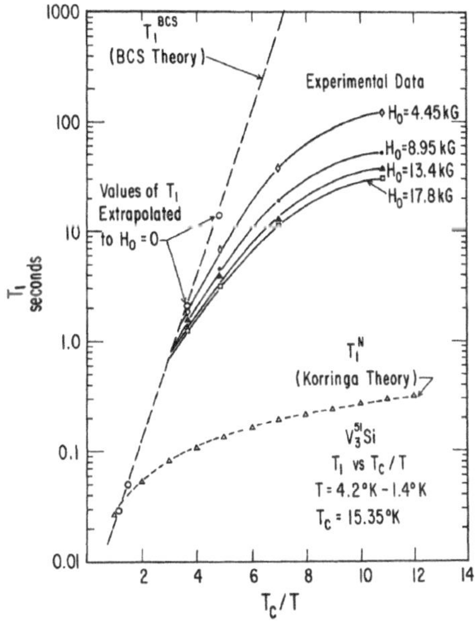

Fig. 2.13 A semilog plot of data for T_1 of ^{51}V in V_3Si versus T_c/T in the superconducting state for four values of applied field as shown from [102]. The data shown exhibit no coherence peaks, but follow exponential BCS theory behavior up to $T_c/T \sim 4$. Beyond that point the mechanism of spin diffusion to the normal fluxoid cores takes over, giving a relaxation rate which is proportional to the applied field (i.e., to the fluxoid density), as expected. See text for further discussion

by a factor 150–200, so that similar behavior cannot be expected on that basis. There seems to be no straightforward explanation for this anomalous result.

Type-II T_1 Behavior for $T \ll T_c$

For values of T well below T_c, an exponential decay $T_{1s}^{-1} \propto \exp(-\Delta_0/k_BT)$ has generally been found to appear. However, as this intrinsic BCS rate weakens, another contribution attributable to the fluxoids is often seen. Since the energy gap near the fluxoids is very small [107], T_{1s} for a region of area $\sim \xi_0^2$ is essentially that of the normal state. When the BCS T_1 process in the bulk of the SC material becomes very slow, then a process whereby the nuclear polarization diffuses to the normal cores becomes predominant. This process was found to be present in three of the four examples given above [99, 101, 102] but was studied most systematically in the case of V_3Si [102].

^{51}V T_{1s} data for V_3Si [102] are plotted semilogarithmically in Fig. 2.13 as a function of T_c/T, where $T_c = 15.35$ K. Results are given for four different field values at temperatures in the SC state which extend down below $T_c/10$. At higher temperatures the data points asymptotically approach the dashed line which represents BCS behavior, i.e. $T_{1s} \propto \exp \Delta_0/k_BT$, where $\Delta_0 = 1.76\,\mathrm{k_BT_c}$. T_{1s} veers away from the BCS curve at $T_c/T \geq 4$, then ranges up to values more than three orders of magnitude longer than its value at T_c. The flattening of these curves is suggested to be caused by spin diffusion to the normal cores, which then becomes the only effective T_1 mechanism at the lowest temperatures. If the spin diffusion time, estimated below, is not an impediment, then at the lowest temperature we should have

$$\frac{T_{1n}}{T_{1s}} = \frac{h\xi_0^2 H}{\Phi_0}. \tag{2.3.19}$$

where h is a numerical factor of order unity. Equation (2.3.19) simply states that the normal T_1 should be diluted by the ratio of $h\xi_0^2$ to the area per fluxoid Φ_0/H. First, we note that $1/T_{1s}$ should be linear in H, as is clearly the case in Fig. 2.13. With the estimate $\xi_0 \simeq 25$ Å for V_3Si given above, we find $h = 1.9$, so that the effective area of normal-state relaxation is a circle of radius slightly smaller than ξ_0 for each fluxoid. The corresponding spin diffusion times for the results in Fig. 2.13 we can estimate with

$$t_{diff} \sim \frac{d^2}{4D_{sp}}, \tag{2.3.20}$$

where d is the distance between fluxoids and D_{sp} is the spin diffusion constant. Taking $d = [2\Phi_0/\sqrt{3}H]^{1/2}$ and $D_{sp} \sim 10^{-12}$ cm^2s [108], we find $t_{diff} \sim 13$ s at H = 4450 G. This is an order of magnitude less than T_1 itself, lending credence to the use of (2.3.19).

While the foregoing scenario gives an appealingly straightforward interpretation of the V_3Si data for T_{1s}, there did arise in the literature a rather knotty problem in connection with this interpretation. Genack and Redfield suggested some time later [109] that the spin diffusion relaxation mechanism in type-II superconductors would not work, because it did not conserve energy. We give the essence of their argument here, then cite further work which provides something of an antidote to this objection.

If spin diffusion takes place along a field gradient, then the simple flip-flop events between neighboring nuclei no longer conserve energy. The difference in Zeeman (field) energies has to be taken up by the spin-spin (usually dipolar) interactions. Viewing this process thermodynamically, the dipolar reservoir then quickly becomes overheated and must, itself, relax to the lattice. The total relaxation process by spin diffusion is thus formulated as a dual process, with Zeeman (magnetization) and dipolar energies both diffusing to the normal cores in the mixed superconducting state. These processes were analyzed in [109] and found to be too slow to provide the observed T_1 processes in V metal [101]. It was concluded that these clearly observed diffusion-like T_1 processes must originate in some other way [109].

The problem is, of course, that no other ideas to explain type-II relaxation have been forthcoming. However, more recently, this problem has been revisited in terms of basically the same ideas, but using more sophisticated mathematical methods (see [110] and references cited therein). These authors have analyzed in detail the relaxation curves of ^{51}V in V metal at $T \ll T_c$ taken from [101] in terms of spatially varying Zeeman and dipolar spin temperatures. They find a good fit to the vanadium data, where the dipolar relaxation time constants are relatively short, thus not obstructing the longer diffusion-mediated Zeeman T_1's.

It seems, then, that the phenomenology of (2.3.19) is basically correct, so that low-temperature, type II T_1 behavior is clearly understood. This is very important for the interpretation of T_1 data for high-T_c systems as well. However, as we shall see, the fundamental behavior of T_1 in high-T_c compounds involves SC gap symmetries which usually have line nodes, giving power-law behavior at low temperatures [111]. Thus, the issues regarding T_1 data analysis differ somewhat from those of d-band SC metals discussed here. Nonetheless, the spin-diffusion-to-normal-cores T_1 mechanism in the mixed state has been found to play a significant role in the phenomenology of cuprate superconducting T_1 effects. The understanding of this effect gained with ordinary 3d metals will be an important resource in the chapters to follow.

Chapter 3
The Superconducting Cuprates: Preliminary Steps in Their Investigation via NMR

The purpose of this chapter is twofold. First, it is partly historical, to recount the somewhat uncertain beginnings and explain how ideas developed the way they did, and secondly, to review the early milestones which laid the foundations for the many years of study which were to follow. These agendas will unfold simultaneously in an essentially chronological recounting of the important first steps in constructing the edifice of high-T_c NMR phenomenology which exists today.

Although the cuprates form a large family of compounds,[1] in recounting the major accomplishments in the field of NMR studies, one's attention is drawn to three compounds, namely $La_{2-x}Sr_xCuO_4$ (LSCO:x), $YBa_2Cu_3O_{7-x}$ (YBCO7-x), and $YBa_2Cu_4O_8$ (Y248). These three have played a major role, both in terms of experimental results and theoretical interpretation. There are other interesting compounds with higher T_c's (e.g. $Tl_2Ba_2CaCu_2O_8$) and with quantum chemistry just as clean ($Hg_2Ba_2Cu_2O_4$). It seems that it was just an historical accident that the three listed above became prominent. Accordingly, they will become our main focus.

In Sect. 3.1 we review the structures, phase diagram, and some of the early theoretical models regarding transport and quasiparticle dynamics. Of course with a research topic as completely new and unfamiliar as the cuprates, one spends a great deal of time and effort at the outset simply establishing the basic experimental features of the system under investigation. Some of this activity is reviewed in Sect. 3.2, where we shall see that the anomalous nature of the spin dynamics (i.e. T_1) was already evident in the earliest high-T_c NMR papers.

A major milestone in understanding all HF effects in the cuprates came with the Mila-Rice-Shastry picture of predominant transferred HF couplings with their quantum-chemical foundations [118, 146]. This work, presented and discussed in Sect. 3.3 for $^{63,65}Cu$, established a model HF Hamiltonian for the principal nuclei,

---

[1] See, for example, the wide variety of cuprate structures reviewed by Hazen [214].

© Springer-Verlag GmbH Germany 2018

R.E. Walstedt, *The NMR Probe of High-T_c Materials and Correlated Electron Systems*, Springer Tracts in Modern Physics 276,
https://doi.org/10.1007/978-3-662-55582-8_3

which has been an important touchstone ever since. The application of this picture to the NMR studies of planar ^{17}O and the ^{89}Y in the YBCO structure is the subject of Sect. 3.4.

The relationship between nuclear spin-lattice relaxation (T_1) and the dynamic susceptibility $\chi(\boldsymbol{q}, \omega)$ via the fluctuation-dissipation theorem has been well known for many years [163]. With the cuprates, however, this quantity quickly became the basis for the expression of static and dynamic NMR, for inelastic neutron scattering (INS) results, as well as for theoretical modeling. In Sect. 3.5 we review the derivation of the basic results and give standard expressions for T_1 at various nuclear sites in cuprates for reference throughout the book.

Finally, in Sect. 3.6 the early and highly significant d-wave interpretation of T_1 in the superconducting state by Monien and Pines [160] is reviewed in some detail. This was based on an RPA model of the dynamic susceptibility which formed a basis for future modeling of normal-state dynamics as well [161, 162]. As we shall see throughout the volume, fundamental d-wave behaviors for T_1 and for the transverse decay constant T_{2g} were firmly established several years in advance of the Josephson junction phase-boundary results that finally turned the tide on the d-wave versus s-wave question.

3.1 Cuprate Structures and Doping Effects: LSCO and YBCO

The first high-T_c materials to be announced [112] left some degree of uncertainty about the actual crystal structures of the superconducting phases. In the weeks that followed, among the first structural refinements to be announced were those of $La_{2-x}Sr_xCuO_4$ (LSCO) [113] and $YBa_2Cu_3O_{7-\delta}$ (YBCO) [114]. For that reason as well as the fact of their ease of synthesis into high-quality specimens and of subsequent characterization, these two compounds and closely related materials have been a strong focus of NMR/NQR studies from the inception down to the present era. In this section we present a brief discussion of the structures, doping, and quasiparticle dynamics in these typical cuprates. Discussions of research studies on the LSCO family of compounds will be given in Chaps. 4 and 6.

3.1.1 *LSCO: A Superconductor with* T_c(max) $\simeq 40K$

LSCO has the K_2NiF_4 crystal structure shown in Fig. 3.1a [115]. The key structural element in all of the cuprate superconductors is the CuO_2 quadratic array (Cu-O_1) in Fig. 3.1a, which has been identified as the region where the hole current is carried in these systems. Note that the O_1 ions also form a quadratic array rotated 45° from that of the Cu, and having a lattice constant of $a/\sqrt{2}$. Cu ions in

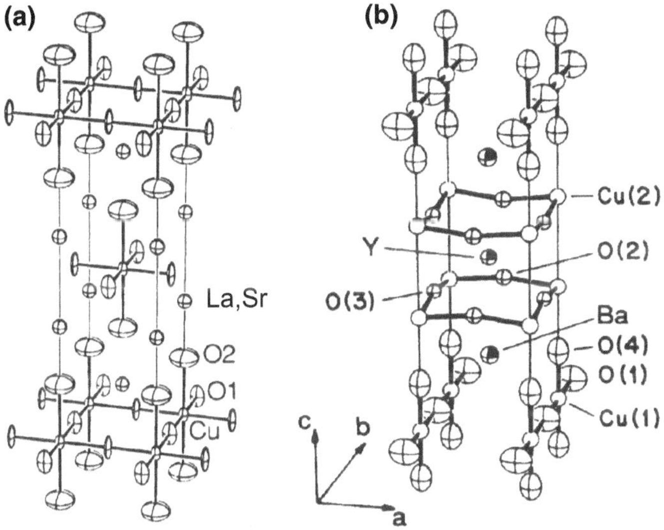

Fig. 3.1 **a** The (K_2NiF_4) structure of $La_{2-x}Sr_xCuO_4$, showing the locus of the CuO_2 conducting plane. The Sr^{2+} dopant ions go substitutionally into the La^{3+} (Ba) sites, leading to a set of mobile holes which run on the square lattice of planar O^{2-} ions. The structure shown has tetragonal symmetry. However, below a few hundred *K* an orthorhombic distortion sets in, where the CuO_6 octahedra tilt through small angles in the [110] planes [116]. **b** The $YBa_2Cu_3O_{7-x}$ structure is shown with the chain layers (Cu(1)) on the top and bottom and a pair of Cu(2)O(2, 3)$_2$ layers in the middle. Note that there is no bonding between adjacent Cu(2) layers except via rather inert Y^{3+} ion cores. The O1 sites are shown filled, forming chains along the *b* axis. The structural symmetry is orthorhombic

adjacent planes are staggered, so that with AFM short–range order (SRO), magnetic couplings between layers are extremely weak. The physics of these structures is strongly two dimensional.

The physics of transport is dramatically different in the cuprates from what it is in an ordinary d-band metal. For LSCO we consider first the case of x = 0, i.e. La_2CuO_4. With the expected valence (La^{3+}, Cu^{2+}, and O^{2-}), the conduction plane consists of Cu^{2+} ions, i.e. with a single hole in the d shell ($d_{x^2-y^2}$) plus O^{2-} closed-shell ion cores. The 3d holes are strongly hybridized with the O^{2-} 2p orbitals, such that over 30% of the localized d-hole density resides on the planar oxygens [119, 120]. In the conventional picture of weakly-interacting fermions, this would simply be a half-filled band like an alkali metal.

On the contrary, however, the hopping process which might lead to hole conduction is strongly inhibited by coulomb repulsion ($U \sim 10\,$eV) on the Cu^{+2} ion complexes in what is known as a "Mott insulator" [122]. Instead, immobile $S = 1/2$ local moments on the Cu^{+2} ions condense to an insulating, antiferromagnetic ground state with $T_N \simeq 325\,$K. In order to generate conductivity in the CuO_2 planes, it is necessary to dope additional holes into the system. This may be done by replacing some of the La^{+3} ions with Sr^{+2} ions as in LSCO. For each such substitution a hole must appear in or near the CuO_2 plane. Above a small threshold concentration these holes become mobile. In the Zhang-Rice singlet picture, they combine with Cu^{2+} holes to form a singlet quasiparticle [125]. After considerable investigation and debate, such singlets have finally been concluded to have negligible HF interactions on their own. It has been suggested that they serve the function of rendering the Cu^{2+} holes itinerant [125].

The LSCO Phase Diagram

The phase diagram of Fig. 1.1 illustrates the resulting behavior as a function of the hole-doping parameter x. First, the antiferromagnetism collapses very quickly with hole doping. At $x \sim 0.02$ there is an insulator-to-metal transition, though conductivity is still rather poor. In this region, spin-glass behavior is thought to occur. At $x \simeq 0.05$ the superconducting phase appears, although the normal-state conductivity is still that of a very bad metal. With an optimum concentration of $x = 0.15$, T_c then peaks at $\sim$40 K. Beyond that point is the *overdoped region*, where the system becomes more and more like a normal metal. T_c then declines to zero at $x \sim 0.35$. The pattern seen with LSCO is repeated with essentially every cuprate superconductor, except that in many cases it is not possible to vary the carrier content all the way from the antiferromagnetic phase through superconductivity and into the overdoped phase.

The line in Fig. 1.1 running from $x = 0.2$ upward to join the $x = 0$ axis at $T \sim$ 500 K is a boundary between the tetragonal and orthorhombic phases in LSCO. However, there is a "crossover" line in approximately that position in various theories of magnetic behavior [240, 244], for which there is growing experimental evidence in other cuprate compounds. In the Barzykin-Pines model this is a boundary between mean-field behavior for the dynamic exponent ($z = 2$) on the high-temperature side and quantum-critical ($z = 1$) behavior on the low-temperature side. There is some NMR evidence for this which will be discussed in Chap. 6.

3.1.2 The 92 K Superconductor $YBa_2Cu_3O_{7-x}$ (YBCO)

Many cuprates have multiple CuO_2 planes neighboring each other with a single layer of cations between them. $YBa_2Cu_3O_{7-x}$ (YBCO7-x) has two such neighboring planes separated by a single layer of Y^{3+} ions, as shown in Fig. 3.1b. This structure also features a "chain" layer, known as the Cu(1) sites, which is a quadratic array of copper ions, but where half of the ligand O^{2-} ions seen in the Cu(2) layers are absent, leaving 1D Cu-O chains running along the b axis of the crystal. The chain sites are a complication in understanding the behavior of YBCO, because as we shall see below, the Cu(1) and Cu(2) sites behave rather differently.

On the other hand, regarding the level of mobile carriers, the YBCO7-x system is *self-doping*. The minimum doping level is effectively zero when x = 1, giving six oxygens per formula unit. With the expected valencies (Y^{3+}, Ba^{2+}, $Cu(1)^{+}$, $Cu(2)^{2+}$, and O^{2-}), $YBa_2Cu_3O_6$ is in a *vacuum state* for conductivity similar to La_2CuO_4. In this structure there are no oxygens at all in the Cu(1) planes, and their corresponding valence state is Cu^{1+}. The maximum doping occurs when x = 0, whereupon the Cu(1) chains are fully oxygenated, and one finds an average charge of +2.33 per Cu ion. This is considered to be slightly overdoped, where the maximum $T_c \simeq 92$ K occurs for $x \simeq 0.1$. It is noteworthy, however, that YBCO7 is *stoichiometric* and therefore has minimal disorder, i.e. far less disorder than optimally doped LSCO.

3.1.3 Physical Models of Itinerant Quasiparticles in Cuprates

At optimum doping levels, then, the cuprate conduction planes then appear to contain a small concentration of mobile holes moving in a sea of $S = 1/2$, exchange-coupled local moments. Because of the coulomb repulsion, these holes are thought to be restricted to the oxygen sites. The description of this mixture as a superconducting quantum fluid is therefore dramatically different from a simple 3d-band of partially-occupied Bloch functions. It has posed a stiff theoretical challenge from the beginning, and even now there is not a clear consensus about it. One thing we understand is that it is not a set of exchange-coupled local moments simply coexisting with a hole band of itinerant fermions. For example, at $T \ll T_c$ all spin magnetism decays to a small fraction of its normal-state level, as we know from NMR shift and relaxation measurements (Sect. 3.2.5). Upon a (low, but) sufficient level of doping, then, all Cu^{2+} holes become effectively itinerant and part of a superconducting quantum fluid at $T < T_c$.

The t-J Model

How this takes place has been the topic of many theoretical papers. We mention here the "t-J" model [123, 124], which is an approximation to the one-band Hubbard model. A number of theory papers have shown that the t-J model Hamiltonian (here simplified)

$$\mathcal{H}_{t-J} = J \sum_{i,j(nn)} \mathbf{S}_\mathrm{i} \cdot \mathbf{S}_\mathrm{j} + t \sum_{i,j(nn)} [c_i^* c_j + c_i c_j^*] \tag{3.1.1}$$

results in a band of itinerant quasiparticles of width $\sim J \sim 0.15$ eV [126].

Zhang-Rice Singlets

A detailed model for the structure of the quasiparticles themselves, known as the "Zhang-Rice singlet model", appeared very early on and is still considered an interesting conceptual picture [125]. In this model the mobile holes form a singlet complex with neighboring Cu^{2+} "localized" holes, rendering them itinerant. In spite of their small numbers, the mobile holes are said to render all of the Cu^{2+} moments effectively itinerant and part of a quantum liquid which condenses into a superconducting spin-paired ground state. That this actually occurs is concluded from decay of the

Cu(2) Knight shift and relaxation rate, and thus the spin paramagnetism of YBCO7, to a value small compared with its normal-state value for $T \ll T_c$ (see Sect. 3.2).

The "Oxygen Hole Band" Issue

Contrary to the Zhang-Rice singlet picture, many early papers considered that the doped holes in cuprates would result in a separate hole band residing on the oxygen lattice with its own independent HF effects. This idea was often advanced when it seemed that the fluctuating $S = \frac{1}{2}$ moments which occupy the Cu^{2+} sites would not give an adequate explanation of NMR shift and relaxation effects at all sites studied. Often the papers concerned made important contributions in spite of maintaining this point of view. Eventually, as will become clear in Chaps. 4 and 5, it was realized that there was simply no need for oxygen hole bands with independent HF effects and no evidence for them either. This issue will be mentioned at appropriate points in reviewing the relevant literature.

3.2 Early NMR/NQR Studies: The Early Predominance of YBCO

It was, of course, realized immediately that NMR studies of the $^{63,65}Cu$, ^{17}O, and in the case of YBCO, ^{89}Y nuclear spins could yield a great deal of detailed information about static and dynamic fermion behavior in these systems. Perhaps because of its very high T_c value, there was an intense focus on YBCO at the outset, which was facilitated by the ease with which samples could be synthesized. Accordingly, there were a great many NMR papers on YBCO, mainly fully doped (YbCO7) during the first year or two. Our focus in this section will be these early studies of YBCO and the issues they raised. LSCO eventually received its share of attention, and will be addressed both in the next chapter and in Chap. 6.

In this section we examine the developments which took place during 1987–88 following the discovery of high-T_c superconductors. It was a time of intensive activity, but at the same time, sample preparation was still in a rather primitive state. There was a definite focus on YBCO7, the 90–92 K superconductor, because of its spectacularly high T_c value, and because it is a stoichiometric compound. Even so, almost all of the samples studied during this period were either random powders or milligram-scale crystals which gave narrow resonance lines, but very small signals. The random powders gave $(-\frac{1}{2} \leftrightarrow \frac{1}{2})$ transition spectra $\sim$0.5 T wide in a typical laboratory (7–8 T) field, and it took a great deal of noise averaging to obtain a single spectrum. Only with NQR was it possible to do a T_1 measurement.

Even so, many basic issues concerning the NMR/NQR study of the cuprates were addressed during this time period, and impressive progress was made toward their resolution. At the end of the 2-year period we are focusing on, there were two events which marked the beginning of a more settled period, namely the appearance of the Mila-Rice-Shastry papers to clarify the quantum chemistry and the transferred–HF–interaction picture [118, 146], and secondly, the advent of the *oriented powder* sample preparation method, which dramatically improved the prospects for precise determination of nearly all experimental NMR parameters. The Mila-Rice-Shastry picture for HF couplings in YBCO is presented in Sect. 3.3.

3.2.1 NMR Shift and Relaxation of the ^{89}Y in YBCO7

Not surprisingly, the first papers on cuprate nuclear spin dynamics to appear featured studies of ^{89}Y T_1 behavior in both the normal and superconducting states [127, 128]. The ^{89}Y nuclei have very small, $I = 1/2$ nuclear spin moments, but are 100% abundant. They are very weakly hybridized with the Cu(2) 3d moments, giving them a long T_1, but a relatively isotropic NMR shift, so that they are easy to study in a random powder sample. The quick appearance of these ^{89}Y studies was a harbinger of a great many fruitful ^{89}Y NMR studies to follow.

The ^{89}Y T_1 data from [127], shown in Fig. 3.2, immediately revealed two striking behaviors. First, T_1 follows a Korringa-like, inverse–T law in the normal state, giving actually a misleading impression of the behavior at its neighboring Cu(2) site to be unveiled soon after. The normal-state data of [128] were in agreement with this finding, and both papers found a normal-state shift $K \simeq 0.02\%$ as well. As for T_1, there is no hint of any coherence peak below T_c; moreover, the decline of $1/T_1$ below T_c is remarkably abrupt. It behaves in a roughly exponential fashion when plotted versus T/T_c, and has a slope much larger that the expected asymptotic BCS slope (plotted). If the BCS curve reflects a gap $2\Delta_{BCS} \simeq 3.5\, k_B T_c$, the ^{89}Y T_1 curve shows a gap value $2\Delta_Y \simeq 11.2\, k_B T_c$. While such a characterization is somewhat superficial at this stage, it shows at least that the superconducting T_1 behavior is dramatically different from that of any ordinary d-band metal (See Sect. 3.6 for further elucidation of T_1 behavior below T_c).

NMR/NQR Studies of ^{63,65}Cu(1, 2): The Quadrupolar Tensor and T_1(T)

Next, many papers appeared in which the four-line NQR spectrum (or part thereof) was reported (^{63,65}Cu(1) and ^{63,65}Cu(2)) in YBCO7 [128, 129, 131, 132]. Of course, for the $I = 3/2$, $^{63,65}Cu$ nuclei the value of η cannot be derived from NQR frequencies alone. The full quadrupolar frequency tensor has been deduced from NMR studies of small single crystals of YBCO7-x, with $x \simeq 0$ [133, 134], yielding essentially equivalent results. After some refinement, these tensors may be given (in MHz) as $\nu_{\alpha\alpha} = (-19.03(7), 19.17(7), -0.16(3))$ for Cu(1) and (15.94(5), 15.56(5), −31.50(5)) for Cu(2) at 100 K [139], where the numbers in parentheses are the uncertainties in the last digit quoted. This quadrupolar tensor has been analyzed by Frank Adrian [142].

NQR Studies of Cu(1, 2) T_1 in YBCO7

NQR was used to measure T_1 for both the ^{63}Cu(1) and ^{63}Cu(2) lines at 22 MHz and 31.5 MHz, respectively, giving the first glimpse of T_1 behavior for the copper sites in the cuprates [128, 129]. The T_1 data from [129] are displayed in Fig. 3.3, where

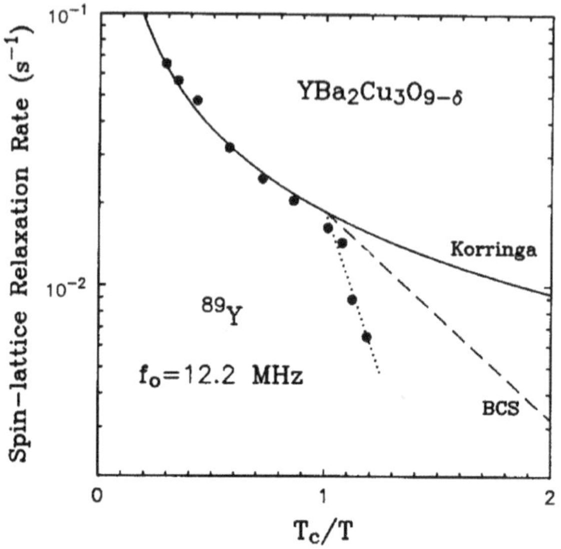

Fig. 3.2 $1/T_1$ of ^{89}Y in $YBa_2Cu_3O_{9-\delta}$ from [127] is plotted versus T_c/T in a semilog plot, where $\delta \simeq 2.1$. In the normal state the data are within error limits of a Korringa curve (solid line, $1/T_1 \propto T$). Below T_c the measured relaxation rate drops much faster than the BCS energy gap would allow (dashed curve—BCS). See text for discussion

markedly different behavior is seen to occur for the two copper sites, both above and below T_c. In the normal state, the Cu(1) site T_1 behavior is rather Korringa-like, similar to the ^{89}Y, whereas the Cu(2) site $1/T_1$ behaves initially like $a + bT$, but then flattening off near room temperature. The normal state data from [128] behave in a similar fashion to that in Fig. 3.3, except that T_1 values appear to be about 15% longer than from [129]. The value quoted in [141] is also slightly longer, but well within error limits of the data in Fig. 3.3. The reader may note that different references cited here give T_1 values which are defined variously, none of which agree with the discussion of T_1 in Sect. 1.1.5. Recall that the NQR value of relaxation time measured for I = 3/2 is 1/3 of the formally defined T_1 (see Sect. 2.1.5). Finally, regarding the data shown in Fig. 3.3, we shall not consider the temperature dependence of T_1 in depth until Chap. 4. We comment briefly on the general features of these data.

Below T_c the Cu(2) relaxation rate declines very steeply, rather like the ^{89}Y data in Fig. 3.2. The data from [128] for $T < T_c$ are also similar, the main difference being that the Cu(2) curve is extended to somewhat lower temperatures, where it is found to level off markedly. Such behavior is very likely sample dependent. Fitting the Cu(2) rate below T_c (Fig. 3.3) to a BCS-style exponential decay, a reasonable fit gives $2\Delta_Y \simeq 8.3\, k_B T_c$ with $T_c = 92\,\mathrm{K}$ [129]. This is a qualitatively similar result to the slope found above for the ^{89}Y, but since the latter studies were done in a field of 50 kG and the present data were taken in zero field, a quantitative comparison may not be warranted. See Sect. 3.6 for further discussion of T_1 below T_c.

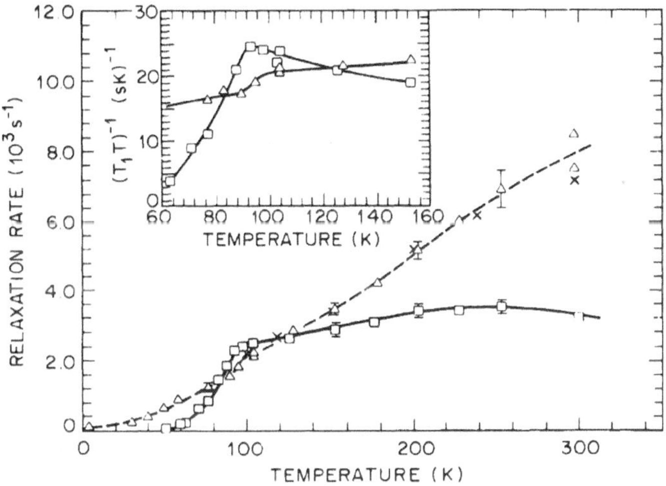

Fig. 3.3 The ^{63}Cu relaxation rate $1/T_1$ is shown plotted versus T for Cu(1) (crosses, triangles) and Cu(2) (squares) in a YBCO7 sample ($T_c \simeq 92$ K) [129]. The NQR rate is shown, which is 3 times the formally defined rate $1/T_1$

3.2.2 Establishing the Site Assignment for the Cu(1) and Cu(2) NQR Spectra

Following the publication of [129] (and [130]) there was a brief period of controversy, caused by the fact that the identification of the NQR frequencies with the Cu(1) and Cu(2) sites in these papers was reversed from the correct assignment stated above. The T_1 results are valid when attributed to the correct sites, but a large portion of the discussion, based on the incorrect identification, would need to be revised [129, 130]. Since a number of discussion points have also been rendered moot by later insights and improved understanding, we shall attempt no such revision here. The incorrect assignment was made on the basis of (i) similarity of normal-state T_1 curves between the ^{63}Cu(1) (Fig. 3.3) and the ^{89}Y (Fig. 3.2) and (ii) on the relative intensities of the 21 and 31.5 MHz NQR lines in [129] as well as in other papers [131, 132]. These intensities were essentially reversed from the 2:1 ratio expected from the Cu(2):Cu(1) site count, showing the general unreliability of NQR intensities recorded at different frequencies.

Following these papers, other papers appeared whose main purpose was to document the correct site assignment. References [133, 134] argued the correct assignment on the basis of data for the EFG tensor. After all, the Cu(2) site has very nearly axial symmetry, while Cu(1) does not, features which are expressed very strikingly in the EFG tensor. Meanwhile, [135] argued the case on the basis of ^{63}Cu relaxation studies conducted on a sample of $GdBa_2Cu_3O_7$. The Gd^{3+} ions each possess a localized moment of nearly $7\mu_B$, which fluctuates as a consequence of exchange couplings with the other Gd^{3+} moments. The nearby Cu(2) nuclei are strongly relaxed by fluctuating dipolar fields from these moments, whereas, the Cu(1) are largely unaffected. These observations confirm the correct assignment.

The $(-\frac{1}{2}, \frac{1}{2})$ Quadrupolar Powder Pattern for Cu(2) in YBCO7

The authors of [129, 130] also conducted NMR intensity studies for the purpose of site verification, partitioning the $(-1/2, 1/2)$ NMR line intensity for YBCO7 into Cu(2) and Cu(1) contributions [137] (see also [133]). The intensity of the NMR spectrum was calibrated by comparison with spectra from a known quantity of Cu_2O, yielding an absolute intensity of 2.7 ± 0.3 Copper ions per unit cell of the YBCO7. The Cu(2):Cu(1) intensity ratio was found to be 2 within experimental error. A byproduct of the latter exercise was a very precise fit to the $\eta = 0$, second-order quadrupolar powder pattern for the $(-1/2, 1/2)$ (central) transition at $T = 105\,\text{K}$ [137], using the same procedures which gave sample powder patterns in Figs. 2.2 and 2.4. This fit is shown in Fig. 3.4, both as spin echo intensity data recorded at pulse separation $\tau = 20\,\mu\text{s}$ and as the extrapolated, $\tau = 0$ curve with its fit to a quadrupolar powder pattern form (see Fig. 2.2) with Gaussian broadening shown as a dashed line. The residual area above the dashed line fit curve is identified as the Cu(1) site intensity. See [137] for a detailed discussion. Shift values for the Cu(2) are also deduced from this fit (see below).

3.2.3 *Variation of the ^{89}Y Shift* $K_{89}(x)$ *with x for YBCO6+x*

A study of the doping dependence of the ^{89}Y NMR shift $K_{89}(x)$ in YBCO6+x by Alloul et al. [136] uncovered some very interesting behavior regarding the Pauli paramagnetism of this system. For $x \geq 0.4$ they found an essentially linear increase $K_{89}(x) \propto (x - 0.4)$, below which K_{89} levelled off to a value which remained constant throughout the insulating phase. It was concluded that the latter value of shift $K_{89} \simeq 2.25 \times 10^{-5}$ relative to the shift of YCl_3 is actually *the zero of* K_{89} *in this system.* The discrepancy with the YCl_3 was attributed to a difference in the chemical shifts of the two systems. However, similar results by Balakrishnan et al. [154] included a study of chemical shifts suggesting that the actual chemical shift in YBCO6+x at low values of x is somewhere above 200 ppm relative to YCl_3, contrary to the foregoing conclusion. Eventually, the latter value gained acceptance as the correct zero for the ^{89}Y spin paramagnetic shift in YBCO.

The behavior of $K_{89}(x)$ with doping is shown in Fig. 3.5. The change in K_{89} above x = 0.4 is seen to be *negative* and is attributed to the core polarization of the Y^{3+} 4*d* orbitals hybridized into the conduction band in the CuO_2 planes of YBCO6+x. The Pauli paramagnetism varies in a similar fashion to K_{89}, so that in a plot of $K_{89}(x)$ versus $\chi_m(x)$ the behavior for $x > 0.4$ is linear with a slope $dK_{89}/d\chi_m = -4.6\,(\text{emu/mol})^{-1}$. This is much smaller than would correspond to the full core polarization HF field, showing that the 4d admixture is very slight. The authors comment that this behavior seems contrary to the proposed separation of spin and charge behavior of the RVB model [123, 124].

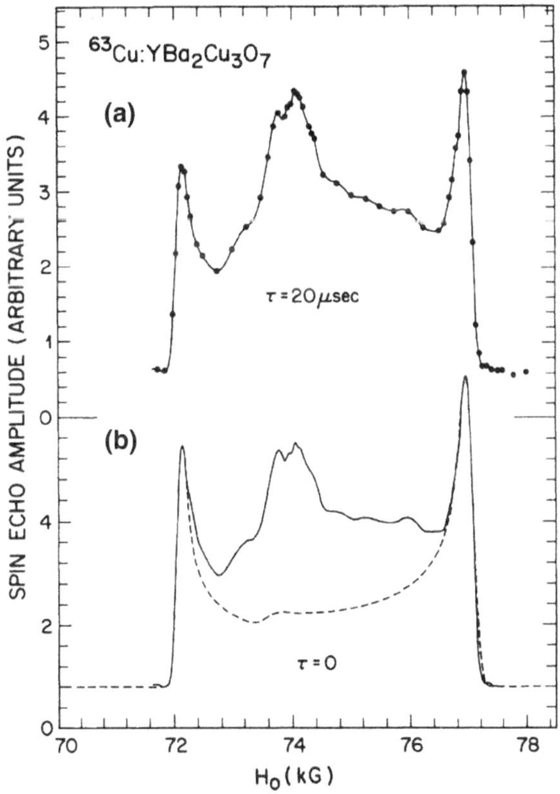

Fig. 3.4 NMR field-sweep spectrum of ^{63}Cu in YBCO7 at $T = 105\,\text{K}$ and an NMR frequency of 84.07 MHz. Part (**a**) is uncorrected data taken with a spin-echo pulse separation of 20 μs. In part (**b**) the signal has been extrapolated back to $\tau = 0$ using measured values of T_2 at various points in the spectrum. The outer singularity peaks are characteristic of an $\eta = 0(-1/2, 1/2)$ quadrupolar powder pattern. Such a powder pattern has been fitted to these peaks (dashed line), and serves to distinguish the Cu(2)-site intensity from that of the Cu(1) sites, which lies above the dashed line. There is a spurious peak to the left of center thought to be caused by accidental powder orientation in this otherwise random sample. The calculated Cu(1) intensity with this scheme is $\simeq$54% of that assigned to the Cu(2) sites

3.2.4 *Physical Models for the Relaxation of ^{63}Cu(2) in YBCO7*

One of the major challenges to the NMR community and to the theorists as well has been the interpretation of the copper-site T_1 data, which is to say the magnitude of T_1, its *anisotropy*, and its temperature dependence. We discuss the temperature dependence of T_1 (for all sites) in Chap. 4, and as of this writing, it is still an unsettled issue. The magnitude and anisotropy of T_1, however, came to light early and were the main influences which brought about the Mila-Rice discussion and formulation

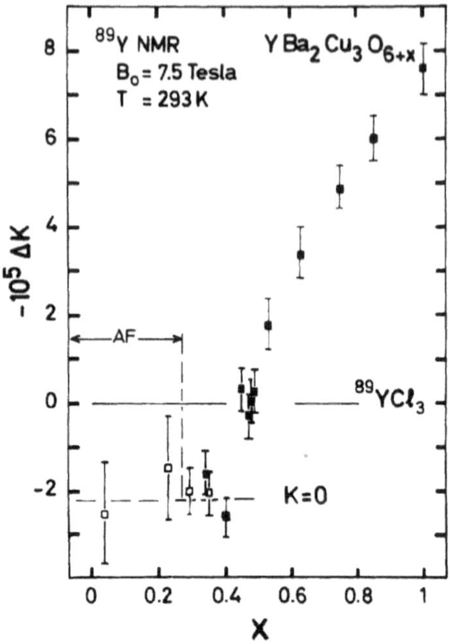

Fig. 3.5 The room temperature ^{89}Y NMR shift in $YBa_2Cu_3O_{6+x}$ is plotted for a series of x value ranging from nearly zero to one. This is interpreted (see text) to mean that the Pauli paramagnetism vanishes below $X = 0.4$ and the shift value below that point is simply the difference in chemical shifts between YBCO6 and YCl_3

of the problem. We review here two background developments which highlighted the situation, namely the model treatments of T_1 in experimental papers by Walstedt et al. [138] and by Pennington et al. [139].

Reference [138] reported the first data on the anisotropy of the ^{63}Cu(2) T_1, finding $T_{1c}/T_{1ab} \simeq 3.5$. This was done with a random powder sample, using NQR ($\theta = 0$), and the up-field and down-field singularities of the $(-1/2, 1/2)$ powder pattern ($\theta \simeq 43°$ and $\theta = 90°$, respectively). The data were not first-rate, but the result for T_{1c}/T_{1ab} held up in later studies with oriented powders [143]. The paper goes on to discuss the feasibility of explaining the foregoing T_1 data using the typical formulation of spin-lattice relaxation in 3d metals given by Yafet and Jaccarino [82] and discussed in detail in Sect. 3.3. The hyperfine constants are well-known for the Cu^{2+}, and the density of states is derived from the susceptibility, where the orbital component is estimated and subtracted off. The result is, if anything, an overestimate for $n(E_F)$. The normally dominant orbital relaxation mechanism is vitiated in YBCO by the nearly pure $d_{x^2-y^2}$ ground state, which has negligible matrix elements for the orbital HF process. As a result, the estimated relaxation rate with this scheme falls far short of the magnitude of the observed rates, and the peak anisotropy available falls well short of the measured ratio. The standard d-band relaxation model was concluded to be incapable of accounting for the cuprate T_1 data.

3.2.5 *Introduction of the* Spin Hamiltonian *Model for the* Cu^{2+}

A different tack was taken by Pennington et al. [139]. The Cu(2) relaxation tensor was measured using NMR on single crystal samples, finding similar numbers, but an anisotropy of 4.2, rather larger than the 3.5 reported in [138]. This discrepancy is most likely a difference in samples. In any case it has never been resolved. For the analysis of T_1 these authors adopt a model of fluctuating Cu^{2+} local moments, adopting a *spin Hamiltonian* form for the spin–HF tensor [144] and simply deducing the anisotropy of the spin–HF field from the corresponding anisotropy of T_1 for Cu(2) given above. Simultaneously, however, Takigawa, Hammel et al. announced their result that the c–axis NMR shift changed very little below T_c, while other components underwent substantial changes [135, 140]. They concluded that the spin paramagnetism declined effectively to zero for $T \ll T_c$. It followed that the net c–axis spin–HF component for Cu(2) is nearly zero! From this observation sprang the Mila-Rice model for Cu(1, 2) HF couplings, presented in Sect. 3.3. Both papers introduced a local moment-based calculation of the orbital susceptibilities on the *Cu*$^{2+}$ sites, which, along with the spin and orbital HF tensors, constitutes the *spin Hamiltonian* picture for the detailed physics of the Cu^{2+} ions [145]. We shall discuss this point of view in detail in Sect. 3.3.

In the meantime, it is interesting to consider the vanishing of the cuprate spin susceptibility $\chi_s(T)$ at temperatures $T \ll T_c$, postulated above, in the context of all of the difficulties over this point which occurred vis-a-vis the *simpler* superconductors discussed in Sect. 2.2.2. We recall that vanishing of $\chi_s(T)$ required that the inverse lifetime from spin-orbit scattering must be small compared with the energy gap [52] (see (2.2.8)). That this must be so for YBCO7 results from the combined facts that (i) Cu is a very light element with relatively small spin-orbit coupling and (ii) energy gaps in the cuprates are roughly an order of magnitude larger than those found for sp-band superconductors. It is reasonable to suppose, then, that $\hbar/\Delta(0)\tau_{so} \ll 1$ (2.2.8a), *even in the mixed state*, from which it follows that $\chi_s(T \ll T_c)/\chi_s(T > T_c) \ll 1$.

3.2.6 *Experimental Breakthrough: Oriented Powder Samples*

Late in 1987 a paper appeared which reported the technique of creating powdered samples of YBCO in which each crystallite is oriented with its c axis parallel to an applied magnetic field [117]. This scheme relies upon the anisotropy of the (orbital) susceptiblity to provide the torque necessary to align each particle with the field. The other necessary condition has to do with the morphology of the polycrystalline material to be oriented. It is assumed that upon reducing sample material synthesized in polycrystalline form to fine powder, each particle of such a powder will be a single crystal, and thus, free to orient with its c axis along the applied field. Any particle consisting of two or more crystallites will not align properly.

The success of this technique is therefore dependent on the particular cuprate under study. YBCO is generally a good candidate for this effect, where one can generally expect the Cu(2) ($-\frac{1}{2} \leftrightarrow \frac{1}{2}$) NMR line to assume a width of 100 Oe or less [143]. Following the original announcement, many NMR studies have appeared using oriented samples [134, 143]. Powder orientation results in an enormous improvement in the NMR signal strength, where signals requiring accumulation of $\sim 10^5$ echo traces for adequate noise resolution [130] become visible on a single oscilloscope trace. The Cu(1) NMR signal is also greatly improved, where the quadrupole splitting along the c axis corresponds to an extraordinarily small component of the quadrupolar tensor, resulting in a weakly split line [143]. The real significance of this method is that one can now measure the anisotropy of T_1 with a conveniently large signal [143], although remarkably good results were obtained with milligram-scale single crystals [139]. To measure oriented samples in a transverse field, one customarily sets the oriented particles in epoxy. Field oriented samples yield a 2D powder pattern in a transverse field, which is also important for studies of the ^{17}O and ^{89}Y NMR parameters.

3.3 The Mila-Rice-Shastry Model: A Universal HF Tensor for the Cuprates

The Mila-Rice-Shastry model [118, 146] for HF coupling with itinerant Cu^{2+} holes in a superconducting cuprate clarified a number of murky points regarding NMR shift and T_1 behavior. Important features were (i) transferred HF couplings and (ii) the ability to account for observed behavior with a single "band" of excitations, and (iii) the quantum chemistry framework for the HF Hamiltonian which emerged. Looking back, this was an important turning point in our understanding of HF interactions and of their dynamics. Previous workers had tried to formulate the NMR shift and relaxation in terms of Cu^{2+} local moment dynamics driving purely on-site HF couplings, but found peculiar results when trying to fit the extraordinarily large T_1 anisotropy at the Cu(2) site [139]. The demise of the local HF coupling picture was sealed by the observation that the Cu(2) *c*-axis shift term attributable to 3d-spin moments was vanishingly small [135, 140], while, in contrast, *c*-axis HF fluctuations dominated the T_1 process [138, 139, 143]. It was clear that there were two or more independent contributions to the c-axis HF coupling which vanished when superimposed to form the net shift, but produced fluctuations which did not cancel and were in fact very large.

The Cu(2)-Cu(2) Transferred HF Coupling

The resolution of the vanishing c axis spin shift question was concluded to be a highly unusual transferred HF coupling between a Cu-site nuclear spin and the four *nn* Cu spin moments, for which it is difficult to find a significant precedent. The putative isotropic transferred *nn* coupling may be written

$$\mathcal{H}_{Tr} = B \sum_{i(nn)} \boldsymbol{I} \cdot \boldsymbol{S}_i, \tag{3.3.1}$$

where the sum is over the four nearest-neighbor Cu(2) sites to the site in question.

Nonetheless, it was argued that the only alternative to this was HF coupling with mobile holes on the planar oxygen lattice. Any such hole state would need to hybridize with the Cu(2) 4s state in order to explain the large, isotropic HF coupling. Suitable oxygen hole states for this purpose were argued to be energetically unfavorable [118]. It was also argued by Mila and Rice that doped holes would be in singlet states which have negligible HF coupling with nuclear moments [157]. Whether this is correct or not is unresolved. To our knowledge, however, no unidentified HF effect, static or dynamic, has ever been reported to contradict this conclusion.

The result is a quantum fluid of Cu^{2+} quasiparticle holes which is expected to be the sole source of HF effects in the normal state and to form spin pairs in the superconducting ground state. Hybridization with the O^{2-} neighbors has been proposed as the source of spin HF effects for the $^{17}O(2, 3)$ sites by Shastry [146], and hybridization with the Y^{3+} 4d orbitals has been demonstrated by Alloul et al. [136]. That this is a correct picture is evidenced by the vanishing of $1/T_1$ at $T \ll T_c$ for the $^{63,65}Cu$, ^{17}O, and ^{89}Y NMR lines for which data have been reported, as well as of the spin-paramagnetic NMR shifts.

3.3.1 General Definition of the HF Tensors

Before entering into the details of extracting the HF tensors from shift and susceptibility data, we set down a general framework for the structure of all HF couplings to be discussed from this point forward. The YBCO HF tensor components are defined by

$$\mathcal{H}_{HF} = \sum_{i,\alpha} I_{i\alpha}^{(1)} \left[A_\alpha^{(1)} S_{i\alpha} + B^{(1)} \sum_{j(nn)} S_{j\alpha} \right] + \sum_{i,\alpha} I_{i\alpha}^{(2)} \left[A_\alpha^{(2)} S_{i\alpha} + B^{(2)} \sum_{j(nn)} S_{j\alpha} \right]$$
$$+ \sum_{i,\alpha} {}^{17}I_{i\alpha} \sum_{j(nn)} C_\alpha S_{j\alpha} + \sum_{i,\alpha} {}^{89}I_{i\alpha} \sum_{j(nn)} D_\alpha S_{j\alpha}. \tag{3.3.2}$$

The four terms in (3.3.2) refer, respectively, to nuclei on the Cu(1), Cu(2), O(2, 3), and Y sites. In the first two terms the *nn* couplings $B^{(1,2)}$ are introduced as defined in the foregoing paragraph. The coefficients $A_\alpha^{(1,2)}$ are composites of the on-site Cu(1, 2) HF couplings which are derived from the spin Hamiltonian model below. For the O(2, 3) and Y, the sums on j cover two and eight *nn* Cu(2) sites, respectively. Omitted are oxygen sites O(1) and O(4) for the sake of simplicity. Their shift and relaxation behavior have been recorded, but are not essential to our understanding of the CuO_2 planes where the superconductivity resides. The ^{63,65}Cu(1, 2) nuclear spins are labelled $\boldsymbol{I}_i^{(1)}$ and $\boldsymbol{I}_i^{(2)}$ in an obvious notation. The ^{17}O(2, 3) spins are labelled $^{17}\boldsymbol{I}_i$ and the ^{89}Y spins are labelled $^{89}\boldsymbol{I}_i$.

Note that all HF effects are assumed to be driven by the Cu^{2+} spin operators $\boldsymbol{S}_j$, which are rendered itinerant by the hole doping as in the Zhang-Rice model [125]. Our viewpoint regarding hole doping is that there are no HF effects which stem directly from the doped holes [157]. Such effects have been discussed extensively in the literature, and we shall review these discussions where appropriate. It is not felt, however, that convincing evidence for oxygen hole-band HF effects has ever been developed for the cuprates.

Evaluation of the HF Coefficients

The clearest and simplest way to evaluate HF coefficients (e.g. $A_\alpha^{(1)}$, $A_\alpha^{(2)}$, C_α, and D_α in (3.3.2)) is by means of analysis of the NMR shifts and the measured susceptibility. In the case of YBCO7-x which we are concerned with here, there is the complication of partitioning the measured susceptibility between the Cu(1, 2) sites and neighbors, principally because modeling the behavior of the Cu(1) sites involves a certain amount of guesswork. The spin-Hamiltonian model is used to estimate the orbital terms. In the end, MR simply partitioned $\chi_s(T)$ equally between the three copper sites of the formula unit. Efforts in the literature to improve on this have only been partially successful [179].

To proceed, then, we first estimate the orbital (Van Vleck) susceptibilities and NMR shifts by means of the spin Hamiltonian model [145]. The core diamagnetic term can be estimated from tables. Subtracting these terms from the measured total then yields the spin susceptibility $\chi_s(T)$. Provisions to be described below will be made for its anisotropy, which for the Cu(2) sites will be $\propto g_\alpha^2$, α = ab, c, according to the spin Hamiltonian.

HF tensor components will then be derived as follows. We take the ^{17}O(2, 3) sites as a typical example, where $\mathcal{H}_{HF17} = \sum_{i,\alpha} C_\alpha^{17} I_{i\alpha} \sum_j S_{j\alpha}$ is the O(2, 3) nuclear spin Hamiltonian term from (3.3.2). Taking the expectation value of this term in the presence of a field H along the β axis and equating this result to a nuclear shift Zeeman term $-\gamma_{17}\hbar H \sum_i I_{i\beta} K_\beta$, we find

$$K_\beta(T) = \frac{2C_\beta}{\hbar\gamma_{17} g_\beta \mu_B N_A} \chi_{s\beta}(T) = \alpha_{17\beta}\chi_{s\beta}(T), \tag{3.3.3}$$

where $\alpha_{17\beta}$ is the shift coefficient and $\chi_{s\beta}(T) = -g_\beta \mu_B N_A \langle S_{j\beta}\rangle / H$ is the molar susceptibility ($\langle S_{j\beta}\rangle < 0$ is independent of j) of a Cu(2) site.

In practice the shift coefficient is determined experimentally by dividing a measured spin-paramagnetic shift by a "best estimate" of the corresponding molar susceptibility. The tensor C_β has the units of energy, so what we will tabulate is the quantity

$$C'_\beta = N_A \mu_B \alpha_{17\beta}/2 = C_\beta/\hbar\gamma_{17} g_\beta, \tag{3.3.4}$$

in units of "kOe/μ_B", since this is the HF field which would be found if there were a magnetization of one μ_B on each Cu(2) site. It is divided by 2, because each O(2, 3) has two Cu(2) neighbors. Appropriate definitions of the HF constant for each case discussed will be given as they come up.

We finally want to emphasize one point, which is that the scale of the HF constants will depend on the value of the $\chi_s(T)$ tensor adopted. It is given in Table 3.4 and corresponds to T = 100 K. No other standard will be used for HF coefficients for YBCO7-x, including for the deoxygenated phases. As we shall see, the shift values for YBCO7 are fairly well established, and if more accurate values for the $\chi_{s\beta}(100\,\text{K})$ are determined, it is a simple matter to scale all of the HF coefficients uniformly. We proceed, then, with the determination of the HF coefficients.

3.3.2 *Extracting the Cu(1, 2) HF Tensors for YBCO7 from Shift and Susceptibility Data Using the Spin Hamiltonian Model*

As a preamble to their quantum chemical calculation of hybridized wavefunctions for the Cu(1, 2) sites, Mila and Rice [118] analyzed the measured shifts and susceptibility for YBCO7 with the aid of a spin Hamiltonian model, yielding both spin and orbital HF tensors for the Cu(1, 2) sites, as well as crystal-field splittings for the Cu(2) site. Although the spin Hamiltonian model is most appropriate for insulators, it appears to provide an accurate description of HF effects in the narrow-bandwidth doped cuprate superconductors.

Since late 1988 when this was first proposed, a good deal of additional data for all the experimental quantities has been reported. Although the original data used served a very important purpose, newer data were published with smaller error bars, and we believe that they lead to a picture with better internal consistency. Accordingly, we discuss here the original K and χ analysis, but re-work the numbers at the same time. The changes are small, but appear to be significant.

The Spin and Orbital NMR Shift Tensors for Cu(1) in YBCO7

We shall use the notation $K_{tot}^{(1,2)}(a, b, c) = K_{orb}^{(1,2)}(a, b, c) + K_s^{(1,2)}(a, b, c)$ to represent the total, orbital, and spin shifts along the a, b, c axes for Cu(1, 2). In Table 3.1 we list the original Cu(1)-site shifts [135, 140] and other data which were published a bit later. The total shift values listed were measured at $T = 100$ K, and the orbital shifts, which are assumed to be temperature independent, were measured at 4.2 K, where it is assumed that $K_s^{(1,2)}(a, b, c) = 0$. The spin components listed then correspond to 100 K. Note that the differences between these two sets of data are not large, but are outside of the stated error bars in a number of cases.

Spin and Orbital NMR Shift Tensors for Cu(2) in YBCO7

For the Cu(2) NMR shift data in Table 3.2 there are additional data sources listed, and we also include susceptibility data. Since it is not feasible to resolve a and b axis data for the Cu(2) shift, nor for the susceptibility, the in-plane data are labelled "ab". Again, the original Mila-Rice data are given on the top line, and other sources of data

Table 3.1 Table comparing Cu(1)-site shift values from the original Mila-Rice data [135, 140] and other data which appeared later. The data from [147] are used in the analysis discussed in the text. The spin components are obtained by subtraction: $K_s = K_{tot} - K_{orb}$. Shift values are in (%). The numbers in parentheses are uncertainties in the last digit quoted

Source	$K_{tot}^{(1)}(a)$	$K_{tot}^{(1)}(b)$	$K_{tot}^{(1)}(c)$	$K_{orb}^{(1)}(a)$	$K_{orb}^{(1)}(b)$	$K_{orb}^{(1)}(c)$	$K_s^{(1)}(a)$	$K_s^{(1)}(b)$	$K_s^{(1)}(c)$
[135, 140]	1.34(2)	0.61(2)	0.60(8)	1.18(2)	0.43(2)	0.31(8)	0.16(2)	0.18(2)	0.29(8)
[139]	1.38(7)	0.55(7)	0.60(4)						
[147]	1.32(2)	0.56(2)	0.59(1)	1.08(4)	0.27(4)	0.25(1)	0.25(4)	0.29(4)	0.33(1)

Table 3.2 Table comparing Cu(2)-site shift values for YBCO7 from the original Mila-Rice data [135, 140] with other sets of data which have subsequently been reported. Data from the bottom line are employed for the analysis discussed here [147]. No distinction is made between the a and b planar axes. The susceptibility components given for [137] are derived by the procedure stated in the text. We note that the spin components are obtained by subtraction: $K_s = K_{tot} - K_{orb}$. Shift values are in (%) and susceptibilities in units of (10^{-4} emu/mol f.u.). The numbers in parentheses are uncertainties in the last digit quoted

Source	$K_{tot}^{(2)}(ab)$	$K_{tot}^{(2)}(c)$	$K_{orb}^{(2)}(ab)$	$K_{orb}^{(2)}(c)$	$K_s^{(2)}(ab)$	$K_s^{(2)}(c)$	$\chi_{tot}(ab)$	$\chi_{tot}(c)$
[135, 140]	0.61(2)	1.27(8)	0.24(2)	1.35(8)	0.37(2)	−0.08(8)	2.50	3.39
[139]	0.59(4)	1.267(1)						
[137]	0.55(5)	1.30(3)					2.24(10)	3.61(10)
[148]			0.25(2)	1.28(2)				
[147]	0.58(2)	1.27(1)	0.28(1)	1.28(2)	0.30(1)	−0.01(1)		

are listed below. There is reasonably good consistency between the last two sources listed. We shall use the bottom line numbers for analysis here. The susceptibility components attributed to [137] were obtained in the following way. The actual data in [137] are measurements on a random powder sample. The 100 K susceptibility from these data (2.70×10^{-4} emu/mol f.u.) has been partitioned between $\chi_{tot}(c)$ and $\chi_{tot}(ab)$ assuming the ratio $\chi_{tot}(c)/\chi_{tot}(ab) = 1.61$, as reported in [149]. This is, to our knowledge, the largest anisotropy ratio which has been reported for an oriented powder sample, and therefore corresponds to an unusually high degree of crystalline alignment. The resulting values for $\chi_{tot}(ab)$ and $\chi_{tot}(c)$ are given in the table and are used in the analysis discussed in the text.

Determination of the Orbital Susceptibility Tensors

Our next step is to use the Cu(1, 2) shift tensors from the bottom lines of Tables 3.1 and 3.2 to determine the orbital susceptibilities, using the relation

$$\chi_{orb}(a, b, c) = K_{orb}(a, b, c)/\alpha_{orb}, \quad (3.3.5)$$

Table 3.3 Table listing orbital susceptibility components derived from the orbital shifts from the bottom lines of Tables 3.1 and 3.2 using (3.3.5). Also, total susceptibilities from second line of Table 3.2 are listed here after correction for atomic core diamagnetism (see text). Susceptibilities are given in units of (10^{-6} emu/mol) for individual sites and (10^{-6} emu/molf.u.) for χ^{tot}_{s+orb}

$\chi^{(1)}_{orb}(a)$	$\chi^{(1)}_{orb}(b)$	$\chi^{(1)}_{orb}(c)$	$\chi^{(2)}_{orb}(ab)$	$\chi^{(2)}_{orb}(c)$	$\chi^{tot}_{s+orb}(ab)$	$\chi^{tot}_{s+orb}(c)$
76.5	19.1	17.7	19.8	90.7	399	536

where $\alpha_{orb} = 2\langle r^{-3}\rangle/N_A$. Equation (3.3.5) may be employed for either Cu site. We take $\langle r^{-3}\rangle = 6.3$ a.u. from [118], yielding $\alpha_{orb} = 141\,(\text{emu/mol})^{-1}$. The resulting orbital susceptibility values for the Cu(1, 2) sites are listed in Table 3.3. Note that the a and b axis values are identical for Cu(2). Also tabulated are susceptibility values labelled "s+orb", which are obtained with the following equation:

$$\chi^{tot}_{s+orb} = \chi^{tot}_{s} + \chi^{tot}_{orb} = \chi_{tot} - \chi_{dia} \tag{3.3.6}$$

where χ_{tot} is the experimental value, and where we use the (isotropic) $\chi_{dia} = -1.75 \times 10^{-6}$ emu/mol from [118]. The measured susceptibilities cannot distinguish between the a and b axes.

Determination of the Spin Hamiltonian Parameters

Using the spin Hamiltonian expressions [145] $\chi_{orb}(c) = 0.70\,(8\mu_B^2/\Delta_0)$ and $\chi_{orb}(ab) = 0.70\,(2\mu_B^2/\Delta_1)$, we may now use the numbers in Table 3.2 to estimate the crystal field energies $\Delta_{0,1}$. The numerical factors 0.70 in these formulas are reduction factors based on measurements [119] and calculations [120] of covalency in the CuO_2 planes.

Inserting the values of $\chi^{(2)}_{orb}(ab, c)$ from Table 3.3 then leads to $\Delta_0 = 2.00$ eV and $\Delta_1 = 2.29$ eV. The calculations of McMahan et al. [150] suggest $\Delta_{0,1}$ to lie in the vicinity of 2.0 eV and to differ by ~10%. We can now evaluate the g-factor expressions $g_\perp = 2 - 2\lambda/\Delta_1$ and $g_\parallel = 2 - 8\lambda/\Delta_0$, where λ is the spin-orbit coupling parameter. The free ion value of $\lambda \simeq -830\,\text{cm}^{-1}$ ([145], Table 7.6). Mila and Rice adopted the value $\lambda = -710\,\text{cm}^{-1} = -0.088$ eV. We suggest that a slightly greater reduction factor is appropriate here, such as the value 0.75 given by Owen and Thornley for Ni^{2+} in MgO [121]. This yields $\lambda = -0.077$ eV, which we employ henceforward. With this value we find

$$g_\perp = 2.067; \quad g_\parallel = 2.308. \tag{3.3.7}$$

The expected value for the anisotropy A_s of the spin susceptibility then becomes $A_s = \chi_{s\parallel}/\chi_{s\perp} = g_\parallel^2/g_\perp^2 = 1.247$. We note that g-factor anisotropies in metals are not unusual, as may be seen in data reported for Cu and other noble metals [33].

Our next task is to partition the measured susceptibility between the Cu(1) and Cu(2) sites. Where Mila and Rice assumed the same isotropic spin susceptibilities for Cu(1) and Cu(2), we shall use the same *axial susceptibility tensors* for both sites.

As discussed in [118] this is an oversimplified viewpoint for the Cu(1) site, for which the error incurred is hopefully not too large.

We equate the "s+orb" values in Table 3.3 to expressions using the same axial spin susceptibility tensor components for the Cu(1) site (a axis) and the Cu(2) sites (c axis). For this purpose we introduce the variable $\chi_{s\perp}$ for a single-site transverse component and $\chi_{s\|} = A\,\chi_{s\perp}$ for the longitudinal component, where A and $\chi_{s\perp}$ are to be determined. This leads to the following equations for A and $\chi_{s\perp}$:

$$\chi^{tot}_{s+orb}(c) = 2\,\chi^{(2)}_{orb}(c) + \chi^{(1)}_{orb}(c) + (2A+1)\,\chi_{s\perp} \qquad (3.3.8a)$$

$$2\,\chi^{tot}_{s+orb}(ab) = 4\,\chi^{(2)}_{orb}(ab) + \chi^{(1)}_{orb}(a) + \chi^{(1)}_{orb}(b) + (A+5)\,\chi_{s\perp}. \qquad (3.3.8b)$$

Inserting the numbers from Table 3.3 we solve (3.3.8), finding $A = 1.166$ and $\chi_{s\perp} = 1.01 \times 10^{-4}$ emu/mol. This "experimental" value for A is somewhat smaller than the spin Hamiltonian g-factor-based estimate of 1.247 calculated in the previous paragraph. The spin susceptibility tensor components deduced here are then 1.01 and 1.18×10^{-4} emu/mol for $\chi_{s\perp}$ and $\chi_{s\|}$, respectively, as compared with the isotropic Mila-Rice value of 1.08×10^{-4} emu/mol.

It is interesting to inquire what the possible source might be for the foregoing discrepancy in values for A. First, it seems likely that the correct value would agree with the spin Hamiltonian model, since other aspects of the model are quite successful. Next, it seems unlikely that approximate treatment of the Cu(1) site behavior is the source of the discrepancy. For example, assuming the Cu(1) site susceptibility to be isotropic would only increase the deduced anistropy for the Cu(2) site by $\sim$1%. On the other hand, the deduced anistropy A is quite sensitive to the experimental anisotropy value, which was taken to be $\chi_t(c)/\chi_t(ab) = 1.61$ [149] in the foregoing analysis. If this were increased by less than 4% to 1.67, (3.3.8) would yield the the spin Hamiltonian value $\chi_{s\|}/\chi_{s\perp} = 1.247$. This result suggests that the correct value for the susceptibility anisotropy may actually be a few percent larger than the largest experimental value which has been reported.

3.3.3 *Quantum Chemistry of YBCO7 and the Hyperfine Tensor*

While it is to be expected that there is strong covalency between the Cu^{2+} moments and the oxygen ligands in the cuprate conducting planes, it seems quite surprising that there is a substantial admixture of the *4s wavefunctions of the nearest-neighbor Cu^{2+} ions* as posited in (3.3.1). Yet this is precisely what the quantum-chemical calculations show [118, 151]. We outline the calculation here and refer the reader to the original papers for the details [118, 151]. The tight-binding quantum-chemical calculations for Cu(2) are based on the diagram shown in Fig. 3.6, where the relevant orbitals are shown at each ionic site. The $d_{x^2-y^2}$ orbital is shown at the central (planar) Cu^{2+} ion, the p_σ orbitals on the oxygen ligands, and the 4s orbitals on the four nn Cu^{2+}

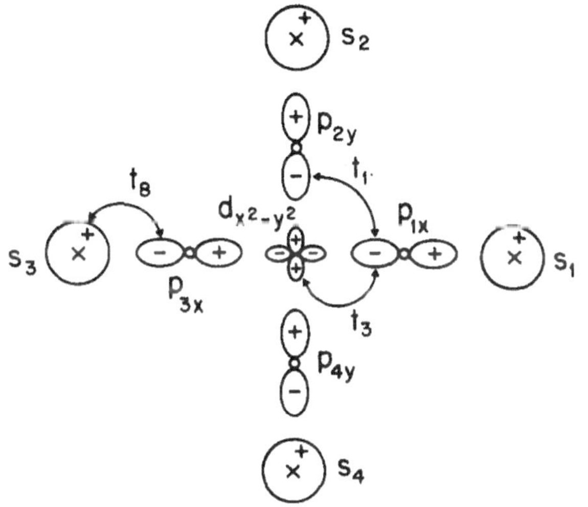

Fig. 3.6 A schematic representation of the $d_{x^2-y^2}$ orbital at a Cu(2) site and the p_σ orbitals of the O^{2-} ligands, as well as the 4s orbitals on the four nn Cu(2) sites, which are admixed to form the total wavefunction of the Cu(2) 3d-holes for the Mila-Rice quantum chemistry calculation (see text). The t_α's show hopping matrix elements, and the signs give phase conventions for admixed orbitals

ions. With the sign conventions given in the figure, the p and s-orbital admixtures which couple to the central $d_{x^2-y^2}$ orbital are p = $(p_{1x} - p_{2y} - p_{3x} + p_{4y})/2$ and s = $(s_1 - s_2 + s_3 - s_4)/2$. The energies are as follows: The 3d-orbital energy is taken as the origin, $\epsilon_d = 0$. The ligand on-site energies are taken to be $\epsilon_p = 4\,\text{eV}$ and $\epsilon_s \sim -6\,\text{eV}$. The hopping energies shown in Fig. 3.6 are $t_1 = 0.33\,\text{eV}$, $t_3 = 1.38\,\text{eV}$, and $t_8 = 3.9\,\text{eV}$. The Hamiltonian is then solved in conjunction with the somewhat more complex Cu(1)-site Hamiltonian.

The Wannier Wavefunction for Cu(2)

The Cu(2) tight-binding (Wannier) wavefunction is written

$$|\psi(2)\rangle = \alpha_d |d_{x^2-y^2}\rangle + \alpha_p |p\rangle + \alpha_s |s\rangle, \tag{3.3.9}$$

where the solution presented gives $\alpha_d = 0.9$, $\alpha_p = -0.34$, and $\alpha_s = -0.26$. In addition there is a coupling to the apical oxygen that is ligand to the nearest Cu(1) site and to the 4s state on that site. This coupling has some effect on the Cu(2) shift, but an essentially negligible effect on the Cu(2) T_1. For simplicity we shall omit these terms. The transferred isotropic HF coupling (3.3.1) from the nn Cu(2) sites may be estimated as follows. The 4s contact HF coupling from EPR data for atomic Cu [152] is

$$\frac{16\pi}{3} |\psi_{4s}(0)|^2 = 0.45 \times 10^{27}\,\text{cm}^{-3}. \tag{3.3.10}$$

Referring to (2.2.2) and (2.2.7), the corresponding 4s contact HF coupling is 2,080 kOe/μ_B. The total transferred coupling is then estimated to be $\alpha_s^2 \times 2080 = 141\,kOe/\mu_B$, as compared with $\sim 160\,kOe/\mu_B$ required to give the nearly vanishing c axis spin-paramagnetic shift. Mila and Rice note that the value stated is for atomic Cu, whereas Cu^{2+} will have an enhanced value of $|\psi_{4s}(0)|^2$ because of the ionic plus charge. There seems little doubt that the *nn* 4*s* admixture effect is the correct answer to the c axis HF anomaly.

The Composite Cu(2) Spin Hyperfine Tensor

Let us now state the remainder of the HF coupling terms for Cu(2) and then formulate the Mila–Rice T_1 relaxation tensor. The full HF tensor may be written [145],

$$\mathcal{H}_{HF}^{(2)} = \sum_{\alpha} A_{\alpha}^{(2)} I_{\alpha} S_{0\alpha} + B(2) \sum_{i(nn)} \boldsymbol{I} \cdot \boldsymbol{S}_{\mathrm{i}}, \tag{3.3.11}$$

with

$$A_{\alpha}^{(2)} = A_{cp}^{(2)} + A_{dip,\alpha}^{(2)} + A_{so,\alpha}^{(2)}, \tag{3.3.12}$$

where the three terms shown are the isotropic core polarization term, the dipolar term, and the *spin-orbit* term, and where $\alpha = (ab, c)$ for the axially symmetric Cu(2) site. The spin-orbit terms, being a second-order perturbation effect to higher crystal field states driven by the spin-orbit coupling, are a special feature of the spin Hamiltonian picture. They are actually an orbital field admixed into the spin HF terms.

The core-polarization term is sometimes given as proportional to $\langle r^{-3} \rangle$, which is a holdover from EPR studies where the origin of this term had not been identified.[2] Here we shall simply appeal to systematic studies, which give $A_{cp}/2\hbar\gamma_n \simeq -127\,kOe/\mu_B$. The dipolar and spin-orbit terms may be written (in energy units)

$$A_{dip,ab} = 2\gamma_n \hbar \mu_B \frac{2}{7} \langle r^{-3} \rangle, \tag{3.3.13a}$$

$$A_{dip,c} = -2\gamma_n \hbar \mu_B \frac{4}{7} \langle r^{-3} \rangle, \tag{3.3.13b}$$

$$A_{so,ab} = 2\gamma_n \hbar \mu_B \langle r^{-3} \rangle \left[-\frac{11\lambda}{7\Delta_1} \right], \tag{3.3.13c}$$

$$A_{so,c} = 2\gamma_n \hbar \mu_B \langle r^{-3} \rangle \left[-\frac{6\lambda}{7\Delta_1} - \frac{8\lambda}{\Delta_0} \right]. \tag{3.3.13d}$$

We use the same parameters used earlier to evaluate the Van Vleck orbital susceptibilities and in (3.3.7) for the anisotropic g-factor to calculate the spin HF tensor. These are $\lambda = -0.077\,\mathrm{eV}$, $\Delta_0 = 2.00\,\mathrm{eV}$, $\Delta_1 = 2.29\,\mathrm{eV}$, and $\langle r^{-3} \rangle = 6.3$ a.u. The resulting values for the Cu(2) spin HF tensor (3.3.11) are listed in Table 3.4 in units of kOe/μ_B (see Sect. 3.3.1). We also employ the reduction factor 0.70 derived from determinations of covalency [119, 120] for these parameters, as well as the core polarization HF field, in the second line of Table 3.4. The value of $B^{(2)}$ shown is adjusted to give minimum error for the Cu(2) spin-paramagnetic shift values $K_s^{(2)}(ab)$ and $K_s^{(2)}(c)$ (bottom line of Table 3.2), with the resulting estimated values $K_{sE}^{(2)}(ab)$ and $K_{sE}^{(2)}(c)$ also listed in the table. A good fit is seen to be obtained. To the far right are listed the components of the composite on-site HF tensor $A_{\alpha}^{(2)}$ defined by (3.3.12).

[2]The work of Heine [35] later revealed that it had no connection with $\langle r^{-3} \rangle$.

Table 3.4 Listing of values of spin HF tensor components for the Cu(2) sites, as derived in the foregoing paragraph. On the second line the free-ion component values have been corrected with a reduction factor 0.70 to account for covalency as mentioned above. A value of $B^{(2)}$ is stated (see (3.3.11)) which minimizes the errors in the spin-paramagnetic shift components given in the bottom line of Table 3.2. This value of $B^{(2)}$ is in reasonable accord with the estimate derived from quantum chemistry [118]. The spin susceptibility tensor associated with each Cu(1, 2) site deduced earlier is shown, along with the Cu(2) site spin-paramagnetic shift tensor corresponding to the HF parameters given. Finally, the composite on-site HF tensor components given by (3.3.12) are listed. All HF tensor components are in units of kOe/μ_B, with susceptibilities in 10^{-4} emu/mol

	$A^{(2)}_{cp}$	$A^{(2)}_{dip,ab}$	$A^{(2)}_{dip,c}$	$A^{(2)}_{so,ab}$	$A^{(2)}_{so,c}$	$B^{(2)}$	$\chi_{s\perp}$	$\chi_{s\parallel}$	$K^{(2)}_{sE}(ab)$	$K^{(2)}_{sE}(c)$	$A^{(2)}_{ab}$	$A^{(2)}_{c}$
Equation (3.3.13)	−127	113	−225	21	131	55						
Corr.	−89	79	−158	15	92	38	1.01	1.18	0.28%	−0.01%	5	−155

3.3.4 T$_1$ Ratios and Electron Spin Correlation Effects

Mila and Rice introduced the concept of electronic spin-spin correlation factors in their T_1 calculation, but did not use them in their discussion of T_1 anisotropy. Subsequent discussions of the T_1 problem in cuprates have made extensive use of this idea. Here, we find it necessary to consider the effect of spin-spin correlations to obtain agreement with experimental data. All HF constants are given in (kG/μ_B). The primed notation for this option introduced with (3.3.3) and (3.3.4) is suppressed in this section for simplicity.[3]

Using expressions similar to MR, we now present results for the Cu(2) relaxation tensor. The $1/T_{1ab,1c}$ have the form $H_{HF}^2(ab, c)\chi_s''(\omega_0)$, where the squared HF field parameters are

$$H_{HF}^2(c) = 2\,[A_{ab}^{(2)\,2} + 4\,B^{(2)\,2} + 8\,A_{ab}^{(2)}B^{(2)}\,\mathcal{K}_{01} + 4\,B^{(2)\,2}\,(2\,\mathcal{K}_{12} + \mathcal{K}_{13})] \quad (3.3.14)$$

and

$$H_{HF}^2(ab) = [A_{ab}^{(2)\,2} + A_c^{(2)\,2} + 8\,B^{(2)\,2} + 8\,(A_{ab}^{(2)} + A_c^{(2)})B^{(2)}\,\mathcal{K}_{01} + 8\,B^{(2)\,2}\,(2\,\mathcal{K}_{12} + \mathcal{K}_{13})], \quad (3.3.15)$$

where $\mathcal{K}_{ij} = 4\langle \boldsymbol{S}_i \cdot \boldsymbol{S}_j \rangle$ is the normalized spin-spin correlation between sites labeled i and j (see Fig. 3.6). The quantity $\chi_s''(\omega_0)$ has the dimensions of a correlation time, but also has in it the density of states and temperature factors that arise out of some specific model of quasiparticle dynamics appropriate to the cuprates. If the system were a Fermi liquid one would have $\chi_s''(\omega_0) \propto T$. Without such modeling this theory cannot give absolute values for T_1, but it can predict the anisotropy of T_1 at the Cu(2) site using the HF tensor data from Table 3.4. Having minimized the error in the spin paramagnetic shift tensor to estimate the value of $B^{(2)}$ given in the table, we insert the parameter values into the ratio of (3.3.15) to (3.3.14) to determine the expression,

$$\frac{T_{1c}}{T_{1ab}} = 2.64\frac{[1 - 1.69\,\mathcal{K}_{01} + 0.38\,(2\,\mathcal{K}_{12} + \mathcal{K}_{13}]}{[1 + 0.29\,\mathcal{K}_{01} + 0.99\,(2\,\mathcal{K}_{12} + \mathcal{K}_{13})]}. \quad (3.3.16)$$

Equation (3.3.16) is to be compared with experimental data giving values of T_{1c}/T_{1ab} of ~4.5 [139], ~3.6 [143], or the later value ~3.4 [153].[4] In any case, it is clear that the spin-spin correlation factors in (3.3.16) are going to have to play a role in order to achieve quantitative consistency with experiment. Regarding (3.3.16), since $\mathcal{K}_{01}$ is a nearest-neighbor correlation it must be negative, while $\mathcal{K}_{12}$ and $\mathcal{K}_{13}$ are

[3]In (3.3.2) and again in (3.3.11)–(3.3.13) hyperfine (HF) terms are introduced, with expressions given in energy units. In (3.3.4) a primed notation is introduced, giving the HF constants in the more convenient kG/μ_B units. In all subsequent discussion, the latter units will be used and the primed notation dropped. See, e.g., Tables 3.4, 3.6, and 4.1, etc.

[4]The difference between these two measurements is outside of error estimates, and can only be attributed to a difference in samples. The lower numbers agree with earlier measurements made on a random powder [138].

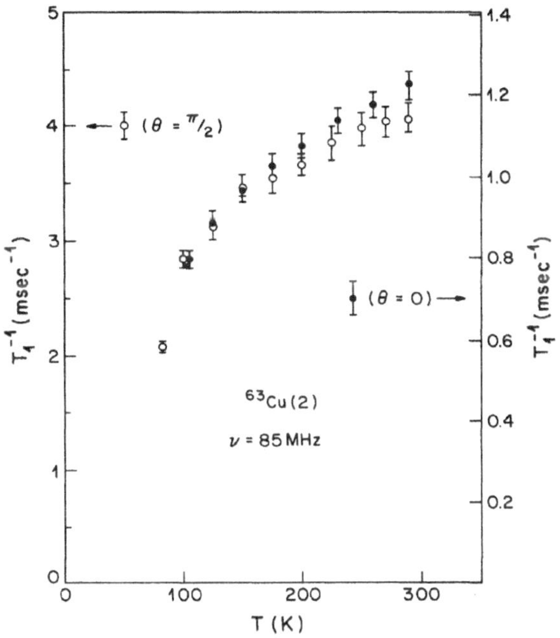

Fig. 3.7 $1/T_{1ab}$ ($\theta = \pi/2$) and $1/T_{1c}$ ($\theta = 0$) are plotted versus temperature over the range $90\,\mathrm{K} \le T \le 300\,\mathrm{K}$ for Cu(2) sites in an oriented powder sample of fully oxygenated YBCO7 [143]. The data are scaled so that T_1 values for the two orientations coincide at $T = 100\,\mathrm{K}$, so as to highlight the decrease of the T_{1c}/T_{1ab} ratio over range of T measured

correlations between neighbors on the same "sublattice", so to speak, so they will be positive and undoubtedly smaller in magnitude that $\mathcal{K}_{01}$. Thus, $\mathcal{K}_{01}$ will have a clear effect of increasing the ratio if its value is negative and appreciable. For example, if we neglect the $\mathcal{K}_{12,13}$ terms for the moment, then a value of $\mathcal{K}_{01} = -0.07$ would yield a T_1 ratio of 4. With only a modest amount of correlation we can obtain a realistic value for T_{1c}/T_{1ab}.

Temperature-Dependent Spin-Spin Correlations

It is not clear from the Mila-Rice presentation, either, whether they intended the correlations to be used as temperature-dependent parameters or not, but a number of subsequent developments have employed them in just that role. For example, temperature-varying correlations give a straightforward explanation for the variation found for the T_1 anisotropy in YBCO7 between 100 and 300 K [143]. These data are shown in Fig. 3.7, where it is seen that T_{1c}/T_{1ab} diminishes by barely 10% over that temperature interval, and is only barely outside of the error bars. Nonetheless, the trend persists over a series of data points and is clearly established.[5]

As we shall see in Chap. 5, extensive use has been made of the $\mathcal{K}_{ij}$'s as temperature-dependent parameters to describe the behavior of T_1. In the Uldry-Meier T_1 analysis, for example [195], the temperature dependence of all T_1's is described in terms of the spin-spin correlations and an effective correlation time. The theoretical basis for this

[5]It would have been desirable to have this effect confirmed by another experimental group. To our knowledge, no such confirmation has been reported.

approach is derived in Sect. 3.5. Meanwhile, we also recall that the first Mila-Rice paper [118] made no attempt to apply their model to the interpretation of the peculiar temperature dependence of the T_1 data. Indeed, theories addressing the behavior of $T_1(T)$ were fairly quick to follow, but are still under discussion, as will be seen in Chaps. 4 and 5.

3.4 Incorporating ^{17}O and ^{89}Y Data into the Mila-Rice-Shastry Picture

Following the Mila-Rice (MR) paper there were many new developments, some going beyond the scope of the present chapter. There were new data on ^{17}O NMR in YBCO, as well as a great deal of discussion regarding the question of a two-fluid picture of quasiparticle dynamics, i.e. whether there is evidence for a separate hole band on the oxygen sites. There were also efforts to treat the temperature dependence of the copper T_1, along with discussion of just how the Cu(1) chain sites in YBCO differ from the Cu(2). In the midst of these many challenging questions and almost immediately following Mila and Rice, reports on the peculiar behavior of the oxygen-deficient phases made a sudden appearance. These effects, known collectively as *the pseudogap* and their interesting history in the high-T_c field are the subject of Chap. 4.

We consider here the first reports on the behavior of the planar ^{17}O NMR, as well as the closely related ^{89}Y. Detailed theories which address the contrasting behavior of ^{63,65}Cu, ^{17}O, and ^{89}Y in YBCO are presented in Chap. 5. Following the discussion of the ^{17}O data for YBCO7, we consider again the possible occurrence of oxygen hole bands along with the Cu^{2+}-related (Zhang-Rice) quasiparticles in the total picture of fermion dynamics.

^{17}O(2, 3) NMR Data and Interpretation for YBCO7

By the middle of 1989, ^{17}O NMR studies on YBCO7 had been reported by several groups [156, 168–170]. Here we shall review and discuss the results reported by Takigawa et al. [156, 168] in some detail. A comparable set of results was reported by Horvatić et al. [169], covering a wider range of temperatures. The high temperature results for the quadrupolar and shift tensors for the O(2, 3) sites from the latter study were found to agree within error bars with those from [168], and the quadrupolar data from [170] is also in accord. The HF tensor is analyzed below and compared with the MR model calculation. There is some disagreement in detail, but this, we suggest, may arise from inaccuracies in the MR quantum chemistry results.

^{17}O spectra were taken on oriented powder samples of YBCO7 [168], for which ^{17}O-enriched oxygen gas was used to exchange ^{17}O into the sample material at a temperature of 900 °C. The resulting (nearly) complete NMR spectrum for all of the oxygen sites is shown in Fig. 3.8 for the case of field oriented along the *c* axis. The identification of all the peaks in this spectrum is a real tour-de-force of NMR shift and quadrupole splitting analysis. Although the *c* axis is expected to be one of the principal axes of the EFG tensors for each oxygen site, the EFG tensors are

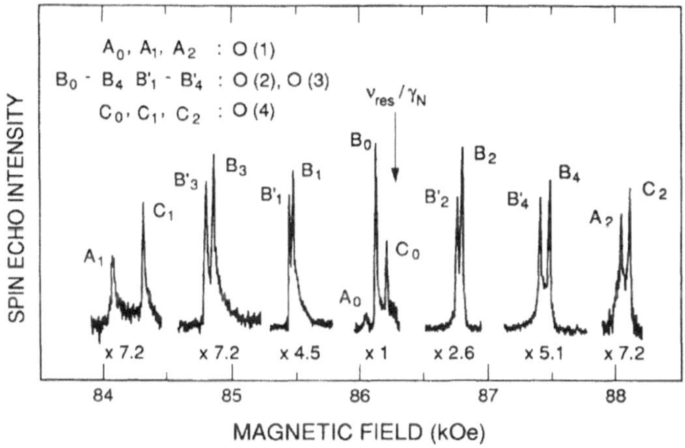

Fig. 3.8 ^{17}O NMR spectrum taken at 49.8 MHz on an oriented powder sample of YBCO7 with enriched ^{17}O content [168]. The field was oriented along the c axis. Sample temperature was 160 K. For ^{17}O $\gamma_n/2\pi$ was taken to be 0.577186 kHz/Oe. The sites were identified as labelled, with the (B_n, B_n') pairs of lines very close together for the O(2, 3) sites. See text for discussion of the quadrupolar and shift tensors for the O(2, 3) sites

decidedly non-axial in character. The peaks have been fitted using the $I = \frac{5}{2}$ second-order perturbation formulas [168]

$$\left[-\frac{1}{2} \leftrightarrow \frac{1}{2}\right] : \ \Delta\nu = K_z + \frac{2}{9}(\nu_x - \nu_y)^2/\nu_0, \tag{3.4.1a}$$

$$\left[\pm\frac{1}{2} \leftrightarrow \pm\frac{3}{2}\right] : \ \Delta\nu = K_z \pm \nu_z + \frac{5}{36}(\nu_x - \nu_y)^2/\nu_0, \tag{3.4.1b}$$

$$\left[\pm\frac{3}{2} \leftrightarrow \pm\frac{5}{2}\right] : \ \Delta\nu = K_z - \frac{1}{9}(\nu_x - \nu_y)^2/\nu_0, \tag{3.4.1c}$$

where $\nu_0 = \gamma_n H_{res}$ and $\Delta\nu = \nu_{res} - \nu_0$. The NMR frequency is ν_{res}, and H_{res} is the value of the field at the peak in question. The quantities $\nu_{x,y,z}$ are quadrupole frequency parameters defined as $\nu_\alpha = \frac{3}{20} eQ\, V_{\alpha\alpha}$. Equation (3.4.1a) will be recognized as (2.1.7) written out with $V_{\alpha\alpha}$ parameters. Note that no relative size or sign relationship is implied here among the $\nu_{x,y,z}$, except, of course, that $\sum_\alpha \nu_\alpha = 0$. The parameters are adjusted for each site so as to give a best fit to the spectrum shown, as well as spectra taken with $\boldsymbol{H} \perp$ the c axis, an exercise of considerable complexity.

For the case of the O(2, 3) sites, it was not clear from the data alone whether the principal axis of the EFG was along the c axis or along the bond axis. To decide this, trial calculations of the EFG tensor were made assuming a point-charge contribution multiplied by a Sternheimer factor $(1 - \gamma)$ [171], where γ was taken to be 9, plus an on-site p-electron contribution of $(\frac{3}{20})(\frac{4}{5}e^2|Q|\langle r^{-3}\rangle)$, where $\langle r^{-3}\rangle$ was taken to be

3.63 a.u. [168]. On this basis, the only reasonable outcome was obtained assuming the principal axis is the bond axis, giving an estimated result (0.97, −0.42, −0.55) MHz as compared with (0.976, −0.598, −0.378) MHz shown in Table 3.5. Note that the transverse axis values are reversed between modeling and experiment, so that the fit is close, but not exact. The principal EFG values for all sites lie along the bond axes.

Our interest in the shift data shown in the table is in comparing them with the MR calculated results and ultimately to see whether the ^{17}O data are compatible with a single-fluid description of the distribution of electronic (i.e. hole) spin density. Shift values are given for all sites at two temperatures, 160 and 30 K. In the latter case, we suppose that the spin susceptibility is essentially zero (see e.g. shift measurements for $T < T_c$ in [140, 147]), so that these residual values may be regarded as orbital shifts. This surmise is verified by a detailed plot of $K_c(T)$ versus T by Horvatić et al. (see Fig. 4 in [169]). Subtracting the 30 K shift values from the 160 K ones, we find estimates of the spin paramagnetic shift components, listed under "K_s(160 K)". The latter values form a tensor for each site, which has been divided into an isotropic component $K_{iso} = (K_{\parallel} + 2K_{\perp})/3$ and an anisotropic (axial) component with $K_{ax} = (K_{\parallel} - K_{\perp})/3$. The latter parameters are tabulated in the final two columns. It is seen for the O(2, 3) that there is a fairly large isotropic component as expected, with quite a sizeable axial shift $K_{ax} = 0.03\%$ as well. It seems that the main hybridization is into the $2p_\sigma$ orbitals as expected.

On the other hand, the MR quantum chemistry calculation predicted a rather smaller value for K_{ax} than what was found. We analyze this point by noting that the axial shift coefficient for a p_σ orbital occupied with a fraction f of the $S = \frac{1}{2}$ moment is given by $\alpha_{ax} = \frac{2}{5} f \langle r^{-3} \rangle / N_A$. According to the MR quantum chemistry (Sect. 3.3), $f = \alpha_p^2/2 = 0.058$. Using the value $\langle r^{-3} \rangle = 3.63$ a.u. [168], we find $\alpha_\sigma = 0.94\,(\mathrm{emu/mol})^{-1}$. With a transverse susceptibility component of 1.01×10^{-4} emu/mol Cu(2) (Table 3.4), we find $K_{ax}(MR) \simeq 0.0095\%$. This is approximately 1/3 the experimental value in Table 3.5. So although the experimental value has a large error bar, it is certainly much larger than the MR calculated value.

Is this evidence for an oxygen hole band? We think not, but unfortunately, there is no trend with doping or even temperature dependence to check such a hypothesis. However, there is another experimental number which comes to mind here, namely the determination of the covalency of La_2CuO_4 at high temperatures using ^{17}O NMR shift measurements [119]. The total fraction of spin density residing on the O(2, 3) with the MR result (foregoing paragraph) is $2f = 11.6\%$. The comparable experimental result for La_2CuO_4 is 31% [119]. Scaling the foregoing estimate by the ratio of these numbers, one finds a result nearly a factor of 3 larger and in much better agreement with the O(2, 3) axial shift in Table 3.5. The La_2CuO_4 measurement is also in good agreement with, but actually slightly less than theoretical values given by Hüsser et al. [120]. We suggest that the value for La_2CuO_4 should not be very different from that for YBCO7. Thus, it is possible that the expected number for the $2p_\sigma$ spin density is quite a bit larger than the MR value, and that the reported $^{17}O(2, 3)$ shift values in [168] are consistent with a one-band picture.

Table 3.5 Experimental quadrupolar tensors and shift data from spectra such as Fig. 3.8 are shown at two temperatures in the left-hand columns, followed by deduced spin components, which are then broken down into isotropic and axial parts [168]. The 30 K values are taken to be of purely orbital origin. Site and orientation labels are shown at the far left. Frequencies are given in MHz and shifts in %

Site	Axis	ν_α	$K_\alpha(160\,\mathrm{K})$	$K_\alpha(30\,\mathrm{K})$	$K_s(160\,\mathrm{K})$	$K_s(iso)$	$K_s(ax)$
O(1)	$\perp c$	−0.484	0.23	...	...	0.10(1)	0.01(1)
	$\perp c$	1.629	0.22	0.13(1)	0.09(1)		
	$\parallel c$	−1.145	0.24	0.13(3)	0.11(3)		
O(2, 3)	$\perp c$	(0.986,0.966)	0.31	0.05(1)	0.26(1)	0.19(1)	0.03(1)
	$\perp c$	−0.598	0.17	0.01(1)	0.16(1)		
	$\parallel c$	(−0.389, −0.369)	0.18	0.02(3)	0.16(1)		
O(4)	$\perp c$	−0.375	−0.02	−0.01(1)	−0.01(1)	0.01(1)	0.02(1)
	$\perp c$	−0.721	0.03	0.04(1)	−0.01(1)		
	$\parallel c$	1.096	0.08	0.03(1)	0.05(1)		

Table 3.6 Spin components of NMR shift are tabulated for the ^{17}O(2, 3) and ^{89}Y, $^{17}K_{s\alpha}$ and $^{89}K_{s\alpha}$, respectively, for YBCO7 at T = 100 K. See text for the origins of these data. The corresponding HF tensors obtained with (3.3.4) are also listed, along with susceptibility tensor from Table 3.4. Note that for the ^{89}Y case the denominator in (3.3.4) is "8" rather than "2". C_α and D_α are given in kOe/μ_B and $\chi_{s\alpha}$ in 10^{-4} emu/mol

Axis	$^{17}K_\alpha$ (%)	$^{89}K_\alpha$ (ppm)	C_α	D_α	$\chi_{s\alpha}$ (100 K)
a	0.26(1)	247	72(3)	1.71(9)	1.01
b	0.16(1)	247	44(2)	1.71(9)	1.01
c	0.16(1)	315	38(2)	1.86(9)	1.18

HF Tensors for ^{17}O(2, 3) and ^{89}Y in YBCO7

In Table 3.6 we collect data for the ^{17}O(2, 3) and ^{89}Y NMR shift tensors and list the corresponding HF tensors defined in (3.3.2) and (3.3.4). For the ^{17}O(2, 3) case we take NMR shift data from Table 3.5. For the ^{89}Y case the data are taken from Alloul et al. [200], which reports the highest numbers available for an oriented powder sample.[6] The HF tensors are then obtained with (3.3.4) using the spin susceptibilities in Table 3.4.

The HF constants for the various sites in and near the CuO_2 in Tables 3.4 and 3.6 are expected to hold for reduced oxygen phases of YBCO7-x as well. A very important test of this idea occurs in Chap. 4, where these values are compared with precisely determined relative values for the YBCO6.63 phase.

3.4.1 *Measurements of* T_1 *for* ^{17}O *in YBCO7*

The foregoing authors also reported T_1 measurements for various oxygen sites in YBCO7 [156]. As found earlier for the ^{89}Y [127, 128], the ^{17}O(2, 3) relaxation in the normal state is very Korringa-like, with $T_1T \simeq 2.7$ sK. The authors chose to give a qualitative discussion of this result, along with their ^{63}Cu(2) data, in terms of the Korringa relation (1.2.8). Following that, however, they discuss the T_1 results in terms of the dynamic susceptibility $\chi''(\boldsymbol{q}, \omega)$, which became the method of choice for analyzing relaxation. In the following section (Sect. 3.5) we present for reference purposes a derivation of the relaxation rates for ^{89}Y, ^{17}O(2, 3), and ^{63}Cu(2) in YBCO in terms of $\chi''(\boldsymbol{q}, \omega)$. We shall use formulas given there for our present discussion of data from [156].

First, let us consider the simple Korringa relation (2.2.8) in relation to the current problem. One could hardly expect it to apply in any strict fashion to YBCO, owing to unknown particle dynamics, q-dependent HF coupling tensors, etc. Nonetheless,

[6] We take larger numbers for the ^{89}Y NMR shift at 100 K to be a sign of full oxygenation. Moreover, the plot reported for $^{89}K_\alpha(T)$ versus T [200] shows a gradual increase from 300 K down to a maximum at $T \sim 120$ K. This is a characteristic of a fully oxygenated sample as well [179], rather than declining curves which then give smaller values of $^{89}K_\alpha$ at T = 100 K [166].

it does serve a sort of qualitative purpose in comparing the behavior of different nuclear sites. The ^{89}Y and ^{17}O(2, 3) T_1 processes are at least Korringa-like in their temperature dependence, though this may be somewhat illusory because of spin-spin correlation effects. Nonetheless, if you take the isotropic shift value from Table 3.5, you find for ^{17}O(2, 3)

$$\frac{1}{K_s(iso)^2 T_1 T} \simeq 1.4 \times \left[\frac{4\pi k_B \gamma_n^2}{\hbar \gamma_e^2}\right]. \tag{3.4.2}$$

To compare the latter result with the Cu(2) site they used T_{1ab}, because it relates to the largest on-site HF term $A_c = -221\ kOe/\mu_B$. Since A_{ab} is very small, however, they divided the expression (3.4.2) by two,[7] finding a multiplier of 11 for the Cu(2) Korringa product. So there is a very clear enhancement of the Cu(2) relaxation, with almost none for the O(2, 3). If we look at the q-dependent expressions for $1/T_1$ in Sect. 3.5, the Cu(2) expression for the field $\perp$ c axis, it is essentially independent of $\boldsymbol{q}$, whereas for both the ^{89}Y and ^{17}O(2, 3) cases, the integrand vanishes at $Q_{AF} = (\pi/a, \pi/a)$. If $\chi''(\boldsymbol{q}, \omega)$ has an AFM enhancement peak, the Cu(2) rate will receive full enhancement, but the ^{89}Y and ^{17}O(2, 3) will have little or none. This is the explanation offered in [156] for the discrepancy in Korringa ratios which they find, but has been put forward by many other observers as well [158]. It was also generally thought to be related to the peculiar temperature dependence of the Cu(2) relaxation rate while the ^{89}Y and ^{17}O(2, 3) rates were Korringa-like.

Reference [156] also presented data on relaxation rates for $T < T_c$ for YBCO7, but discussion of those results will be combined with consideration of the Cu(2) data in Sect. 3.6.

3.4.2 *The One-Band, Two-Band Debate: ^{89}Y and ^{17}O(2, 3) NMR in YBCO*

The MR model theory purported to explain the Cu(1, 2) NMR (in YBCO7) behavior at 100 K with a single spin fluid, where all HF couplings were assumed to originate with the $S = 1/2$ coupled moments on the Cu^{2+} sites. At the same time or slightly later, other interpretive papers appeared as well as experimental papers with more data on the ^{89}Y and ^{17}O NMR shift and relaxation properties.

In this section we consider arguments presented on the question of one quantum fluid versus two separate fluids needed to account for the T_1 data. Monien et al. [164] presented an analysis similar to MR, including the transferred HF term for the Cu(2) sites, but also considering a possible "oxygen hole band" characterized by susceptibility χ_h as well as a Cu(2) HF coupling A_h. In similar fashion to Sect. 3.3, they concluded that the one-fluid model could not explain the data unless there were spin-spin correlations. However, they also pointed out the possibility that the (χ_h, A_h)

[7] Thus, only one axis is effectively contributing to the T_1 rate.

oxygen band, if present, could be the explanation. The question was how to decide whether the hole band was necessary.

A substantive discussion of this point was given by Mila and Rice themselves in a follow-up paper [157]. We shall briefly recap their main points here. First, they note that while one could consider the possibility of a hole band on the O(2, 3) sites, the Zhang-Rice singlet picture essentially rules this out, since the singlets are magnetically inert.[8] Thus, one would expect that HF parameters should not depend on the doping level. In this connection they noted that in the large ^{89}Y study by Alloul et al. [166], the slope of K versus χ plots is invariant from full doping all the way down into the insulating regime. Thus, the ^{89}Y HF coupling is independent of doping, consistent with the Zhang-Rice singlet picture.

Next, they consider what one would expect if there actually were a hole band. On the basis of quantum chemistry, MR argue that such a hole band would have a much stronger HF coupling with the ^{89}Y nuclei than the Cu^{2+} moments have. The reasoning is as follows. First, the Cu^{2+} are too far from the ^{89}Y for direct coupling, and for the (strongly) hybridized $2p_\sigma$ orbitals on the O(2, 3), the MR Wannier wavefunction (3.3.9) has the wrong symmetry to couple to the Y *s* or *d* orbitals. The answer, they suggest, is a possible double hybridization path through $2p_\pi$ orbitals to the Y 4*d*. This is, of course, very weak, and they argue that any hole band, which starts with either the $2p_\sigma$ or $2p_\pi$ orbitals, will be at least 5 times larger. Because of this, the ^{89}Y HF parameter in such a case would vary sharply with hole concentration. Since it does not, they conclude that an oxygen hole band is essentially ruled out.

There was further discussion of oxygen bands in a few additional papers [147, 160], but no really cogent evidence confirming their existence. In Chap. 5 a discussion is presented of several models for calculation and interpretation of T_1 data for various nuclear species in the YBCO family of compounds. With the development of various theoretical models and schemes for interpretation of relaxation effects, there seemed less and less motivation to include oxygen hole bands. There was no evidence to support their suggested existence once it was realized that $\boldsymbol{q}$-dependent susceptibilities and HF couplings could account for the observed behavior. There, however, we shall also reconsider the possibility of an orbital relaxation term.

3.5 Formulation of T_1 in Terms of $\chi''(\mathbf{q}, \omega)$ and q-Dependent HF Couplings

Even though transferred HF couplings quite obviously underlie the shift and relaxation effects of the ^{89}Y nuclei in YBCO [127, 128, 136, 154], their presence—and the concomitant effects of antiferromagnetic (AFM) spin-spin correlations—were brought to our attention by the MR [118] analysis. However, the spin-spin correlation picture was not destined to become the language of choice for this problem, but

[8]There were, however, two-band models proposed at the time, such as that given by Emery and Reiter [167].

rather that of the well-known dynamic susceptibility $\chi(\boldsymbol{q}, \omega)$ pioneered by Moriya [58] (see also the discussion by Jaccarino [155]). While these pictures are, as will be seen below, completely equivalent, the $\boldsymbol{q}$-space methodology is perhaps a bit more convenient, as well as tying in directly with INS measurements. While the potential of this approach to distinguish the behavior of ^{63,65}Cu, ^{17}O, and ^{89}Y in YBCO was noted early on [137], its dramatic consequence of cancelling AFM fluctuations at the planar oxygen was first articulated by the analysis of Shastry [146]. Subsequently, it has been widely invoked for discussion of T_1 effects [111, 156–158, 161, 162].

3.5.1 *Derivation of the Dynamic Susceptibility* $\mathbf{T_1}$ *Equation*

The formula for T_1 in terms of $\chi''(\boldsymbol{q}, \omega)$ and q-dependent HF couplings is widely quoted. For reference, we give here a simple derivation of this result, starting from the master equation for the interaction representation density matrix for nuclear spins $\rho^*(t)$ [5],

$$\frac{\mathrm{d}\rho^*}{\mathrm{d}t} = -\int_0^\infty \mathrm{d}\tau\, \overline{[\mathcal{H}_1^*(t),\, [\mathcal{H}_1^*(t-\tau), \rho^*(t)]]}, \tag{3.5.1}$$

where $\mathcal{H}_1$ is the spin-lattice interaction being treated here as a perturbation, and the overline represents an average over t which is long compared with the correlation time scale of τ, but short compared with the time for the nuclear spin polarization to change. The star indicates operators which have been transformed into the interaction representation as in $\mathcal{O}^*(t) = \exp(\mathrm{i}(\mathcal{F} + \mathcal{H}_0)t)\mathcal{O}\exp(-\mathrm{i}(\mathcal{F} + \mathcal{H}_0)t)$, where $\mathcal{H}_0$ is the unperturbed nuclear spin Hamiltonian, and $\mathcal{F}$ is the Hamiltonian for the rest of the dynamic variables in the system. Here, for simplicity we take $\mathcal{H}_0$ to be a Zeeman term with a field H applied along the cartesian axis labelled α,

$$\mathcal{H}_0 = -\omega_0 \sum_i I_{\alpha i}, \tag{3.5.2}$$

where $\omega_0 = \gamma_n H$. The density matrix for spin-lattice relaxation in such a case is simply $\rho^*(t) = \rho(t) = a(t)\sum_i I_{\alpha i}$, where $a(t)$ is a variable which represents the changing nuclear magnetization during a T_1 experiment.

We illustrate the $\boldsymbol{q}$-dependent HF coupling effects with $\mathcal{H}_1$ for the Cu(2) sites,

$$\mathcal{H}_1 = \sum_{j,\beta} I_{j\beta}\left[A_\alpha S_{j\beta} + B\sum_{k=nn} S_{k\beta}\right], \tag{3.5.3}$$

where the B term is an isotropic coupling between $\boldsymbol{I}_j$ and the four nn spins $\mathbf{S_k}$ to the j site. It will simplify the procedure if we neglect the imaginary part of the time factors $\exp(\pm\mathrm{i}\omega_0 t)$ which occur for each of the operators $I_{j\beta}$ in the interaction representation, and simply represent them as a factor $cos\omega_0\tau$ times the correlation functions. The correlation functions are assumed to be even in the variable τ, and the $sin\omega_0\tau$ terms will always vanish in the applications considered here.

A second simplification occurs if we operate on both sides of (3.5.1) with $\overline{Tr(I_{i\alpha}\ldots)}$. Since the spin correlation functions are even functions of τ so that $\overline{S_{j\beta}(t)S_{k\beta}(t+\tau)} = \overline{S_{j\beta}(t+\tau)S_{k\beta}(t)}$, we may write (3.5.1), in part, as

$$a^{-1}\frac{da}{dt} = -\sum_i^{(N)}\int_0^\infty d\tau cos\omega_0\tau\overline{[A_xS_{ix}(t)+B\sum_{k=nn}S_{kx}(t)][A_xS_{ix}(t+\tau)+B\sum_{k=nn}S_{kx}(t+\tau)]}, \quad (3.5.4)$$

where we are displaying only the x axis fluctuation term as an example.

The next step is to Fourier transform the spin operators with

$$S_{i\beta}(t) = \sum_q^{(N)} e^{i\boldsymbol{q}\cdot\boldsymbol{r}_i} S_{\boldsymbol{q}}(t)\,, \quad \beta = x, y, z, \quad (3.5.5)$$

where the summations are normalized so that $\sum_q^{(N)}\exp[i\boldsymbol{q}\cdot(\boldsymbol{r}-\boldsymbol{r}_i)] = \delta_{\boldsymbol{r},\boldsymbol{r}_i}$, etc.[9] In (3.5.5) $S_{\boldsymbol{q}}(t)$ is defined to be an isotropic, $\boldsymbol{q}$-dependent spin fluctuation operator. Using (3.5.5) in both factors in (3.5.4) and noting that $\sum_i^{(N)}\exp(i(\boldsymbol{q}+\boldsymbol{q}')\cdot\boldsymbol{r}_i) = \delta_{\boldsymbol{q},-\boldsymbol{q}'}$, (3.5.4) becomes

$$a^{-1}\frac{da}{dt} = -\sum_q^{(N)} A_x(\boldsymbol{q})^2\int_0^\infty d\tau\overline{S_{\boldsymbol{q}}(t)S_{-\boldsymbol{q}}(t+\tau)}cos\omega_0\tau, \quad (3.5.6)$$

where $A_x(\boldsymbol{q}) = [A_x + 2B(cos(q_xa)+cos(q_ya)]$. In (3.5.6) the $A_x(\boldsymbol{q})$ are in circular frequency units per unit of spin. We now proceed to replace the time average with an ensemble average

$$\int_0^\infty d\tau\overline{S_{\boldsymbol{q}}(t)S_{-\boldsymbol{q}}(t+\tau)}\cos\omega_0\tau = \int_0^\infty d\tau\langle S_{\boldsymbol{q}}(0)S_{\boldsymbol{q}}(\tau)\rangle_{Ens}\cos\omega_0\tau = S(\boldsymbol{q},\omega_0), \quad (3.5.7)$$

where we have also assumed that $S_{\boldsymbol{q}}(t)$ has inversion symmetry. The quantity on the far right of (3.5.7) is known as the *dynamic structure factor.*

3.5.2 *The Fluctuation-Dissipation Theorem*

The dynamic structure factor may be related to $\chi(\boldsymbol{q},\omega_0)$ using the *fluctuation-dissipation theorem* [163], which says

[9] In this and chapters to follow, the summation $\sum_q^{(N)}$ indicates an integration over the 2D cuprate Brillouin Zone (BZ). A convenient equivalent operator may be written $(a^2/4\pi^2)\int_{-\pi/a}^{\pi/a}dq_x\int_{-\pi/a}^{\pi/a}dq_y$. For a $\boldsymbol{q}$-independent integrand, this BZ integral is normalized to unity.

$$Im\,[\chi_s(\boldsymbol{q}, \omega_0)] = \chi_s''(\boldsymbol{q}, \omega_0) = \frac{(e^{i\hbar\omega_0\beta} - 1)}{\hbar e^{i\hbar\omega_0\beta}} S(\boldsymbol{q}, \omega_0), \tag{3.5.8}$$

where $\beta = 1/k_B T$ and we note that $\hbar\omega_0\beta \ll 1$. The subscript "s" is a reminder that the spin-only susceptibility defined by (3.5.8) has a dimension of inverse energy. Inserting this into (3.5.6) and combining terms for x and y-axis fluctuations yields for $-a^{-1}da/dt$,

$$\frac{1}{T_{1\alpha}} = \frac{\gamma^2 k_B T}{\mu_B^2} \sum_{\boldsymbol{q}}^{(N)} \sum_{\beta} A_\beta(\boldsymbol{q})^2 \chi_\beta''(\boldsymbol{q}, \omega_0)/(g_\beta^2 \omega_0). \tag{3.5.9}$$

The latter expression for $1/T_{1\alpha}$ has $A_\beta(\boldsymbol{q})$ in Gauss per unit of spin and a conventional susceptibility per formula unit of CuO_2. The sum on β is over the two directions normal to α.

Next, we introduce a simplification just for convenience. One could use (3.5.9) by absorbing the g-factor denominator into the HF coefficients and work with the primed coefficients via (3.3.4). Following Auler et al., we find it more convenient to define an *isotropic susceptibility* via $\chi_{is}''(\boldsymbol{q}, \omega) = 4\,\chi_\beta''(\boldsymbol{q}, \omega)/g_\beta^2$, whereupon (3.5.9) becomes

$$\frac{1}{T_{1\alpha}} = \frac{\gamma^2 k_B T}{4\,\mu_B^2} \sum_{\boldsymbol{q}}^{(N)} \sum_{\beta} A_\beta(\boldsymbol{q})^2 \chi_{is}''(\boldsymbol{q}, \omega_0)/\omega_0. \tag{3.5.10}$$

In this way only the natural anisotropy of the HF couplings from (3.3.2) comes into the T_1 formula.

Next, we present expressions for $1/T_1$ for the various nuclear species in YBCO, using the HF tensors developed in the text. For the c axis relaxation of Cu(2), $A_\beta = A_{ab}$, for both terms, so (3.5.9) becomes

$$\left.\frac{1}{T_{1c}}\right|_{Cu(2)} = \frac{\gamma^2 k_B T}{2\,\mu_B^2} \sum_{\boldsymbol{q}}^{(N)} [A_{ab} + 2B\,(\cos q_x a + \cos q_y a)]^2 \chi_{is}''(\boldsymbol{q}, \omega_0)/\omega_0\,, \tag{3.5.11}$$

where the susceptibility is per CuO_2. For the field in the ab plane, we have

$$\left.\frac{1}{T_{1ab}}\right|_{Cu(2)} = \left.\frac{1}{2T_{1c}}\right|_{Cu(2)} + \frac{\gamma^2 k_B T}{4\,\mu_B^2} \sum_{\boldsymbol{q}}^{(N)} [A_c + 2B\,(\cos q_x a + \cos q_y a)]^2 \chi_{is}''(\boldsymbol{q}, \omega_0)/\omega_0\,. \tag{3.5.12}$$

However, for the Cu(2) HF couplings given in Table 3.4 A_c is the dominant term, so that for $1/T_{1ab}$ the contribution of $1/2T_{1c}$ is essentially negligible. Unless the correlations are known to be large, a good estimate can be obtained by dropping the "B" term as well, so that the HF coupling becomes essentially independent of $\boldsymbol{q}$.

For the planar oxygen sites and the yttrium sites in YBCO the HF couplings are dominated by hybridization effects with the 2s [146] and 4d [136] holes, giving isotropic HF coupling terms

$$A_O(\boldsymbol{q}) = 2C\,cos(q_x a/2) \quad \text{and} \quad A_Y(\boldsymbol{q}) = 8D\,cos(q_x a/2)\,cos(q_y a/2), \qquad (3.5.13)$$

respectively, where it is noted that the HF coupling for the Y^{3+} is thought to be dominated by the $4d$ core-polarization HF field [136].

The corresponding expression for the ^{17}O relaxation rate is, e.g.,

$$\left.\frac{1}{T_{1c}}\right|_O = \frac{\gamma^2 k_B T}{2\,\mu_B^2}\sum_{\boldsymbol{q}}^{(N)}[C_x^2(1+cos(q_y a)) + C_y^2(1+cos(q_x a))]\chi''_{is}(\boldsymbol{q},\omega_0)/\omega_0\,, \qquad (3.5.14)$$

where T_1 along other axes is analogous to the case for Cu(2).

For the case of ^{89}Y, we give here an expression in which interplanar correlations are neglected. Thus,

$$\left.\frac{1}{T_{1c}}\right|_Y = \frac{4\gamma^2 k_B T}{\hbar\mu_B^2}\sum_{\boldsymbol{q}}^{(N)}[D_x^2 + D_y^2](1+cosq_x a)(1+cosq_y a)\chi''_{is}(\boldsymbol{q},\omega_0)/\omega_0. \qquad (3.5.15)$$

The essential physics of the T_1 process is more clearly evident if we recast (3.3.14) and (3.3.15) in terms of the T-dependent parameters $\mathcal{K}_{ij}$ and τ_e, to be defined presently.[10]

$\mathcal{K}_{ij}$ and $ø_e$ Expressed in Terms of $Ø''(\boldsymbol{q},\omega)$

Dynamic AFM spin-spin correlation effects in these relaxation processes are embodied in the factors such as $(1 + cos\, q_{x,y}a)$, the mean value of which will approach zero in the event of strong correlations. Such correlation factors were introduced by Mila and Rice [118], where in Sect. 3.3 they took the form $\mathcal{K}_{ij} = 4\langle S_{i\alpha}S_{j\alpha}\rangle$. It is useful to introduce an equivalent definition of the $\mathcal{K}_{ij}$ in terms of $\chi''(\boldsymbol{q},\omega)$. For example, the sum on $\boldsymbol{q}$ in (3.3.14) can be put into the form

$$\sum_{\boldsymbol{q}}^{(N)}[\;] = \frac{\mu_B^2}{k_B T}[C_x^2 + C_y^2][1+\mathcal{K}_{01}]\tau_e\,, \qquad (3.5.16)$$

where

$$\tau_e = \frac{k_B T}{\mu_B^2}\sum_{\boldsymbol{q}}^{(N)}\frac{\chi''_{is}(\boldsymbol{q},\omega_0)}{\omega_0} \quad \text{and} \quad \mathcal{K}_{01} = \frac{k_B T}{\mu_B^2\tau_e}\sum_{\boldsymbol{q}}^{(N)} cos\,q_{x,y}a\,\frac{\chi''_{is}(\boldsymbol{q},\omega_0)}{\omega_0}. \qquad (3.5.17)$$

The quantity $\tau_e(T)$ is a fundamental correlation time for itinerant spin fluctuations in the CuO_2 planes. Furthermore, $\tau_e \sim \hbar k_B T N(0)^2$ gives the correlation time along with the statistical weight of the T_1 process, where $N(0)$ is the density of states/atom at the Fermi surface for one direction of spin in an energy band environment. The other

---

[10]It is noteworthy that the parameter $\mathcal{X}_{at}$, used in the first edition of this monograph, is being replaced here by the more intuitive quantity $(\mu_B^2/k_B T)\tau_e$.

factor $(1 + \mathcal{K}_{01}(T))$ gives the temperature variation of T_1 arising from dynamic correlations for the case of transferred HF interactions. Equations (3.5.16) and (3.5.17) therefore effect a separation of these two fundamental—and very different—factors in the temperature dependence of T_1. This result will be used extensively in subsequent chapters.

To complete the picture of spin–spin correlations used for the YBCO family of compounds, we have, in analogy with (3.5.17),

$$\mathcal{K}_{12} = \frac{k_B T}{\mu_B^2 \tau_e} \sum_{\boldsymbol{q}}^{(N)} \cos q_x a \cos q_y a \frac{\chi''_{is}(\boldsymbol{q}, \omega_0)}{\omega_0} \text{ and } \mathcal{K}_{13} = \frac{k_B T}{\mu_B^2 \tau_e} \sum_{\boldsymbol{q}}^{(N)} \cos 2q_{x,y} a \frac{\chi''_{is}(\boldsymbol{q}, \omega_0)}{\omega_0}. \tag{3.5.18}$$

We may now introduce these forms into the T_1 expressions (3.5.11)–(3.3.15). As an example we consider the expression (3.5.11), where the HF coefficient factor $[\,]^2$ may be expanded as

$$[A_\alpha^2 + 4B^2 + 2B^2(\cos 2q_x a + \cos 2q_y a) + 8B^2 \cos q_x a \cos q_y a + 4A_\alpha B\,(\cos q_x a + \cos q_y a)]. \tag{3.5.19}$$

Using the definitions (3.5.17) and (3.5.18), the expression for $1/T_1$ becomes

$$\left.\frac{1}{T_{1c}^{mf}}\right|_{Cu(2)} = \frac{\gamma_{63}^2}{2} \tau_e \{A_{ab}^2 + 4B^2[1 + 2\,\mathcal{K}_{12} + \mathcal{K}_{13}] + 8A_{ab}B\mathcal{K}_{01}\}, \tag{3.5.20}$$

where we recognize the same expression for the HF constants and correlation functions as in (3.3.14). Expressions such as (3.5.20) give a more direct idea of the dependence of T_1 on the spin-spin correlations than expressions given earlier in terms of $\chi''(\boldsymbol{q}, \omega)$.

The corresponding equation for $^{17}O(2, 3)$ is

$$\left.\frac{1}{T_{1z}}\right|_{O} = \frac{\gamma_{17}^2}{2} \tau_e [C_x^2 + C_y^2][1 + \mathcal{K}_{01}], \tag{3.5.21}$$

and for ^{89}Y is

$$\left.\frac{1}{T_{1z}}\right|_{Y} = 4\,\gamma_{89}^2 \tau_e [D_x^2 + D_y^2][1 + 2\mathcal{K}_{01} + \mathcal{K}_{12}], \tag{3.5.22}$$

for any permutation of axes (x, y, z). Equations (3.5.20)–(3.5.22) are quite general and are useful for discussing cases where $\tau_e(T)$ and the $\mathcal{K}_{ij}(T)$'s are both changing substantially with temperature. We emphasize that the behavior of $\tau_e(T)$ is independent of any HF coupling tensor, being a fundamental property of the quasiparticles via $\chi''_{is}(\boldsymbol{q}, \omega)$.

For examples of the application of the foregoing results to the pseudogap problem in high T_c, see Sects. 4.1 and 4.2.

3.6 A d-Wave Model for the NMR Shift and T_1 at $T < T_c$

Following Mila and Rice, a number of relaxation model calculations were developed which made use of the random phase approximation (RPA) to include electron–electron interaction effects in the calculations. Several of these model calculations are reviewed in Chap. 5, as well as a proposed orbital relaxation term for the ^{63}Cu(2) based on the marginal Fermi liquid hypothesis [159].

Among the RPA schemes which appeared early on was an interacting Fermi liquid theory with strong AFM exchange couplings put forward by Monien and Pines (MP) [160], where they used the RPA to construct a model for $\chi''(\boldsymbol{q}, \omega)$ with an adjustable enhancement level in the vicinity of Q_{AF}. An important innovation in this work was the introduction of an energy gap with d-wave symmetry for T_1 and NMR shift calculations for $T < T_c$. Even though there were theoretical developments from very early on suggesting the suitability of d-wave superconductivity for the cuprates [175–177], up to this point no one had tried to interpret NMR data with the d-wave picture. In spite of the apparent success of this interpretation as seen in [160], widespread acceptance of d-wave symmetry did not occur until the dramatic confirmation of d-wave superconductivity in the cuprates by means of phase jumps at Josephson junction boundaries, which was still several years away [172, 173]. The NMR modeling described here followed closely upon a successful application of d-wave theory to Raman scattering experiments [174].

The RPA theory of [160] was applied to the Cu sites in YBCO7, giving a realistic account of the shift and of T_1 versus T at for $T < T_c$. This first NMR test of d-wave calculations was an important step along the path of eventual adoption of the singlet d-wave interpretation of cuprate superconductivity. By this point in time, there was ample T_1 data below T_c on which to apply calculations such as this. As we shall see, however, *both the theoretical details and some of the data* were destined to undergo revision in later years. Nonetheless, these preliminary exercises in understanding the cuprate NMR data were both very encouraging and highly suggestive.

3.6.1 The NMR Shift Behavior for $T < T_c$

To discuss the Cu(1, 2) NMR shift below T_c, we must first address the experimental situation. We therefore digress briefly here to discuss the NMR shift measurements conducted by Barrett et al. [147] and Takigawa et al. [178]. Both papers appear to be very careful pieces of work. The biggest problem involved in such measurements is accounting for the diamagnetic shift of the internal field in the superconducting vortex state. This was dealt with in sharply contrasting ways in the two papers cited, leading, unfortunately, to slightly different results. Here we suggest a compromise interpretation.

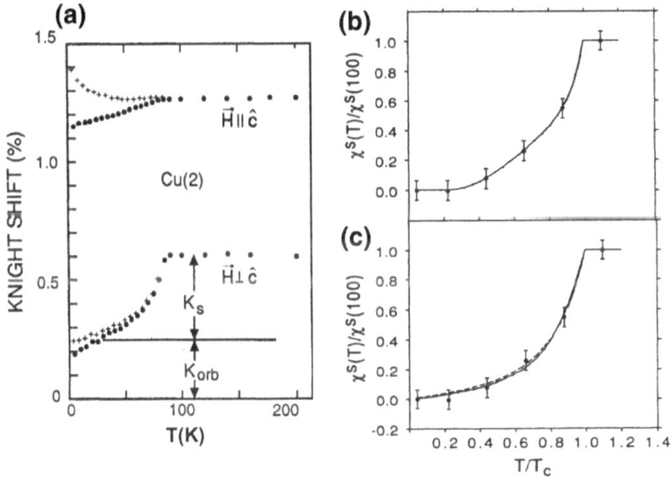

Fig. 3.9 **a** Corrected and uncorrected NMR shift data for Cu(2) in YBCO7 from [178] are shown plotted as a function of temperature for field parallel (top) and perpendicular (bottom) to the c axis. The correction scheme is described in the text. Cu(2) NMR shift data for YBCO7 from [147], i.e. $K_s(c) - K_s(ab)$ normalized to 1 at 100 K, are plotted versus T/T_c in (**b**) and (**c**). Yosida function fits to the data are shown using s-wave symmetry (**b**) and d-wave (d′-wave) symmetries (**c**) from Fig. 3.10a as a solid (dashed) line. Fit parameters are shown in Table 3.7 and discussed in the text

Correcting for Diamagnetism via Magnetization

In [178] the internal field was corrected for using an effective demagnetizing factor, which is applied to the measured magnetization of the oriented powder sample. The demagnetizing factor was determined with a careful calibration procedure, which is too lengthy to discuss here. However, the risk with such a procedure lies perhaps not in the demagnetizing factors, but in using magnetization measurements on a powder sample set in epoxy. There is always a risk of non-intrinsic Curie-Weiss-like anomalies affecting such a procedure. Indeed, the c-axis shift for Cu(2) has an upturn below ~50 K which we suggest may be spurious. This is shown in Fig. 3.9a where the Cu(2)-site shift results are plotted with corrections. Interestingly, no such feature is found for the ab-plane result. This same kind of anisotropic Curie-Weiss background effect was also reported later in an attempt to determine the intrinsic normal-state susceptibility of YBCO [148]. In the present case we note that the shift corrections become nearly 0.25% as $T \to 0$. The size and temperature dependence of this effect do not seem compatible with superconducting diamagnetism.

Calibration of Internal Field with the ^{89}Y Shift

On the other hand, in [147] the ^{89}Y NMR line was used as a reference for the internal field at low temperatures, which is a very sound method in principle, as long as the ^{89}Y NMR shift is reliably known. If, for example, the resonance frequencies ν_{63} and ν_{89} were measured in the same field, then we have

$$K_{Cu} = \left[\frac{\nu_{63}\gamma_{89}}{\nu_{89}\gamma_{63}}\right] + K_Y - 1. \tag{3.6.1}$$

In practice, $K_Y \simeq -0.03\%$ at T = 100 K, and it was assumed [147] that $K_Y \propto Y_s(T)$, $Y_s(T)$ being the Yosida function [49] for s-wave gap symmetry. Since one may be looking here for a d-wave Yosida function effect, the foregoing assumption may cause a problem, but in fact the difference is small enough to be within the error bars. The Cu(2) shift results are shown in Fig. 3.9b and c. It is noteworthy that the maximum c-axis internal field correction (for $K_s(c)$) is ~0.05% or only ~20% of that found at low T in Fig. 3.9a. Although the latter correction cannot be arbitrarily curtailed, it seems likely that it is primarily responsible for the discrepancy between the c-axis results. The ab-plane results are also slightly outside error bars (0.37(2)% for [178] and 0.30(2)% for [147]). However, this difference is as much a result of disparate normal-state shifts as it is of low-T corrections, which are similar.

We suggest the following composite picture based on the combined results of [178] and [147]. The c-axis result of [147] is to be accepted as is, which says that $K_s(c)$ is negligibly small on the scale of $K_s(ab)$. The axial and isotropic shifts thus behave in a very similar fashion, since they both reflect the behavior of $K_s(ab)$. Thus, we suggest that the contrasting behaviors of $K_{ax}(T)$ and $K_{iso}(T)$ highlighted in [178] may have been generated, in part, by subtracting and adding the strongly corrected c-axis shift, respectively. It is suggested that the sharp contrast which results is unreliable. On the other hand, the corrected $K_{ab}(T)$ curve in Fig. 3.9a shows a very interesting linear-T behavior below 50 K which is clearly not a result of the correction process. This may therefore be a sign of d-wave symmetry behavior, as reflected in the MP interpretation below. The behavior of $K_{ax}(T)$ in Fig. 3.9b and c is also compatible with d-wave, as we shall see below. Both of these papers considered the isotropic component of shift as heavily contaminated with "oxygen hole band" effects. Following MR and MR2 [118, 157], we maintain their occurrence to be unlikely.

3.6.2 The Monien-Pines Calculations of NMR Shifts Below T_c

MP began by applying different pairing symmetries to the interpretation of NMR shift variation below T_c. For comparison purposes they applied the pairing symmetries shown in Fig. 3.10a. The pure s-wave and d-wave are flat and $\propto cos\,\phi$, respectively, where ϕ is the azimuthal angle for the cylindrical 2D Fermi surface for the CuO_2 planes. They argue in detail that triplet pairing states are inappropriate for this case [160]. Then, still concerned about the possibility of an oxygen hole band, following [178] they consider only the axial component of the shift, which they suggest would reflect only the spin paramagnetism of the $Cu(2)^{2+}$ sites[11] The data which were fitted are shown in Fig. 3.10b [178] and in Fig. 3.9b and c.

[11]The essential assumption here is that any oxygen hole-band NMR shift would be isotropic.

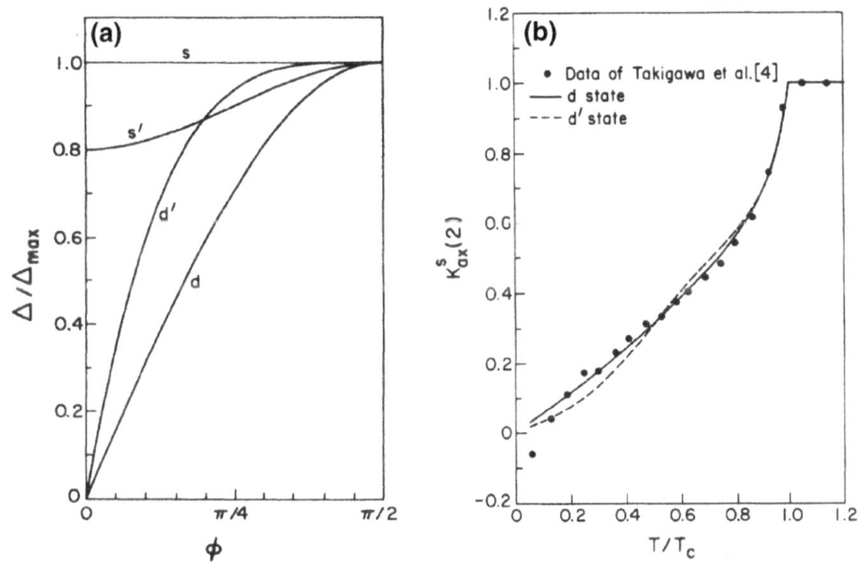

Fig. 3.10 **a** Gap energy variation for several choices of gap symmetry from [160] shown as a function of azimuthal angle ϕ over one quadrant. **b** Plot of $K_s(ax)$ from [178] as a function of T/T_c (dots) for temperatures in the superconducting state. The solid and dashed lines are calculated fits for gap functions labelled "d" and "d′" in the (**a**) portion [160], as described in the text. Noteworthy is the d-wave-like linear T behavior at the lowest temperatures

Calculating the Yosida Function

To compare with NMR shift data, the Yosida function Y(T) is calculated using the weak-coupling BCS formula for an anisotropic, singlet superconductor,

$$Y(T) = \int_{-\infty}^{\infty} \mathrm{d}E \left[-\frac{\partial f}{\partial E} \right] N(E), \tag{3.6.2}$$

where $N(E)$ is the density of states, and f is the Fermi distribution function. The behavior of $Y(T)$ for an s-state gap is $Y_s(T) \propto \exp(-\Delta_{min}/k_B T)$ for $T \ll T_c$, Δ_{min} being the minimum gap energy as $T \to 0$. In contrast, for the case of a d-state gap $N(E) \propto E$ for $E \ll \Delta(0)$, leading to a linear variation of $Y_d(T)$ with T for $T \ll T_c$. MP used the maximum gap value $\Delta(0)/k_B T_c$ and the specific heat jump at T_c

$$\Delta C/C \propto \left[\frac{\partial \Delta^2}{\partial T} \right]_{T_c} \tag{3.6.3}$$

as parameters for calculating $Y_d(T)$ which is plotted in Fig. 3.10b. The d-wave result is seen to be in good agreement with the Cu(2)-site K_{ax} data from [178]. Values of the parameters for the cases of (i) weak-coupling BCS theory and (ii), (3.6.2) fitted to the NMR shift data in Figs. 3.9b, c and 3.10b are given in Table 3.7. Strong coupling

Table 3.7 Table of fitting parameters Δ_0/k_BT_c and $\Delta C/C$ from [160] for the NMR shift data in Fig. 3.10b [178] (top part) and for the data in Fig. 3.9b and c (bottom). The values of these parameters for weak-coupling BCS theory are shown on the left-hand side for comparison. The pairing states shown in Fig. 3.10a were used for these calculations

Data plot	Pairing state	Weak coupling Δ_0/k_BT_c	Weak coupling $\Delta C/C$	Optimal fit Δ_0/k_BT_c	Optimal fit $\Delta C/C$
Figure 3.10b	d	2.13	0.95	2.44	1.97
Figure 3.10b	d′	1.93	1.17	1.74	2.36
Figure 3.9b	s	1.76	1.43	1.90	2.77
Figure 3.9b	s′	2.10	1.27	2.16	2.49
Figure 3.9c	d	2.13	0.95	14.7	1.76
Figure 3.9c	d′	1.93	1.17	3.13	2.14

effects are seen to be appreciable, but also, the d-wave calculation is quite successful in reproducing the linear temperature behavior of the shift data.

3.6.3 *Calculating the Relaxation Decay Curve*

By fitting the relaxation, the viability of using d-wave gap symmetry to interpret NMR data was tested. We begin with a brief synopsis of the calculation of T_1 [160]. In the presence of an AFM interaction $J_{eff}(\boldsymbol{q}, T)$ the RPA expression for the dynamic susceptibility becomes

$$\chi(\boldsymbol{q}, \omega, T) = \frac{\chi_0(\boldsymbol{q}, \omega)}{1 - N(0)J_{eff}(\boldsymbol{q}, T)\chi_0(\boldsymbol{q}, \omega)}, \tag{3.6.4}$$

where $\chi_0(\boldsymbol{q}, \omega)$ is the noninteracting susceptibility. For a T_1 calculation such as (3.5.9) from the previous section, one only needs the imaginary or dissipative part $\chi''(\boldsymbol{q}, \omega)$. Monien and Pines make the following approximations to treat T_1 behavior below T_c. Assuming AFM enhancement effects are large (such as for Cu(2) in YBCO), the enhancement denominator in (3.6.4) is first replaced by its average, temperature-dependent value at Q_{AF}, so that (3.6.4) becomes

$$\frac{1}{T_1} = \frac{1}{(T_1)_{free}} \frac{1}{[1 - \lambda(T)]^2}, \tag{3.6.5}$$

where

$$\lambda(T) = \langle N(0)J_{eff}(\boldsymbol{Q}_{AF}, T)\chi_0'(\boldsymbol{Q}_{AF}, T)\rangle. \tag{3.6.6}$$

It was then further assumed that the real part of the dynamic susceptibility at Q_{AF}, $\chi'_0(Q_{AF}, T)$, scales with temperature like the noninteracting uniform susceptibility $\chi'_0(q = 0, T)$. T_1 (3.6.5) can then be expressed in terms of the Yosida function (see Sect. 2.2.2)

$$Y(T) = \chi'_0(q = 0, T)/\chi'_0(q = 0, T_c) \tag{3.6.7}$$

as

$$\frac{1}{T_1} = \frac{1}{(T_1)_{BCS}} \frac{1}{[1 - \lambda(T_c)Y(T)]^2}, \tag{3.6.8}$$

where $1/(T_1)_{BCS}$ may be calculated for any pairing symmetry desired in the absence of quasiparticle interactions. In this simple approximation the falloff of $1/T_1$ below T_c will be enhanced by the decline of the enhancement factor, which in turn falls off according to how the Yosida function diminishes below T_c.

For other than s-wave gap symmetries, we may write the BCS relaxation expression (cf. Sect. 2.2.2)

$$\frac{1}{(T_1T)_{BCS}} \propto \int_{-\infty}^{\infty} \mathrm{d}E \left[-\frac{\partial f}{\partial E}\right] [N^2(E) + M^2(E)], \tag{3.6.9}$$

where $N(E)$ and $M(E)$, which account for the BCS coherence factor, are defined by [9]

$$N(E) = \left\langle Re\left[\frac{E}{(E^2 - \Delta_k^2)^{1/2}}\right]\right\rangle_{FS} \quad and \quad M(E) = \left\langle Re\left[\frac{\Delta_k}{(E^2 - \Delta_k^2)^{1/2}}\right]\right\rangle_{FS}. \tag{3.6.10}$$

The latter definitions allow for the possibility of a complex gap function. Combining (3.6.8) with (3.6.10) yields an expression for the relaxation rate which will decline below T_c, both because of the growth of the superconducting energy gap and because of the decline of the enhancement factor owing to the decrease in $Y(T)$. Because of the d-wave gap function described above, the linear-E density of states will lead to asymptotic $1/T_1 \propto T^3$ behavior for $T \ll T_c$.

Application of MP T_1 Calculations to Cu(2) Data for YBCO7

MP illustrated the above formalism by using it to fit the NQR T_1 data on ^{63}Cu(2) in YBCO7 by Imai et al. [180]. The fitting function is shown as a solid line in Fig. 3.11a, and is seen to give an excellent representation of the data over two full orders of magnitude of the decline of $1/T_1$ below T_c. We make a number of explanatory remarks regarding this T_1 fit. First, for the d-wave gap function (see Fig. 3.10a), there is no vestige of the coherence peak, unlike what is found for the s-wave cases [160]. Moreover, the very steep decline found just below T_c can be achieved in two different ways. This is illustrated by the fit parameters shown in Table 3.8. The actual curve plotted in Fig. 3.11 does not allow the enhancement factor to change below T_c, for reasons to be discussed presently. As a consequence, a very large gap parameter of $5.12k_BT_c$ is needed to cause the steep fall-off near T_c. However, MP

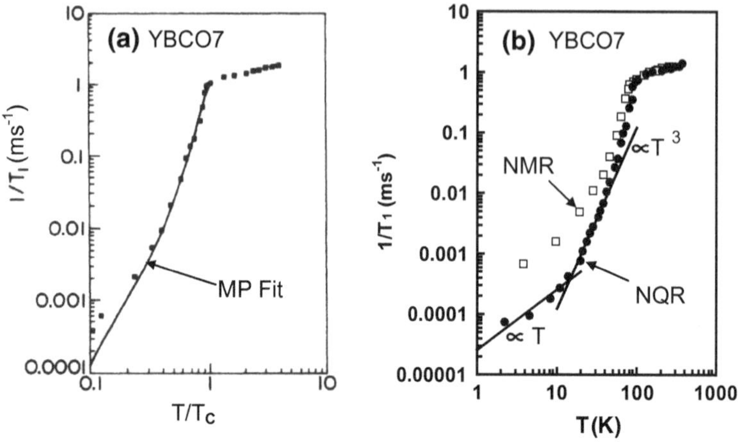

Fig. 3.11 **a** NQR T_1 data for Cu(2) in YBCO [180] is plotted versus T/T_c for temperatures above and below T_c. The MP theoretical fit described in the text is shown as a solid line. **b** The same data as in part (**a**) are plotted as dots on a scale of absolute temperature. Also plotted for comparison are a set of NMR data (open squares) for the same nominal compound taken in a field of 7.4 T [156]. For the NQR data the regions of data which vary as T^3 and as T are indicated with straight lines (see text for discussion)

Table 3.8 Table of fitting parameters for the T_1 data of Fig. 3.11a using a d-wave gap function [160]. The top line shows parameter values for a fit assuming a constant enhancement below T_c. In this case the effective value of λ is zero. The second line shows an essentially equivalent fit in which λ is allowed to decay below T_c according to the d-wave Yosida function. The remaining parameters are those used to fit the NMR shift data, as shown in Fig. 3.9c

Weak coupling Δ_0/k_BT_c	Weak coupling $\Delta C/C$	Optimal fit Δ_0/k_BT_c	Optimal fit $\Delta C/C$	$\lambda(T_c)$
2.13	0.95	5.12	1.89	0.0
2.13	0.95	2.44	1.97	0.5

have calculated another fit curve for the data shown, corresponding to the second row of parameters in the table, which is almost indistinguishable from the first ([160], Fig. 3d). In the latter case the enhancement parameter $\lambda \simeq 0.5$ (see [160]) contributes to the decline of $1/T_1$, but the gap parameter $\Delta_0/k_BT_c = 2.44$ is only half as great as for the calculation shown. This may be a more physically reasonable scenario. The corresponding (second row) parameters from Table 3.8 is in fact the same set used to fit the shift data in Fig. 3.9c.

Referring to Fig. 3.11b, the solid dots plot the same data as in the (a) panel. There we compare the decline of $1/T_1$ with a line of slope $\propto T^3$ in the middle region and a linear T curve at the lowest temperatures. The $\propto T^3$ behavior is expected at low T as mentioned earlier, because of the linear-E character of $N(E)$ which results from the d-wave gap having a line of nodes. Further, the suggested crossover to linear T behavior often results from disorder in the sample causing quasiparticle states to occur in the gap region. The progression of $T^3 \rightarrow T$ behavior has been found to occur below T_c in many non-cuprate unconventional superconductors. A few cases we could mention are ^{115}In in $CeIrIn_5$ [181], ^{69}Ga in $PuCoGa_5$ [182], ^{69}Ga in $PuRhGa_5$ [183], and ^{23}Na in $Na_{0.35}CoO_2 \cdot yH_2O$ [184].

Does AFM Enhancement Decay Below T_c?

Next, we discuss the difference between the NMR and NQR data shown in Fig. 3.11b. Along with the NMR data shown, Hammel et al. also reported a T_1 curve for the (planar) O(2, 3) sites over the same temperature range. Comparison of the NMR data for the Cu(2) and the O(2, 3) created something of a sensation when *their variation below T_c turned out to be precisely the same over two decades*. In terms of the MP theory outlined above the implication is that even though the Cu(2) undergoes substantial enhancement in the normal state, *any such enhancement becomes temperature independent below T_c*. This was taken to be a constraint on the theory, and is the reason why MP calculated the solid curve fit in Fig. 3.11a without allowing the enhancement to vary with temperature.

However, the contrast between the NQR and NMR measurements suggests very strongly that there is an additional T_1 mechanism operating in a high magnetic field. Detailed studies which came later showed this to be the case, namely relaxation to the normal cores in the vortex state just as occurred in V_3Si (see Fig. 2.1.3). For that reason, the Cu(2) and O(2, 3) relaxation curves in [156] are not really intrinsic superconducting behavior. Later studies, as we shall see in Chap. 4, show that the AFM enhancement in the SC state really does decay away to zero at low temperatures. Therefore, for the data of Fig. 3.11a, the fit calculated including enhancement decay is probably the more realistic one.

In sum, we find that the d-wave description of the YBCO7 NMR effects for $T < T_c$ according to the MP model theory is successful in an essentially quantitative way. Fitted parameter values suggest that strong-coupling effects are quite important. It remains to be seen whether this kind of RPA model for the dynamic susceptibility can be applied to interpret the disparate temperature dependence which occurs in the normal state. This is the theme of a series of theory papers which came soon after MP and which are expounded in some detail in Chap. 5. We shall see that the latter papers uphold the continued success of the single quantum fluid picture of these systems that was initiated by the papers by Mila and Rice. Meanwhile, the pseudogap effect for underdoped materials came to light, and as we shall see in the next chapter, became a major sub-theme of cuprate physics.

Chapter 4
Pseudogap Effects in Cuprate NMR Studies

Up to this point the discussion has focused on fully oxygenated YBCO7 and nearby compositions, for which T_c is very near its peak value $\simeq 92\,K$ for this compound. In 1989, detailed studies began to appear for *underdoped* YBCO7-x, $0.3 \leq x \leq 0.6$, as well as for $La_{2-x}Sr_xCuO_4$ over its entire range of superconducting behavior $0.05 \leq x \leq 0.25$. These phases brought the new and surprising *pseudogap* effect to light, an effect that is unique to the cuprates and is most pronounced in the underdoped phases. However, over a long–time perspective it seems that nearly all superconducting cuprates are subject to anomalous behavior that is referred to as the *pseudogap* effect. Thus, the affected properties that depend on excitations near the Fermi surface behave as though parts of the FS were acquiring an energy gap and effectively disappearing.

First observed via NMR properties [25, 166, 185, 186], the essential pseudogap effect was a gapped appearance to the NMR shifts, susceptibility, and low-energy excitations via $1/T_1T$. Examples of this type of behavior, observed in T_1 studies on oxygen–depleted YBCO6.64, are shown in Fig. 4.1. In part (a) of the figure we see that the rise in $1/T_1T$ as T descends is diminished as compared with YBCO7 (dashed line), and there is no longer any noticeable feature at $T_c(\sim 64\,\mathrm{K})$, with a marked roll–off of $1/T_1T$ above T_c. In part (b) similar behavior is seen for ^{89}Y with data taken on the same specimen as in Fig. 4.1a. In this case, however, the roll–off of $1/T_1T$ starts at temperatures well above 200 K, with the magnitude at $T = T_c$ down to less than a quarter of its value near room temperature.

A short time later, a comprehensive study appeared of both NMR shift and T_1 for ^{89}Y in a series of oxygen–deficient phases of YBCO (Fig. 4.2), showing that both the spin paramagnetism (i.e., the shift $K_s(T)$) and the Korringa relaxation gradually decay away at low temperatures in progressive fashion as the oxygen doping in this system is reduced [166]. While the NMR pseudogap results were first interpreted with a Fermi-liquid model, that interpretation was later revised [200]. Early results appeared to establish another important high–T_c result contrary to the

© Springer-Verlag GmbH Germany 2018

R.E. Walstedt, *The NMR Probe of High-T_c Materials and Correlated Electron Systems*, Springer Tracts in Modern Physics 276,
https://doi.org/10.1007/978-3-662-55582-8_4

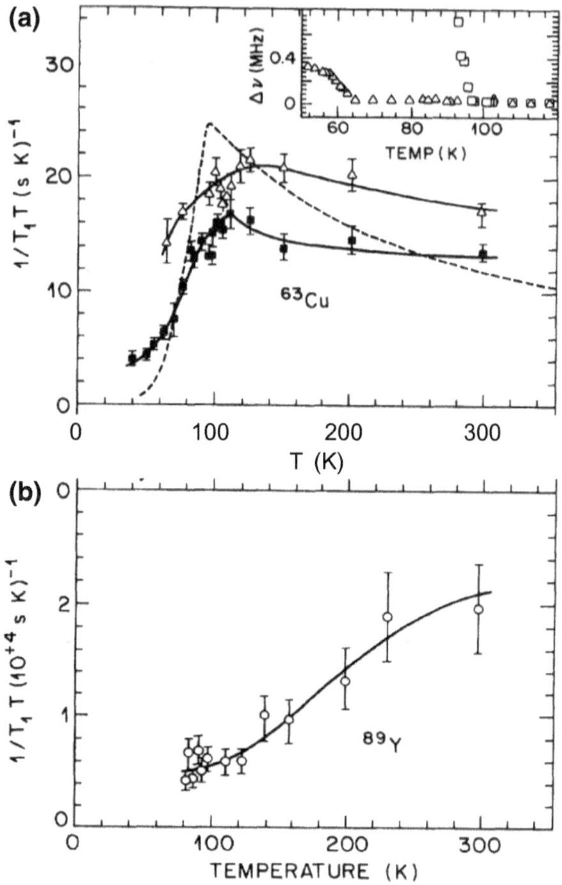

Fig. 4.1 **a** $1/T_1T$ is plotted versus T over the interval $40\,\mathrm{K} \leq \mathrm{T} \leq 300\,\mathrm{K}$ for two NQR peaks associated with ^{63}Cu(2) sites in YBCO6.64 ($T_c \simeq 64\,\mathrm{K}$) [185]. For comparison, the dashed line shows the behavior recorded earlier for the YBCO7 phase ($T_c \simeq 90\,\mathrm{K}$). The inset shows the shift in the resonance frequency of the NQR coil which occurs as the sample goes superconducting at $T \simeq 64\,\mathrm{K}$. **b** $1/T_1T$ is plotted versus T for ^{89}Y NMR in the same sample used as in (**a**). On account of the pseudogap, these data show a sharp, characteristic deviation from Korringa–like behavior seen in the YBCO7 phase

usual Korringa behavior: The relaxation rate product $1/T_1T$ in some instances is very nearly proportional to the spin susceptibility $\chi_s(T)$ (cf. 2.2.8). In Sect. 4.2 we will discuss the physical meaning of this along with a theoretical model which yields this result. It was to be several years before techniques such as ARPES and specific heat would demonstrate that actual gapping of the Fermi surface was primarily responsible for the behaviors in Figs. 4.1 and 4.2. Those results are reviewed briefly below.

However, it was noted at the time [166] that Friedel had suggested that "pseudogaps" might occur as a result of AFM correlations in the band theory of these materials [187]. The detailed physics of this mechanism was not forthcoming, but the name stuck: these effects were ever–after attributed to a "pseudogap" or to a "spin–gap" by those who thought prematurely that the effects were limited to spin–dynamical effects of the mobile carriers.

In this chapter we review and analyze pseudogap phenomena for both the $La_{2-x}Sr_xCuO_4$ Sect. 4.1 and $YBa_2Cu_3O_{7-x}$ Sect. 4.2 series of compounds. For this purpose we use data derived from INS studies of these systems as well as from

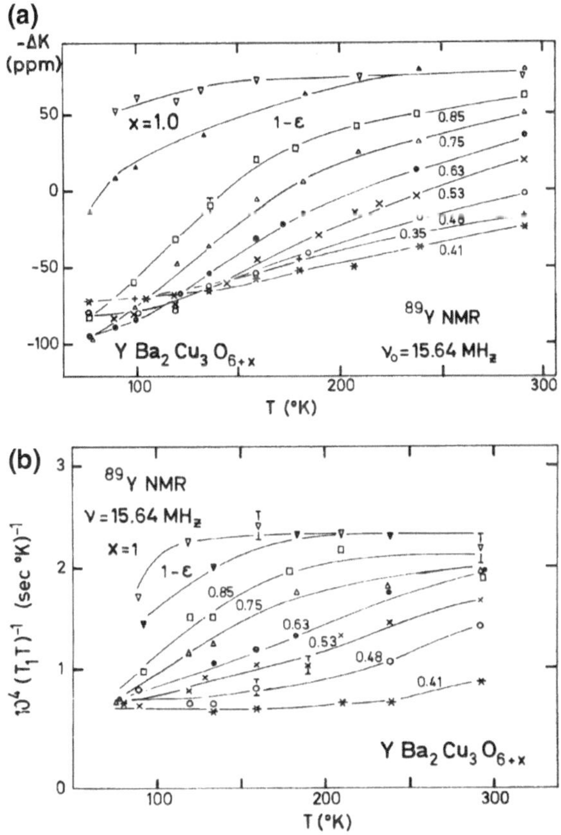

Fig. 4.2 **a** The negative of the ^{89}Y NMR shift ΔK is plotted as a function of temperature for a series of compositions YBCO6+x, for $0.4 \leq x \leq 1.0$ [166]. ΔK is measured relative to the ^{89}Y gyromagnetic ratio found in a solution of YCl_3. In Sect. 4.2 the chemical shift is determined to be $^{89}K_{orb,\alpha} \simeq 160$ ppm, where $\alpha =$ ab, c. **b** Plots of $1/T_1T$ versus T for the same series of compositions used in **a** [166]. In both cases the data show the progressive development of a pseudogap effect in which they exhibit a maximum at progressively higher T for increasing *dilution* of hole carriers. These data can also be plotted to show the relation $1/T_1T \propto (^{89}K_{orb,\alpha} - \Delta K_\alpha(T))$ characteristic of the pseudogap region (see text)

NMR shift and relaxation measurements. At this point there are, for certain doping levels, INS data for both of these compounds that provide an absolute, quantitative calibration of $\chi''(\mathbf{q}, \omega)$. Such data may be employed in the spin–lattice relaxation formulation of Sect. 3.5.2 to estimate the variation of T_1 over a wide range of temperatures. For both YBCO and LSCO there is quantitative agreement between INS and NMR data at low temperatures. However, INS intensity that represents the bulk of the relaxation rates $1/T_1$ at higher temperatures *as well as the pseudogap effect* is absent from the available INS data for both systems. This type of analysis, the first of its kind, will be spelled out in detail for LSCO and YBCO in Sects. 4.1 and 4.2, respectively.

We present a summary of ARPES and specific heat data that revealed for the first time that pseudogap effects actually correspond to progressive gap formation in the Fermi surface. These studies launched a theoretical investigation of the physical origin of the pseudogap effects that has been ongoing. Recent theoretical work will be summarized briefly. We also review the interesting case of the YBCO–related compound $YBa_2Cu_4O_8$, which in its stoichiometric form constitutes an underdoped cuprate with a pronounced pseudogap.

We also highlight in this chapter a very interesting susceptibility scaling effect for LSCO that was originally reported by D. C. Johnston [189]. It was found that $\chi_s(T)$ for a series of underdoped samples of $La_{2-x}Sr_xCuO_4$ not only exhibited a gap-like feature, but were found to scale rather precisely using a reduced temperature variable T/T^*. Although this effect was interpreted at the time in terms of a theoretical model for insulating antiferromagnets, it seems clear in retrospect that it is linked with the pseudogap effect. Recently this effect has also received an interpretation with a first-principles theory. The data and interpretation are presented briefly in Sect. 4.1.

An excellent review of both theory and experimental results up to 2006 on many aspects of "doped Mott insulators", including pseudogap effects, has been given by Lee, Nagaosa and Wen (LNW) [188]. These authors seek to characterize the pseudogap systematically, noting that this is best done with the spin paramagnetic NMR shift or susceptibility. They also point out that the behavior of $T_1(T)$ for the Cu(2) site is not a suitable indicator of the strength of the pseudogap effect, where it sometimes agrees with the NMR shift behavior and sometimes not. While this is a well–taken point, we believe that the origin of this problem lies in the dynamical spin–spin correlation factors $\mathcal{K}_{ij}$ that vary strongly with temperature and could mask the behavior of the basic quasiparticle dynamics in the T_1 formula (see 3.5.20). As will be seen in the discussion below, the pseudogap effect for the T_1 process is much simpler and clearer when the dynamic spin–spin correlation effects are taken account of.

4.1 Pseudogap Effects and Dynamic Susceptibility in $La_{2-x}Sr_xCuO_4$ (LSCO)

Apart from introducing the LSCO structure in Sect. 3.1, there has not been up to this point much discussion of LSCO. In spite of its rather low T_c relative to other cuprates, LSCO has played a major role in high-T_c physics research. This was because, (i) it was one of the first cuprate structures identified [113], (ii) it has a very interesting and accessible phase diagram (Fig. 1.1), and (iii) it is relatively easy to make large single crystals of LSCO. For the latter reason a great many neutron scattering measurements have been conducted on this system, making its dynamic susceptibility perhaps the best characterized of all of the cuprates.

We take up the case of LSCO first, because of clearer INS data and because of its simplicity as a single–layer cuprate. Further, it illustrates in a clear–cut way the complexities of the normal–state T_1 process in cuprates. From the standpoint of NMR technique, LSCO is a far simpler system to work with than YBCO. First, there are only one Cu^{2+} site and two ^{17}O sites, making spectra and relaxation studies generally easier to interpret. For the purposes of conducting ^{17}O NMR studies in crystals, LSCO is also a favorable case, because it has very high oxygen mobility at relatively modest temperatures. Thus, it is easy to exchange ^{17}O into a crystal and homogenize it with simple annealing procedures.

The foregoing circumstances notwithstanding, we shall see that the pseudogap effect in LSCO is not quite the same as in the other cuprates. One rarely even finds the word pseudogap applied to LSCO. While the NMR shift and susceptibility behavior is pseudogap-like (with important differences), the relaxation behavior of the ^{63}Cu is quite different from that of YBCO6.63. One reason for this is the important role played by the dynamical spin-spin correlation effects on the T_1 process. The dynamic susceptibility analysis we present below is the first one in which the latter effects were taken into account [204]. However, see also the work of Uldry and Meier in Sect. 5.1.

Before addressing the INS results and relaxation effects, we recount a unique and interesting scaling effect in the static susceptibility exhibited by LSCO.

4.1.1 *Spin Susceptibility Scaling for $La_{2-x}Sr_xCuO_4$*

Around the time that the first studies of underdoped YBCO7-x appeared, it was revealed by D. C. Johnston that the static susceptibility of $La_{2-x}Sr_xCuO_4$ (LSCO) obeys a simple scaling law that can be summarized by a single equation [189]. Over essentially the entire range of superconducting compositions, one has

$$\chi(x,T) = \chi_0(x) + \chi_{s,max}(x)\, F_s(T/T_{max}(x)) + C(x)/T\,, \qquad (4.1.1)$$

where $\chi_0(x)$ is a temperature-independent background susceptibility, $\chi_{s,max}$ is the maximum spin susceptibility for composition x, which occurs at $T_{max}(x)$, and $F_s(T/T_{max}(x))$ is a scaling function valid for $0.05 \leq x \leq 0.30$ [189]. C(x) is a spurious Curie term which occurs at higher doping levels [232]. By definition $F_s(1) = 1$. The term $\chi_0(x)$ represents Van Vleck orbital and diamagnetic contributions (see Sect. 2.3.2). The function $F_s(T/T_{max}(x))$ is such that $\chi(x,T)$ is qualitatively similar to the YBCO6.63 result in Fig. 4.16.

To illustrate the behavior of $\chi(x,T)$ for LSCO we display the somewhat more complete data set reported later by Nakano et al. [232] in Fig. 4.3 for $0.08 \leq x \leq 0.26$, all superconducting samples. While the latter data extend up to $\sim 550\,K$, data from another study extending up to $\sim 1000\,K$ show that $\chi(x,T)$ for $x \leq 0.09$ had no (accessible) maximum value, but all data plots for $x > 0.09$ exhibited a maximum at some temperature [233]. *This Curie-Weiss-like rise in susceptibilities at low temperatures for the higher doping levels is one factor which sets LSCO apart from the pseudogap behavior of other cuprates.*

Nonetheless, there is a clear gap feature in $\chi(x,T)$, which may be seen in Fig. 4.4, where $\chi_s(T)/\chi_{s,max} = (\chi(x,T) - \chi_0)/\chi_{s,max}$ is plotted versus T/T_{max} [232]. The data for the eleven compounds shown follow this scaling curve to a remarkable degree. It seems highly plausible that this behavior is related to the pseudogap effect identified in both YBCO6.63 and Y1248. This scaling of data may not reflect a simple

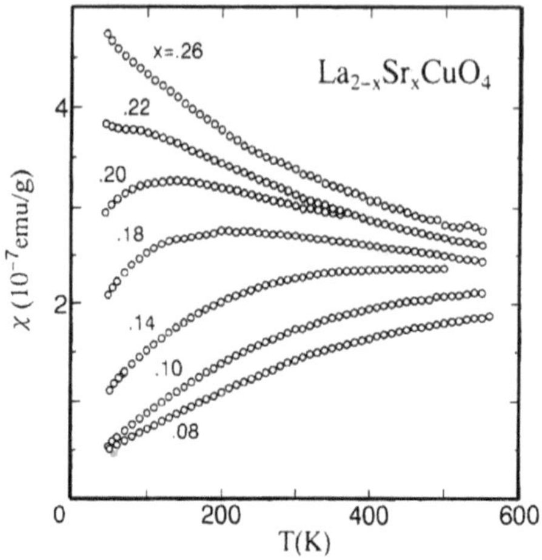

Fig. 4.3 Susceptibility data for $La_{2-x}Sr_xCuO_4$ are plotted as a function of temperature for a range of random powder samples with $0.08 \leq x \leq 0.26$ [232]. These data are found to scale as described in the text and in Fig. 4.4

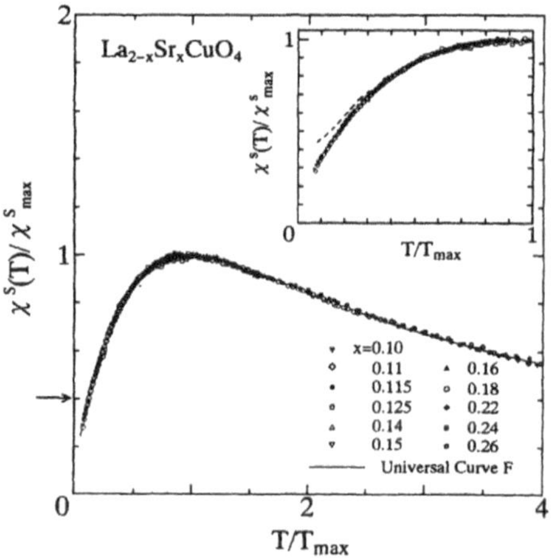

Fig. 4.4 The data from Fig. 4.3 are plotted as $\chi_s(T)/\chi_{s,max}$ versus T/T_{max} as defined in (4.1.1). The resulting scaling curve defines the function $F_s(T/T_{max}(x))$ from the latter equation as well, identified as "Universal Curve F" [232]. The inset shows that compared with previous scaling curves (e.g. [189]) represented by the dashed line, the present data decline more sharply as $T \rightarrow 0$. As noted in the text, the theory by Varma [222] predicts $\chi_s(T) \propto T^{1/2}$ near $T = 0$, very near to what is observed

energy gap parameter as in Fig. 4.16. The limiting, low-T behavior $\chi_s(T) \propto T^{\beta}$, $\beta \sim 0.5$, is actually one of the predictions of the pseudogap theory given by Varma.

It is interesting to examine the behavior of the other parameters in the fit to (4.1.1). The values obtained for $T_{max}(x)$, $\chi_0(x)$, and $C(x)$ are displayed as a function of x

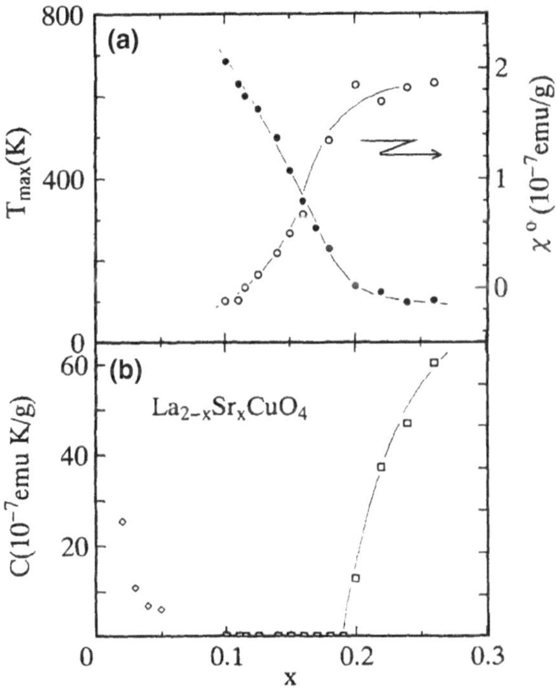

Fig. 4.5 **a** The fitting parameters $T_{max}(x)$ and $\chi_0(x)$, and **b** $C(x)$ from (4.1.1) are plotted as a function of x for the series of fits shown in Fig. 4.4. The behavior of these parameters clearly divides this doping range into regions I and II, where $x \simeq 0.20$ is the boundary line. See text for more discussion

in Fig. 4.5a. It is seen that there is a peculiar sort of transition point between data fits for $0.10 \le x < 0.20$ (Region I) and $0.20 \le x \le 0.28$ (Region II). In Region I, T_{max} declines in a roughly linear fashion[1] from $T_{max} \sim 700$ K for $x = 0.10$ down to $T_{max} \sim 100$ K at $x = 0.20$. In region II, $T_{max} \sim 100$ K. In Region I, χ_0 rises from a slightly negative value at $x = 0.10$ to $\chi_0 \sim 1.8 \times 10^{-7}$ emu/g at $x = 0.20$ and remains constant above that point. Finally, C(x) = 0 in Region I, rising rapidly in Region II to where it becomes a major feature of the susceptibility data. Although there is no result given which identifies $\chi_0(x)$ as the residual susceptibility at $T \ll T_c$, we regard $\chi_s(T)$ as the spin susceptibility. The magnetic behavior then distinguishes Regions I and II very clearly as "underdoped" and "overdoped", respectively. Or in a more speculative vein, one could term these *subcritical* and *supercritical* relative to a putative quantum critical point at $x \simeq 0.20$.

4.1.2 Behavior of the ^{17}O and ^{63}Cu NMR Shifts in LSCO

Scaled ^{17}O NMR shift values are plotted as a function of temperature in Fig. 4.6 for x = 0.15 and 0.24 [232]. The shift data used for these plots are from Ishida et al. for

[1]There were no fits to data below x = 0.10.

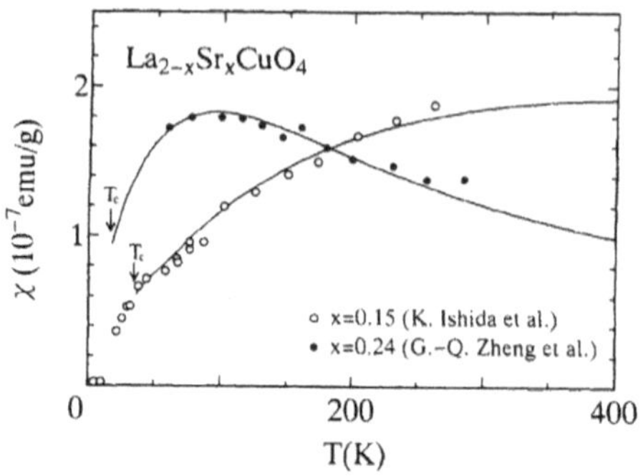

Fig. 4.6 The ^{17}O NMR shifts measured for $La_{2-x}Sr_xCuO_4$ with $x = 0.15$ [237] and 0.24 [181] are fitted to susceptibility curves such as those shown above in Fig. 4.3 [232] and plotted versus temperature. The scale of shift values is such that C_c =50-60 kOe/μ_B

x = 0.15 [237] and from Zheng et al. for x = 0.24 [181]. The lines drawn represent susceptibility data from [232], but the shift scales have been omitted from the figure. We note that the original data reflect *c*-axis shift coefficients 50-60 kOe/μ_B, which is somewhat larger than the values of C_α for YBCO6.63 in Table 4.2. In both cases, the scaled ^{17}O shift values are a good quantitative match to the susceptibility data.

Data for the ^{63}Cu *c*-axis NMR shifts in LSCO (not shown) have been found to have the following features [234]. With field along the *c* axis, compositions in the superconducting range show a very nearly constant shift $K_c \simeq 1.22\%$ independent of temperature. The latter shift is driven mainly by Van Vleck orbital magnetism. For fields in the *ab* plane, shift data appear to follow the same trends as the susceptibility data in Fig. 4.3. Although to our knowledge they have not been scaled with susceptibility data as in Fig. 4.6, a simple estimate of $\Delta K_{ab}/\Delta\chi$ gives a value $A_{ab} \simeq 149$ kOe/μ_B, within 2–3% or so of the values given below (Table 4.2) for the composite ^{63}Cu(2) shift coefficients in YBCO. For the Cu^{2+} at least, the Mila-Rice HF Hamiltonian from Sect. 3.3.3 appears to have nearly universal validity in the cuprates.

^{63}*Cu NQR Spectra and* Sr^{2+} *Doping in LSCO*

Whereas oxygen-deficient YBCO7-x shows line-broadening caused by crystalline disorder, LSCO exhibits nearly–resolved ^{63}Cu NQR and NMR "impurity" lines attributable to sites neighboring the Sr^{2+} dopant ions. NQR spectra are shown in Fig. 4.7a for three Sr^{2+} doping levels [238]. The spectra have been analyzed into pairs of ^{63,65}Cu NQR lines representing sites neighboring the impurities and "bulk" Cu^{2+} sites away from the Sr^{2+} sites. Since the charge environment near a dopant site is quite different from that of a bulk site, it is not surprising to find a different EFG at such sites. For a sample with $x = 0.30$ there is a very high probability that a Cu^{2+} site will have a Sr^{2+} neighbor, and the corresponding impurity line is now larger than the bulk line. As may be seen in Fig. 4.8, the T_1 behavior of these well-resolved NQR lines is nearly indistinguishable. Perhaps it is because of this high level of impurity-induced disorder in LSCO that T_c peaks at $\sim 1/2$ the peak level of hole doping for YBCO7-x.

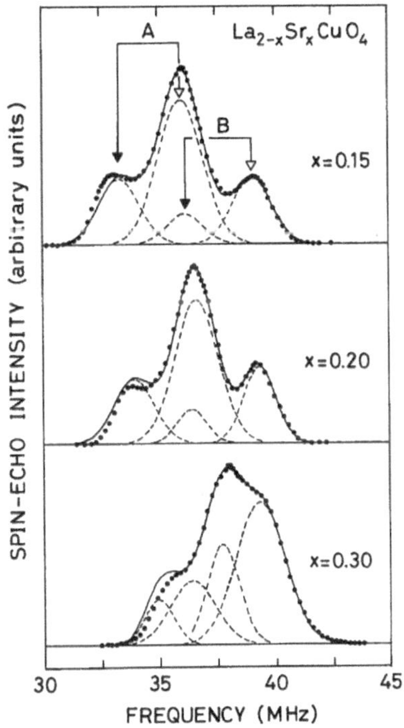

Fig. 4.7 NQR spectra for $^{63,65}Cu$ in LSCO are shown for three typical values of x. the spectra have been analyzed into a pair of "A" lines, which represent bulk material away from the doping centers, and a pair of "B" lines representing sites which have at least one Sr^{2+} neighbor. For each case there are two lines representing the two isotopes with their roughly 7:3 (63:65) abundance ratio. For $x = 0.30$ the B lines have well over half of the intensity. The dashed lines shown are Gaussian components in the spectrum which sum up to the solid lines drawn, giving a reasonable facsimile of the experimental spectra

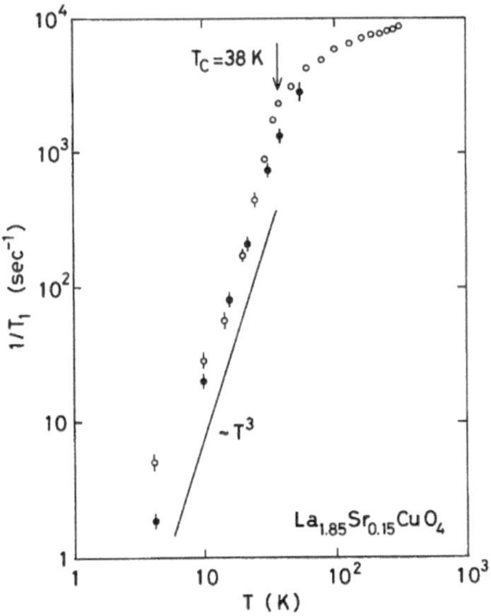

Fig. 4.8 Data are plotted for $1/T_1$ as a function of temperature for both the A (open circles) and B (closed circles) sites identified in Fig. 4.7 in a sample with $x = 0.15$. The two data sets, where comparable, are quite similar in character. The A-site measurements show a typical flat region in the normal state, with a steep drop below T_c, then a T^3 region. Very little evidence is seen of the pseudogap feature found in the susceptibility for this composition

4.1.3 T_1 Results for Planar ^{63}Cu and ^{17}O

We now begin our discussion and analysis of the T_1 processes of planar ^{63}Cu and ^{17}O nuclear spins in an optimally doped crystal of LSCO. This work was completed relatively early (1994), as the INS data on $La_{1.85}Sr_{0.15}CuO_4$ was the first to become available for any cuprate [274, 275]. The relaxation data shown in Fig. 4.9 are, to our knowledge, the only relatively complete data set for any doping level of LSCO. [239] Experimental (c-axis) data are plotted as $(T_1T)^{-1}$ for both ^{63}Cu and the apical ^{17}O in part (a), where ^{63}Cu NQR data are plotted as filled squares, while some earlier data from Imai et al. [283] are plotted as "+" symbols. These data sets are in very good accord where they overlap. In addition, data for the apical ^{17}O scaled to coincide with the ^{63}Cu data between 100 K and 200 K are shown as dots with error bars. Since the apical ^{17}O is relaxed entirely by local HF contact with one Cu-site spin, it is expected to behave roughly like the ^{63,65}Cu, as is seen to be the case.

Figure 4.9b presents $(T_1T)^{-1}$ versus T for the planar ^{17}O in the same crystal. Along with the T_1 data (dots) are plotted measurements of the NMR shift (triangles) scaled

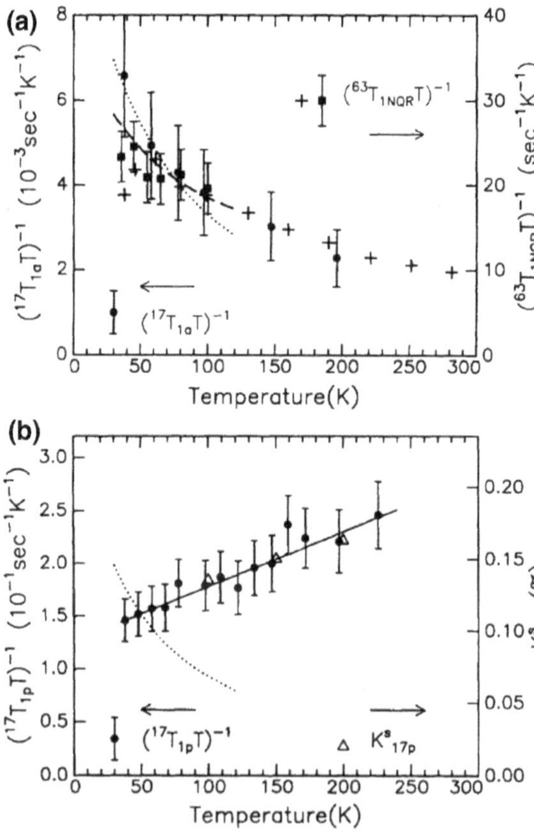

Fig. 4.9 **a** Data for $1/T_1T$ is plotted versus T for ^{63}Cu (filled squares, +'s: right-hand scale) and for apical ^{17}O (dots: left-hand scale) in LSCO:0.15 [239]. The data plotted as +'s are from Imai et al. [283]. The dotted curve is a plot of the calculated rate based on INS data given with (4.1.3) and HF constants from [239]. See text for discussion of expected behavior with HF constants from Table 4.1. **b** $1/T_1T$ data (dots) are plotted versus T for planar ^{17}O in LSCO:0.15 [239]. Also plotted are data points for the c-axis NMR shift K^s_{17p} (open triangles), which have been scaled to coincide with the relaxation data (right-hand scale). The calculated behavior for $1/T_1T$ based on INS data given with (4.1.3) and HF constants from Table 4.1 is shown as a dotted line. See text for discussion of these results

to coincide with the relaxation data. In retrospect we suggest that the coincidence of the shift and relaxation data is probably accidental. The striking feature of this plot is the sharp contrast between the planar ^{17}O and ^{63}Cu relaxation curves. In terms of the Eqs. (3.5.20) and (3.5.21) that relate them to $\chi''(\mathbf{q}, \omega)$, it is seen that they are both driven by the same particle dynamics (i.e., $\tau_e(T)$), so that any contrast between them can be ascribed to the spin–spin correlation factors $\mathcal{K}_{ij}$. The discussion of these data in [239] omitted any consideration of the $\mathcal{K}_{ij}$'s, leaving some important effects to be discovered.

Again, the INS data cited were used to estimate values of $1/T_1T$ for ^{63}Cu using (3.5.20), but ignoring the correlation effects. The results are shown as a dotted line cutting through the data plot at a rather steep angle [239]. With very limited parameter adjustability, the agreement between INS data and the ^{63}Cu relaxation rate is quite satisfactory. However, our current estimate of the HF parameters, i.e. B in particular in Table 4.1, is somewhat smaller than that used in [239]. The current values would therefore lead to estimates of $1/T_1T \sim 30\%$ smaller than those plotted in the figure. We suggest that this would give a somewhat more realistic result, although it would clearly show that some other mechanism—presumably from an unidentified contribution to $\chi''(\mathbf{q}, \omega)$—becomes important at higher temperatures. This conjecture was borne out in a later analysis of these data [204]. The particulars of that analysis are the focus of the next subsection.

4.1.4 Analysis of NMR Data for LSCO Using INS Data for $\chi''(\mathbf{q}, \omega)$

In the foregoing subsections we discussed data for the NMR shifts, uniform susceptibility, and spin-lattice relaxation times T_1 for ^{63}Cu and ^{17}O in LSCO for a doping level $x \sim 0.15$ near the peak value of $T_c \sim 40$ K. This system was seen to exhibit a peculiar sort of pseudogap effect, which strongly affects the susceptibility even at the peak of the T_c curve, but allows the ^{63}Cu T_1 curve to follow a Curie-Weiss-like behavior down to temperatures very close to T_c. Here, we gain further insight into the behavior of quasiparticle dynamics in LSCO by conducting a joint analysis of INS data for the dynamic susceptibility of LSCO and the T_1 behavior of both planar ^{63}Cu and ^{17}O.

This system is unique in having received early on a very thorough characterization of $\chi''(\mathbf{q}, \omega)$ by means of INS. The observable part of $\chi''(\mathbf{q}, \omega)$ consists of four mildly incommensurate peaks near (π, π); these have been scanned in great detail over a wide range of frequencies. As a consequence, LSCO became a classic "test bed" for the theoretical relationship between NMR and $\chi''(\mathbf{q}, \omega)$. We first review the theoretical basis for such an investigation. The fluctuation-dissipation theorem relation between spin-fluctuation-generated T_1 processes in a metallic environment and the dissipative term $\chi''(\mathbf{q}, \omega)$ of the dynamical susceptibility, as recounted in Sect. 3.5, had been known for many decades [58] before it was tested using experimental INS data for

$\chi''(\mathbf{q}, \omega)$. The latter test was conducted via T_1 measurements for ^{63}Cu and ^{17}O in LSCO:0.15 [239] as described in the previous subsection. This test did not include the effects of dynamical spin–spin correlations, which are evidently important. The test did establish that the expected quantitative correspondence occurs as expected at temperatures near T_c, though it did show divergence at higher temperatures.

Our goal then is to correlate two pieces of experimental data using an exact theoretical relationship.[2] This relationship, embodied in (3.5.20)–(3.5.22), was given its first serious experimental test in 2011 [204], yielding limited success and pointing up a new set of problems.

4.1.4.1 Review of INS Data for Optimally-Doped LSCO

We first set the stage with a brief review of relevant INS data for $La_{2-x}Sr_xCuO_4$, with x$\sim$ 0.15. It seems that of all the cuprate compounds, LSCO alone was available in large, uniform crystals with homogeneous oxygen content. Conveniently enough, $\chi''(\mathbf{q}, \omega)$ is a quantity which can be measured directly by means of INS. Although INS measurements can only be made at frequencies several orders of magnitude higher than those at which NMR data are recorded, it is not apparent that there are any low–frequency physical processes that would invalidate an extrapolation of $\chi''(\mathbf{q}, \omega)$ data to frequencies near zero.

Some low-frequency INS data for $\chi''(\mathbf{q}, \omega)$ are shown in Fig. 4.10. In part (a) we see data for scans through two of the four incommensurate peaks located at $(\pi, \pi) \pm \delta(0, \pi)$ and $(\pi, \pi) \pm \delta(\pi, 0)$, where $\delta = 0.245$ for $La_{1.86}Sr_{0.14}CuO_4$ [275]. The peaks are seen to be well resolved, with relatively little difference between behavior near $T_c = 35$ K and that at 4.6 K. There is, however, an increase in intensity midway between the peaks at higher energy. That trend is also visible in the lower-energy scans from Thurston et al. [276] in part (b) of Fig. 4.10, where at $\omega = 1.5$ meV the peaks are completely resolved, while at 4.0 meV there is already some filling in of the space between peaks. The incommensuration varies with composition, $\delta \simeq 2x$ [274], but it shows no measurable variation with energy at the lowest energies scanned [276].

From available INS results for LSCO, then, it does not appear that incommensurability will vanish at lower (i.e. at NMR) frequencies. At higher frequencies, however, early work showed the incommensurability fading into a single broad peak by $\omega = 15$ meV [274]. That this trend would continue to higher energies was verified by studies to above 200 meV carried out at ISIS [277]. In that study, results for $La_{2-x}Sr_xCuO_4$(x = 0.14), were compared with the undoped antiferromagnet La_2CuO_4, both $\mathbf{q}$-scans and energy spectra. For La_2CuO_4, incommensurability is not visible in the lowest energy scans (25–50 meV) reported [277], nor any above

[2]This relationship is very nearly exact under the assumption that spin hyperfine processes are predominant. However, it has not yet been determined how important the orbital relaxation rates are for the cuprates (see Sect. 5.6).

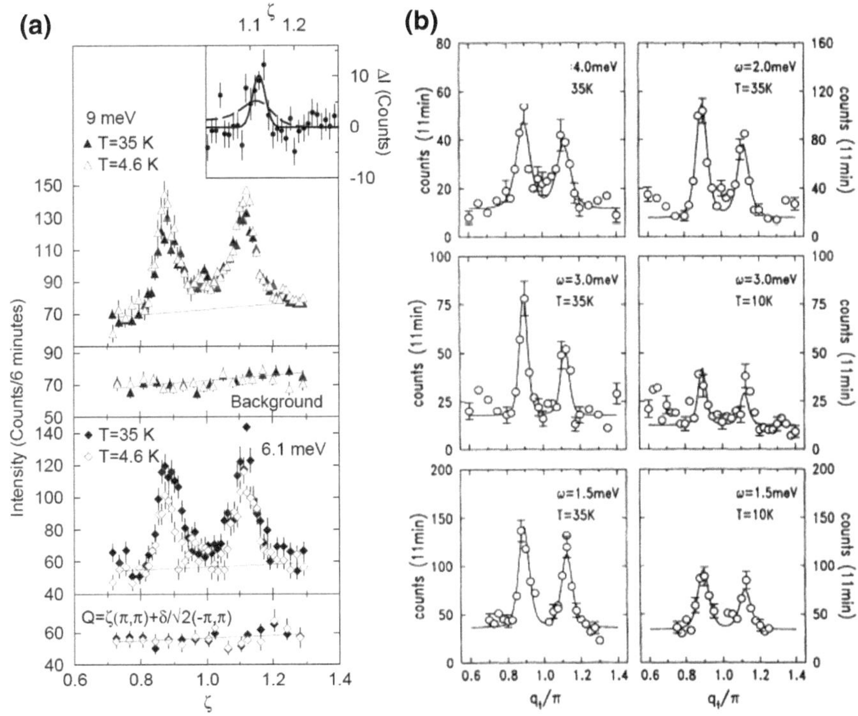

Fig. 4.10 **a** INS **q**-scans along the path shown at the bottom of the figure are plotted for $T \sim T_c$ and for $T \ll T_c$ for the two scattering energies shown [275]. These data show well-resolved incommensurate peaks and a relatively small difference in scattering intensity for the two temperatures. The inset shows the difference in intensity measured for the two temperatures at 9 meV. Note the small, but definite increase in intensity between the peaks between 6 meV and 9 meV. **b** INS data at energies $\omega \leq 4.0$ meV, showing very well-resolved incommensurate peaks with no measurable variation of the incommensuration δ with ω. These authors find $\delta = 0.22$ for $La_{2-x}Sr_xCuO_4$ (x = 0.15). They also find a relatively small variation of intensity with temperature below T_c

that. The resolution in **q**-space may be somewhat marginal for this purpose, although the AFM (Bragg?) peak at 295 K for La_2CuO_4 showed a width $\Delta\mathbf{q} \sim 0.25$ r.l.u.

We display three composite data plots from Hayden et al. [277], namely the equal-time spin correlation function $S(\mathbf{q}) = \int_{-\infty}^{\infty} d\omega\, \chi''(\mathbf{q}, \omega)$, the zero-frequency susceptibility $\chi'(\mathbf{q}) = (2/\pi) \int_0^{\infty} d\omega\, \chi''(\mathbf{q}, \omega)/\omega$, and the local susceptibility $\chi''(\omega) = \sum_{\mathbf{q}}' \chi''(\mathbf{q}, \omega)$. These quantities have been computed from the results of a spallation–source study [277]. The local susceptibilities are shown in Fig. 4.11a, b, where it is noteworthy that the results are plotted in absolute units, and where in (b), reactor-based results are included below $\sim 25\, meV$. Below, we shall compare these data with the absolute susceptibility results which were used to calculate T_1 in [239]. The results in Fig. 4.11a, b show that the effect of doping is to concentrate $\chi''(\mathbf{q}, \omega)$ at energies below 50 meV (as compared with La_2CuO_4), and that the peak seen at 22 meV is a "new energy scale" for this superconductor [277]. This energy is also a rough upper limit for the appearance of incommensurability. The solid line plotted in panel (b) is

$$\chi'' = \frac{\Gamma\omega}{(\Gamma^2 + \omega^2)}, \tag{4.1.2}$$

with $\Gamma = 22 \pm 5\,\text{meV}$.

The equal-time correlation function $S(\mathbf{q})$ and the zero-frequency susceptibility $\chi'(\mathbf{q})$ obtained with the Kramers–Kronig relation in [277] are shown in Fig. 4.11 for the undoped ((c) and (d)) and the doped ((e) and (f)) cases, respectively. Owing to the near-constant width of $\chi''(\mathbf{q}, \omega)$ in $\mathbf{q}$-space for $La_{2-x}Sr_xCuO_4$(x $= 0.14$) [277], there is very little difference between $S(\mathbf{q})$ and $\chi'(\mathbf{q})$. The associated energy integrals have been carried out for ω ranging from 15 to 150 meV. The central peak in Fig. 4.11e, f has an HWHM of $\kappa = 0.26(5)\,\text{Å}^{-1}$, which is just slightly larger than the incommensurate splitting. Since the Kramers-Kronig integral favors low energies, one wonders whether there would be visible incommensurate peaks in $\chi'(\mathbf{q})$ if the reactor-based intensity (i.e., that which gives the open circles in Fig. 4.11b) were included in the calculation of $\chi'(\mathbf{q})$. As noted below, such a contribution might be appreciable for indirect spin-spin coupling with the planar ^{17}O.

The Calibrated Susceptibility for T_1 Calculations in LSCO

In order to discuss T_1 behavior in LSCO we need two main ingredients. The first is a calibrated susceptiblity function $\chi''(\mathbf{q}, \omega)$ for vanishing ω, with detailed $\mathbf{q}$-dependence, and the second is a measured HF tensor. We begin with the calibrated susceptibility that was originally used in [239]. It originated with the measurements of Mason et al. [275]. The absolute calibration for these was presented in [278], but had also been used, it seems, in the paper by Hayden et al. [277]. We shall recount here the calibration used in [239]. Some careful checking finds that the latter result leads to a value for the initial slope of $\chi''(\omega)$ in Fig. 4.11b within 10% or less of the plotted value.

The low-energy INS data for $\chi''(\mathbf{q}, \omega)$, part of which is displayed in Fig. 4.10a, may be represented in the normal state by the function

$$\left.\frac{\chi''(\mathbf{q}, \omega)}{\omega}\right]_{\omega\to 0} = \frac{K_0 \mathcal{A}^2\, \theta[\Gamma_c - \Gamma_n(\mathbf{q})]}{\Gamma_n(\mathbf{q})^2}, \tag{4.1.3}$$

where $\Gamma_n(\mathbf{q}) = \mathcal{A}^2[\kappa^2 + R(\mathbf{q})]$, with

$$R(\mathbf{q}) = \frac{1}{8\pi^2 a_L^2 \delta^2}\{[(q_x - q_y)^2 - \pi^2\delta^2]^2 - [(q_x + q_y)^2 - \pi^2\delta^2]^2\}. \tag{4.1.4}$$

Here $a_L = 3.8\,\text{Å}$ is the CuO_2 lattice constant and $\delta = 0.245$ is the discommensuration [275]. The other quantities in (4.1.3) are

- The constant $\mathcal{A}^2 = 3.858\,\text{eVÅ}^2$ comes from a form used to fit INS data at finite values of ω [275].

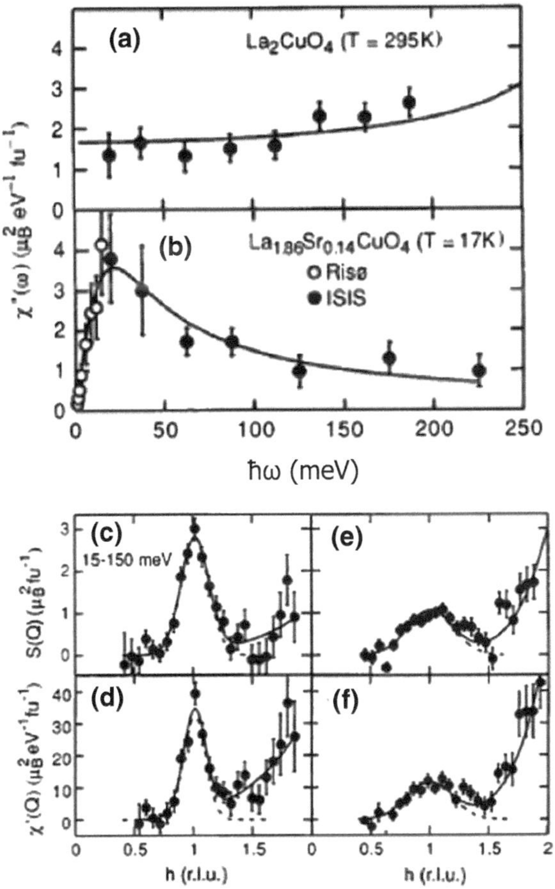

Fig. 4.11 **a** and **b** The local susceptibilities $\chi''(\omega)$ for La_2CuO_4 (295 K) and $La_{2-x}Sr_xCuO_4$(x = 0.14) (17 K), derived from the data presented in [277], are plotted as a function of energy $\hbar\omega$. As compared with La_2CuO_4, the effect of doping is seen to concentrate the amplitude at low energies, giving a peak at $\sim$ 22 meV. Below that point are plotted data from earlier reactor-based studies [275], which serve to determine the initial slope used for T_1 calculations. **c** and **e** are the equal-time spin correlation functions $S(\mathbf{q})$ defined in the text, for La_2CuO_4 and $La_{2-x}Sr_xCuO_4$(x = 0.14), respectively. **d** and **f** are zero-frequency susceptibilities $\chi'(\mathbf{q})$ obtained with (6.3.1) in a similar fashion. For LSCO:0.14 (**e** and **f**) these quantities are nearly indistinguishable, since the width of $\chi''(\mathbf{q}, \omega)$ shows very little dependence on energy [277]. In **c**–**f** the rise in intensity to the right is caused by phonons, where the suggested behavior of the quasiparticle properties is indicated by dashed lines. Although the spallation source resolution width is not as good as in reactor-based studies, it is clearly well below the peak widths found for LSCO:0.14 in **e** and **f**

- The peak width parameter $\kappa = [(0.034)^2 + 31.35(k_B T)^2]^{1/2} \text{Å}^{-1}$ is also from an expression used to fit data over a range of ω values, but is stated here for $\omega = 0$. In this expression $k_B T$ is to be entered in units of eV. We note that this representation of κ has been updated [278] from what was used in [239]. The two expressions give very similar values for temperatures of 100 K or less.
- The calibration obtained using acoustic phonons [278] gives $K_0 \mathcal{A}^2 \simeq 2.5\mu_B^2$.
- The function $\theta[p] = 1$ if $p > 0$ and 0 otherwise. In (4.1.3) Γ_c is taken to be an estimated maximum magnon energy $\sim$0.3 eV.

The foregoing function describes $\chi''(\mathbf{q}, \omega)$ in $\mathbf{q}$-space as having four incommensurate peaks located at $(\pi, \pi) \pm (0, \pi)\delta$ and $(\pi, \pi) \pm (\pi, 0)\delta$. These peaks have adjustable width given by κ. As stated, the function has units of μ_B^2/eV^2 per formula unit. This is the initial slope of $\chi''(\mathbf{q}, \omega)$, determined by INS data at energies of just a few meV. From the standpoint of INS methodology, it is supposed that this function would apply in the energy range in which it was determined and to all energies below that as well.

As we shall see below, however, there is a problem with this interpretation in connection with the T_1 data for the planar ^{17}O, which will be discussed in Sect. 6.3. We note here that there is also a second problem with using INS measurements as a calibration of $\chi''(\mathbf{q}, \omega)$ for NMR frequencies. This has to do with the behavior in the superconducting state. We know from T_1 measurements for ^{63}Cu in LSCO (as well as those for other superconductors at $T < T_c$) that the effective value of $\chi''(\mathbf{q}, \omega)$ declines typically by a large factor. For LSCO $1/T_1$ drops by a factor $\sim$ 100 at a temperature of 4.2 K relative to its value in the normal state [279]. The INS data for $\chi''(\mathbf{q}, \omega)$ as given by Thurston et al., for example, shows no decline whatever below T_c, particularly at the lowest energies tested $\hbar\omega \sim 1.5$ meV. As the latter authors have commented [276], this is a situation for which "further theory is required". It seems, then, that we do not always understand when the INS results will apply at NMR frequencies and when they do not.

Hyperfine Constants for LSCO

Below, we apply the calibrated expression for $\chi''(\mathbf{q}, \omega)$ to T_1 data for ^{63}Cu and planar ^{17}O in LSCO. For this purpose, we first review the analysis of the appropriate NMR shifts and uniform susceptibility data for $La_{2-x}Sr_xCuO_4$(x = 0.15). The numbers are similar to those for YBCO given in Chap. 3, but need, of course, to be established independently. Working from the ^{63}Cu shift results of Ohsugi et al. [279], it is first noted that $K_c(T)$ is constant at $\sim$1.20% well into the superconducting state. Similar to YBCO, then, we have $A_c + 4B \simeq 0$. Next, K_c^{vv} is slightly smaller than that ($\simeq$ 1.28%) for YBCO. Thus, the crystal-field energies for LSCO are a bit higher and the g-factors slightly smaller as well. Estimated values are given in Table 4.1. From the K_α^{vv} tensor we use the effective orbital HF shift coefficient $\alpha_{orb} = 141$ (emu/mol)$^{-1}$ given with (3.3.5) for YBCO. This leads to the Van Vleck susceptibility components given in Table 4.1.

The Van Vleck susceptibility tensor is combined with the estimate $\chi_{dia} \simeq -99 \times 10^{-6}$ emu/mol [235, 280] to extract the spin susceptibility tensor from

Table 4.1 Compilation of parameters related to the analysis of ^{63}Cu and planar ^{17}O NMR shifts and susceptibilities for $La_{2-x}Sr_xCuO_4$(x = 0.15) at $T = 300$ K. These quantities are defined and discussed in the text. The units are as follows: NMR shifts: %; susceptibilities: 10^{-6} emu/mol; HF constants: kOe/spin. The HF constants are defined by (2.3.2). They will have their proper energy units when multiplied by $10^3 \hbar\gamma_{17,63}$

g_{ab}	g_c	K_{ab}^{vv}	K_c^{vv}	K_{ab}^s	χ_{ab}^{vv}	χ_c^{vv}	χ_{ab}^s	χ_c^s	B	A_{ab}	A_c	K_a^{17}	K_b^{17}	K_c^{17}	C_a	C_b	C_c
2.06	2.27	0.27	1.20	0.41	19.1	85.1	145	177	76	19	−304	0.274	0.162	0.214	108	64	77

$$\chi_\alpha^{expt} = \chi_\alpha^s + \chi_\alpha^{vv} + \chi_{dia}. \tag{4.1.5}$$

The powder average susceptiblity was found to be 98×10^{-6} emu/mol (300 K) [239], a value within 5% of those given by Ishida et al. [237] and Johnston [281]. Since $\chi_\alpha^s \propto g_\alpha^2$ [58], we define a quantity χ_0^s with the relation $\chi_\alpha^s = (g_\alpha^2/4)\chi_0^s$. This leads to the expression for the powder average spin susceptibility $\frac{1}{3}(\chi_c^s + 2\,\chi_{ab}^s) = 1.137\,\chi_0^s$ using g-values from the table. With (4.1.5) and the numbers given, we find $\chi_0^s = 137.1 \times 10^{-6}$ emu/mol. The values for χ_α^s in Table 4.1 follow immediately. These are essentially the susceptibility results given in [239]. Note that the spin susceptibilities are substantially larger than those found for YBCO7 (Table 3.4).

We now can extract the ^{63}Cu spin HF constants for $La_{2-x}Sr_xCuO_4$(x = 0.15). The shift coefficient for $K_{ab}^s \simeq 0.41\%$ (300 K) [279] is $K_{ab}^s/\chi_{ab}^s = 28.3\,(\text{emu/mol})^{-1}$. Using (3.3.4) this gives $A_{ab} + 4B = 325$ kG/spin. From studies of undoped La_2CuO_4 [282] we take the result $A_{ab} - 4B = -286$ kG/spin. Combining the latter two relations leads to A_{ab} and B as given in Table 4.1, as well as the estimate $A_c = -4B = -304$ kG/spin. All results are summarized in Table 4.1. We believe that these slightly smaller numbers than used in [26] give a more realistic outcome.

The corresponding values C_α for the planar ^{17}O were derived from room-temperature NMR shift measurements and reported in [239]. The shift measurements, which have been corrected for orbital background shifts $K_{a,b,c}^{orb} = (0.046, 0.008, -0.027)$ (in %, from Takigawa et al. [190]), and their corresponding HF constant values are listed in Table 4.1.

Evaluation of T_1 *for* ^{63}*Cu and* ^{17}*O from INS Data*

With HF couplings from Table 4.1, we can now evaluate T_1 for ^{63}Cu and ^{17}O in $La_{2-x}Sr_xCuO_4$(x = 0.15) using the INS susceptibility data embodied in (4.1.3). Relaxation rates are calculated for ^{63}Cu with (3.5.20) and for ^{17}O with (3.5.21) [239]. We note that the T_1 process for any nucleus in this system consists of two principal ingredients, namely the dynamical spin–spin correlation coefficients for the first three neighbor pairs[3] $\mathcal{K}_{1a}$, $\mathcal{K}_{1b}$, $\mathcal{K}_2$, $\mathcal{K}_{3a}$, and $\mathcal{K}_{3b}$, and the correlation time parameter τ_e that represents the quasiparticle dynamics of the system. Here we have[4]

[3]Note the change in notation for the $\mathcal{K}_{ij}$'s from that in Sect. 3.5. This will be used throughout this chapter and the next.

[4]We drop the "iso" subscript on $\chi''(\mathbf{q}, \omega)$ and insert "I" to indicate the INS value given by (4.1.3) and (4.1.4). The isotropic "iso" value is, however, used throughout (see discussion with (3.5.10).

$$\mathcal{K}_{n\alpha I} = \frac{\int_N d\mathbf{q}\, g_{n\alpha}(\mathbf{q})[\chi_I''(\mathbf{q},\omega)/\omega]}{[\int_N d\mathbf{q}\, \chi_I''(\mathbf{q},\omega)/\omega]}, \tag{4.1.6}$$

where n $= 1, 2, 3$ is the neighbor index and $\alpha = a, b$. Here a and b designate in–plane axis parallel and perpendicular to the incommensuration axis, respectively, while $g_{1\alpha} = \cos(q_\alpha a)$, $g_2 = cos(q_a a)cos(q_b a)$, and $g_{3\alpha} = cos(2q_\alpha a)$. Finally, we have

$$\tau_{eI}(T) = \frac{k_B T}{\mu_B^2} \int_N d\mathbf{q}\, \chi_I''(\mathbf{q},\omega)/\omega \tag{4.1.7}$$

for the quasiparticle correlation time. Note that the $\mathcal{K}_{n\alpha I}$ correlation factors do not depend on the scale of $\chi_I''(\mathbf{q},\omega)$, but only on its shape and width in $\mathbf{q}$ space. $\tau_{eI}(T)$ on the other hand represents the time scale and T–dependence of the quasiparticle dynamics.

Using the above components, the nuclear relaxation rates may then be written

$$\frac{1}{^{63}T_{1c}} = \frac{\gamma_{63}^2}{2}\,[A_{ab}^2 + 4B^2 + 2B^2(4\mathcal{K}_2 + \mathcal{K}_{3a} + \mathcal{K}_{3b}) + 4A_{ab}B(\mathcal{K}_{1a} + \mathcal{K}_{1b})]\,\tau_e \tag{4.1.8}$$

and

$$\frac{1}{^{17}T_{1c}} = \frac{\gamma_{17}^2}{4}[C_a^2 + C_b^2](2 + \mathcal{K}_{1a} + \mathcal{K}_{1b})\,\tau_e \tag{4.1.9}$$

for planar ^{63}Cu and ^{17}O, respectively, the two nuclear species of interest. In these equations A_{ab}, B, C_a and C_b are hyperfine tensor components in units of Gauss per spin. Note that because NMR frequencies are very small on the scale of INS energies, INS data used here are extrapolated to $\omega = 0$.

Using the functional form for $\chi_I''(\mathbf{q},\omega)$ represented by (4.1.3) and (4.1.4) and other data given, we may evaluate the relevant $\mathcal{K}_{n\alpha I}$'s for LSCO over temperatures ranging from T_c up to 300 K. The results are shown in Fig. 4.12. The temperature dependence exhibited is modest, except that the factor $(2 + \mathcal{K}_{1aI} + \mathcal{K}_{1bI})$ that multiplies the ^{17}O relaxation is very small and tends to diminish at low temperatures.

NMR and INS data that form the basis for our discussion of T_1 and pseudogap effects in LSCO are analyzed and displayed in Fig. 4.13 as values of $\tau_e(T)$ and $\tau_{eI}(T)$. There are three separate sets of data points shown, namely values of τ_{eI} derived from INS data for $\chi_I''(\mathbf{q},\omega)$, then τ_e values from T_1 data for both ^{17}O (circles) and ^{63}Cu (squares), which are obtained using (4.1.9) and (4.1.8), respectively and the T_1 data in Fig. 4.9. Values of τ_{eI} are seen to decline only very slowly with T in the normal state. Thus, $\chi_I''(\mathbf{q},\omega)$is seen to obey an ω/T scaling law to quite a good approximation. Such behavior has been attributed to the formation of stripes in the normal state of LSCO [193, 194]. Finally, the most important thing to note in this figure is that if $\chi_I''(\mathbf{q},\omega)$ represented the entire dynamic susceptibility of the quasiparticles in LSCO, then all three curves for τ_{eI} and τ_e would coincide and the problem would be solved. In fact, we find three disparate curves, so there is quite a bit to discuss.

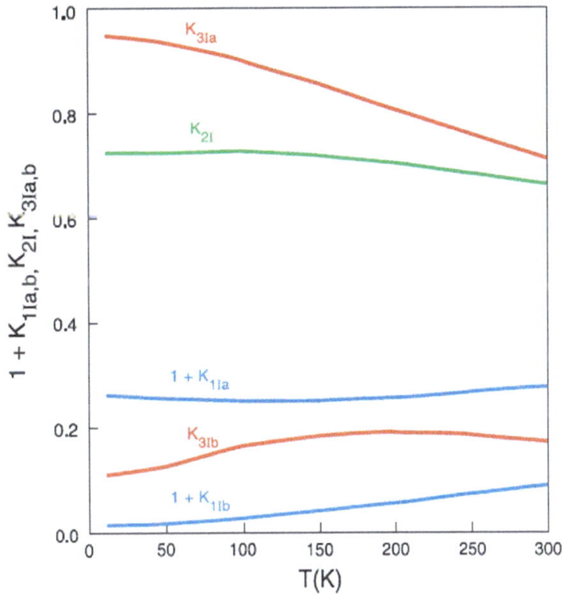

Fig. 4.12 The spin–spin correlation factors $\mathcal{K}_{n\alpha}$ defined in (4.1.6) have been evaluated using the data in (4.1.3) and (4.1.4) and are plotted here as a function of temperature. It is to be emphasized that these plots are based on INS data for $La_{1.85}Sr_{0.15}CuO_4$. Accordingly, they are labeled $\mathcal{K}_{nI\alpha}$. The temperature dependence is fairly modest, except to note that the quantity $(2 + \mathcal{K}_{1Ia} + \mathcal{K}_{1Ib})$ that modulates the ^{17}O relaxation rate (see 4.1.9) is very small compared with unity

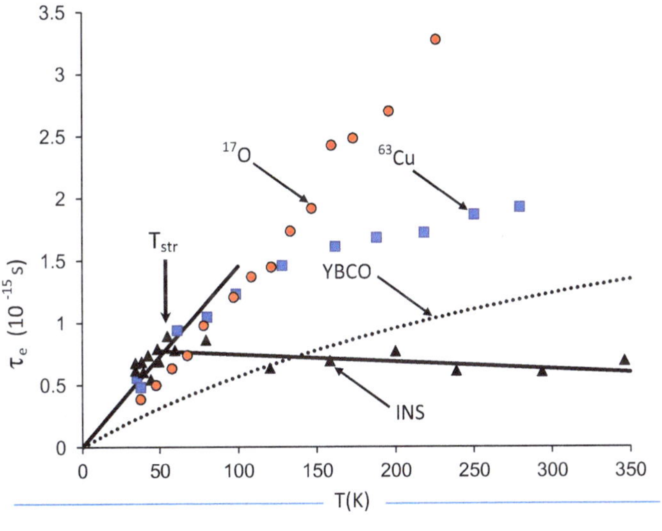

Fig. 4.13 Plots of τ_e and τ_{eI}, as defined in the text, as a function of temperature T. The red dots (^{17}O) and blue squares (^{63}Cu) are show values of $\tau_e(T)$ derived from T_1 data in Fig. 4.9 using $\mathcal{K}_{nI}$'s obtained from Fig. 4.12. See text for details. An estimate of $\tau_{eI}(T)$ based on (4.1.7) and INS data for $\chi_I''(\mathbf{q}, \omega)$ (black triangles) shows a very slight decline with T. Then, below the onset of stripes labelled T_{str}, $\tau_{eI}(T)$ executes a Korringa–like downward ramp that is echoed by the behavior of $\tau_e(T)$ derived from T_1 data, a correspondence that occurs with no adjustable parameters. The dotted line labeled "YBCO" shows $\tau_e(T)$ behavior reported by Uldry and Meier [195] with a method of analysis based only on experimental data. This illustrates the weaker T_1 process found in YBCO7

Below a cutoff temperature that we have labeled T_{str}, whatever is causing the ω/T scaling ceases to operate, and values of τ_{eI} begin falling toward zero in a roughly linear–T fashion. Whatever gives rise to T_{str}, we want to emphasize that it is definitely higher than T_c, which for LSCO never exceeds a value $T_c \sim 40\,\mathrm{K}$. What is also notable here is that values of τ_e deduced from the ^{63}Cu T_1 data descend with declining T in good accord with τ_{eI} below T_{str}. The quantitative agreement here is quite remarkable, because there are no adjustable parameters. The formulation of the T_1 process via the fluctuation–dissipation theorem in Chap. 3 is quantitatively verified in LSCO at temperatures $T \leq T_{str}$.

There are, on the other hand, some clear surprises in Fig. 4.13. The main surprise is that above T_{str} both plots of τ_e rise high above that for τ_{eI}, clearly indicating that there is another source of magnetic fluctuations in LSCO that is not represented by $\chi''_I(\mathbf{q}, \omega)$. The most salient possibility for this contribution is another term in $\chi''(\mathbf{q}, \omega)$ that has a low amplitude, but occupies a broad region in the BZ so that it has statistical weight equal to if not greater than that of $\chi''_I(\mathbf{q}, \omega)$. The amplitude of such a term could be as small as just a few percent of the $\chi''_I(\mathbf{q}, \omega)$ peak amplitude, yet have equal weight. Finding the baseline for such a term could require polarized neutrons, rendering such an enterprise extraordinarily difficult.

Introducing the Pseudogap Term $\chi''_P(\mathbf{q}, \omega)$.

Thus, we assume the existence of another term $\chi''_P(\mathbf{q}, \omega)$, so that the total dynamic susceptibility becomes

$$\chi''(\mathbf{q}, \omega)_{total} = \chi''_I(\mathbf{q}, \omega) + \chi''_P(\mathbf{q}, \omega). \tag{4.1.10}$$

The contributions of these two terms to NMR dynamics are independent and additive. The reason that it is called the pseudogap term is because any pseudogap effect in LSCO will be a property of $\chi''_P(\mathbf{q}, \omega)$. In the earlier paper on this analysis, it appears that such a pseudogap is present [204]. This presents rather unusual behavior, where there are two disparate terms in $\chi''(\mathbf{q}, \omega)$, one having a pseudogap and the other not. Below we discuss the introduction of parameters to characterize $\chi''_P(\mathbf{q}, \omega)$. Before that, we note that (4.1.10) will be incorporated into (3.5.20) and (3.5.21), yielding two sets of terms of the form of $(1/^{63}T_{1cI} + 1/^{63}T_{1cP})$ (4.1.8) and of $(1/^{17}T_{1cI} + 1/^{17}T_{1cP})$ (4.1.9). Terms subscripted '1cI' will incorporate $\mathcal{K}_{n\alpha I}$ and $\chi''_I(\mathbf{q}, \omega)$ from (4.1.6) and (4.1.7), respectively, and those subscripted '1cP' will have $\mathcal{K}_{n\alpha P}$ and $\chi''_P(\mathbf{q}, \omega)$. The INS data as described above determine the values of $1/T_{1cI}$ for both nuclear species. These values are then subtracted from the experimental data (Fig. 4.9) to determine values of $1/^{63}T_{1cP}$ and $1/^{17}T_{1cP}$ at each temperature plotted (with some interpolation). Those are then the input data which determine the width and amplitude parameters of $\chi''_P(\mathbf{q}, \omega)(T)$.

We adopt a squared Lorentzian profile for $\chi''_P(\mathbf{q}, \omega)$ in $\mathbf{q}$ space, which then has fourfold symmetry in the basal plane. Following [278], this gives the form $\chi''_P(\mathbf{q}, \omega)/\chi''_P(peak) = q_w^4/(q_w^2 + q_x^2 + q_y^2)^2$ with the origin at $\mathbf{Q} = (\pi, \pi)$. With fourfold symmetry there are only three correlation coefficients $\mathcal{K}_{nP}$, n = 1, 2, 3 (4.1.6). With this formulation in mind, the goal is to find values of $\chi''(peak)$and

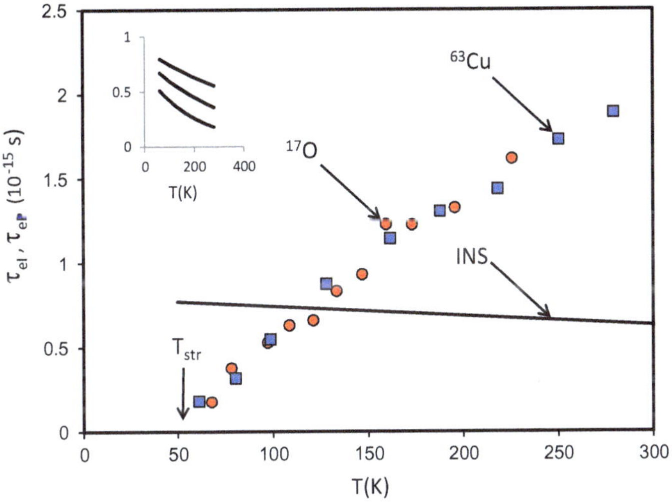

Fig. 4.14 Values of τ_{eP} deduced from T_1 data and the squared Lorentzian model for $\chi''_P(\mathbf{q},\omega)$ are plotted for both ^{17}O (circles) and for ^{63}Cu (squares). By adjusting the $\mathcal{K}_{nP}$'s as described in the text, the two curves for τ_{eP} versus T are made to coincide as they must. This process yields the variation of the $\mathcal{K}_{nP}$'s with T (see inset) and estimates of the width parameter for $\chi''_P(\mathbf{q},\omega)$. The resulting curve for $\mathcal{K}_{1P}$ versus T is represented quite accurately by the temperature function given in the text. For comparison, the INS–derived value of $\tau_{eI}(T)$ is shown as a solid line

q_w at each temperature. Although there are two parameters and two unknowns, this is a somewhat nonlinear procedure that requires a bit of interpolation. In practice, as one progresses along the temperature scale, the coefficient of τ_{eP} that determines $1/^{63}T_{1cP}$ changes very slowly, so $1/^{63}T_{1cP}$ can be used to estimate τ_{eP}. Then the rapidly changing coefficient for $1/^{17}T_{1cP}$ can be used to estimate q_w. Iterating this process, one finds that it converges quickly, giving essentially coincident curves of τ_{eP} versus T for the two nuclear species, as is shown in Fig. 4.14. There, values determined for τ_{eI} and τ_{eP} are plotted separately versus T. This result suggests that the somewhat hypothetical form adopted for $\chi''_P(\mathbf{q},\omega)$ is physically realistic. The curves of the $\mathcal{K}_{nP}$, (n = 1, 2, 3) are shown in the inset to Fig. 4.14.

The key correlation parameter $\mathcal{K}_{1P}$ can be satisfactorily represented here as

$$\mathcal{K}_{1P}(T) = -0.81\, exp[-(T-50)/600]. \tag{4.1.11}$$

The small value exhibited by the quantity $1+\mathcal{K}_{1P}$ at low temperature and its steep increase with T serve to explain the dramatic difference between the temperature profile of $1/^{17}T_{1cP}$ and $1/^{63}T_{1cP}$ seen in Fig. 4.9.

At a more general level, the predominance of $1/T_{1P}$ over $1/T_{1I}$ at the higher temperatures is clear evidence for a hitherto unresolved term in $\chi''(\mathbf{q},\omega)$ in LSCO. Moreover, the resulting coincident curves for $1/^{17}T_{1cP}$ and $1/^{63}T_{1cP}$ and their joint extrapolation to zero at a temperature near T_{str} gives evidence for a consistent interpretation in terms of the squared Lorentzian model for $\chi''_P(\mathbf{q}, \omega)$ and for an apparent pseudogap effect in LSCO. It does, however, require a substantial amount of analysis to detect the presence of this effect, and it depends quite strongly on the apparent vanishing of τ_{eP} (i.e. of $1/^{17}T_{1cP}$ and $1/^{63}T_{1cP}$) in the vicinity of $1/T_{str}$. Whether the presence of $\chi''_P(\mathbf{q}, \omega)$ can be resolved by INS techniques is an important question. Some discussion of this point is given with the presentation of similar results for the case of YBCO6.5 in the following subsection. Again, the dotted curve in Fig. 4.13 shows values of τ_e for YBCO7 determined in a purely empirical scheme described by Uldry and Meier [195]. It illustrates the distinctly stronger T_1 processes in LSCO as compared with YBCO7.

4.2 The Pseudogap in Oxygen–Deficient YBCO7-x

Since pseudogap effects in the NMR data first came to light in studies of YBCO7-x with some depletion of the oxygen content (and thus of hole doping), it was for this system that the most precise and comprehensive bodies of data were first developed. In this section we discuss a set of results on both the NMR shifts and relaxation rates for ^{89}Y, ^{17}O(2, 3), and ^{63}Cu(2) for a representative composition YBCO6.63 [148, 190, 191]. This compound is a well-characterised example of a high-T_c pseudogap system.[5] Relaxation data for ^{63}Cu(2) are shown in Fig. 4.1, where the low-frequency excitation spectrum ($\sum_q A_q^2 \chi''(q, \omega)/\omega$) is seen to decline at low temperatures as though an energy gap had set in below $T \sim 300\,K$. At $T_c \simeq 64$ K there is no break or other feature in the Cu(2) T_1 data (Fig. 4.1a), in striking contrast with the sharp drop which takes place for the YBCO7 phase (see the dashed line in Fig. 4.1a as well as data in Fig. 3.11). Nonetheless, monitoring the rf coil susceptibility (Fig. 4.1a, inset) gives good evidence that the sample material undergoes a transition to bulk superconductivity at that temperature. This behavior is confirmed throughout for this cuprate.

In this subsection we first analyze some very elegantly presented NMR shift data to obtain values for hyperfine coefficients for the 60 K phase of YBCO, which are then compared with those for YBCO7. Here, the case of ^{89}Y presents special problems in that, first, dipolar HF fields are appreciable, but do not have the Spin–Hamiltonian form exhibited by (3.3.2) that was adopted for subsequent analyses of T_1 processes. Second, ^{89}Y also has the slightly unfortunate property that it couples with Cu(2) spins in two neighboring CuO_2 planes of the YBCO structure, and thus, its HF fluctuations are subject to interplanar spin correlations, which have been found to be appreciable. For these reasons, we will discuss ^{89}Y HF effects in the next two subsections on HF constants and early relaxation models for the pseudogap systems, but they will not be included in the joint analysis of T_1 and INS data for $\chi''(\mathbf{q}, \omega)$ in the final subsection.

[5] Although a continuum of T_c values has been reported for YBCO7-x, the "*60 K phase*" is the most homogeneous underdoped phase, since dT_c/dx is quite small in that region.

We present and discuss the calibrated measurements of $\chi''_I(\mathbf{q},\omega)$ by Stock, et al. [201] in the third subsection, and in the fourth, a joint analysis of NMR and INS data along the lines of that for LSCO in Sect. 4.1.4 is presented for YBCO6.5.

4.2.1 *HF Parameters for YBCO6.63 from Scaling of Shift Data*

We first discuss NMR shift and susceptibility data for YBCO6.63 with the goal in mind of extracting HF parameters to use in fitting the corresponding T_1 data. In Fig. 4.15a NMR shift data for ^{63}Cu(2) in an oriented powder specimen are plotted for field orientation both parallel and perpendicular to the c axis [185]. The c axis–data show only a slight increase below T_c, so that $^{63}K_{orb,c} = 1.28(1)\%$ and $^{63}K_{s,c} = -0.01(1)\%$ (near 300 K). This result for $^{63}K_{orb,c}$ is indistinguishable from that for YBCO7.0 [147]. The zero–temperature intercept for $^{63}K_{ab}(T)$ is $^{63}K_{orb,ab} = 0.25(2)\%$, which is also unchanged from YBCO7 [147]. The downward–curving spin paramagnetism is the essence of spin pseudogap behavior.

In Fig. 4.15b are plotted ^{17}O NMR shift values ($^{17}K_{\|}$, $^{17}K_{\perp}$, $^{17}K_c$) as functions of temperature for the three principal axes of the O(2, 3) sites [190]. The quadrupole tensor has also been evaluated and is virtually indistinguishable from what was determined for YBCO7 (see Fig. 3.8 and discussion). The shift $^{17}K_{\|}(T)$ corresponding to the Cu(2)-Cu(2) bond axis is seen to be more than half again as large as the transverse components. The latter ($^{17}K_{\perp}(T)$ and $^{17}K_c(T)$) have slightly different chemical (i.e. orbital) shifts, but their temperature variation is essentially indistinguishable. The ^{17}O(2, 3) shift can therefore be characterized by axial and isotropic components defined by

$$^{17}K_{ax} = (^{17}K_{\|} - {}^{17}K_{\perp})/3, \tag{4.2.1a}$$

$$^{17}K_{iso} = (^{17}K_{\|} + {}^{17}K_{\perp} + {}^{17}K_c)/3. \tag{4.2.1b}$$

As with YBCO7 (Sect. 3.4) there is a dominant isotropic HF effect from 3d-2s hybridization [146] and a lesser axial shift term corresponding to dipolar coupling from the 3d-2p$_\sigma$ hybridization, although the latter admixture is much greater than the former.

The ^{17}O NMR shifts $^{17}K_{\|}$ and $^{17}K_{\perp}$, plus that for ^{63}Cu(2) [190] have been scaled to coincide with $^{17}K_c(T)$, the most accurate data set, between T_c and 300 K, and are plotted with separate vertical axes in Fig. 4.16. All shifts are seen to follow a common temperature variation very closely, i.e. that of the spin susceptibility. The scaling fits then yield the following relations:

$$^{17}K_{ax}(T) = 0.1886 \times {}^{17}K_c(T) + 0.0151\%, \tag{4.2.2a}$$

$$^{17}K_{iso}(T) = 1.057 \times {}^{17}K_c(T) + 0.039\%, \tag{4.2.2b}$$

$$^{63}K_{ab}(T) = 1.522 \times {}^{17}K_c(T) + 0.32\%. \tag{4.2.2c}$$

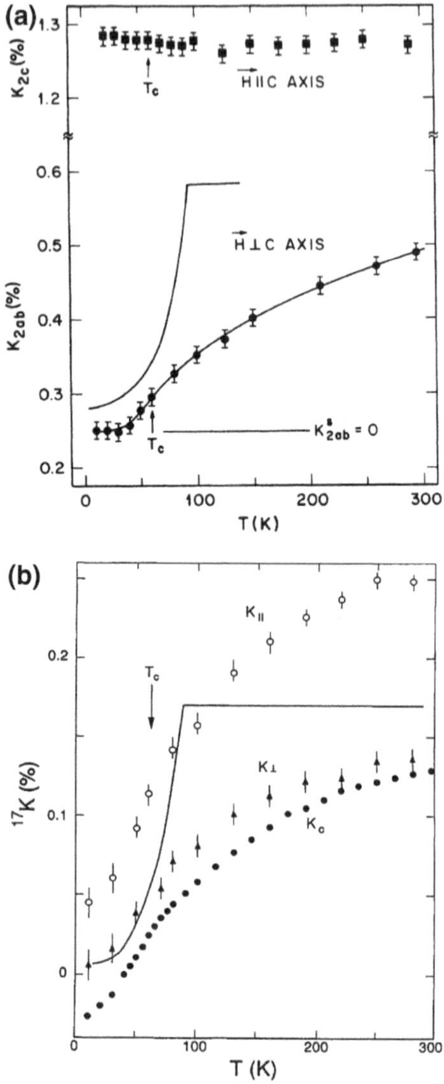

Fig. 4.15 **a** The NMR shift for ^{63}Cu(2) in an oriented-powder sample of YBCO6.64 [185] is plotted versus temperature for field orientation parallel and perpendicular to the c axis. The upper solid line is a representation of data for YBCO7, while the lower one is a guide to the eye. As with YBCO7 it is assumed that the orbital shifts are independent of temperature and that the spin susceptibility goes to zero at $T \ll T_c$. The breakdown of both shifts into spin and orbital components is given in the text. **b** Data for the ^{17}O(2, 3) NMR shift is plotted as a function of temperature up to 300 K for an oriented powder sample with field oriented along three mutually perpendicular axes, where the axis for $^{17}K_\parallel$ is the Cu(2)–Cu(2) bond axis [190]. The solid line is the variation of $^{17}K_c(T)$ for YBCO7 [147]. Analysis of these results is given in the text

This result gives additional strong evidence for a one-component quasiparticle state, since multiple bands would have to have susceptibilities with precisely the same spin-gapped temperature variation to produce this result. Such an effect seems quite implausible.

The scaled curves in Fig. 4.16 come apart a bit below T_c, an effect that is attributed to varying levels of superconducting diamagnetic shielding for the different cases. However, because $^{17}K_\parallel$ and $^{17}K_\perp$ are both measured with field $\perp$ c axis, there is no diamagnetic correction to $^{17}K_{ax}$ (see (4.2.1a)). Extrapolating the curves in

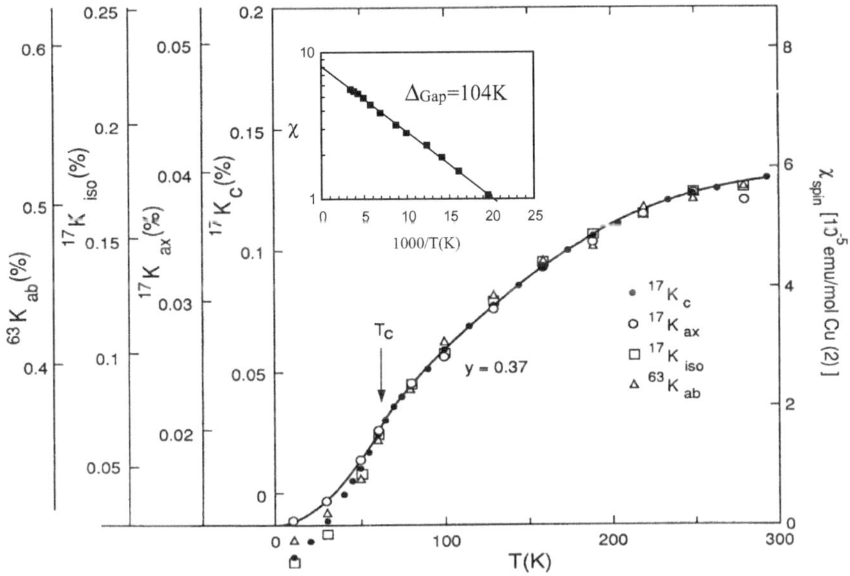

Fig. 4.16 **a** Data for $^{63}K_{ab}(T)$ for the Cu(2) sites [190] (similar to Fig. 4.15a) as well as for $^{17}K_{\parallel}(T)$ and $^{17}K_{\perp}(T)$ have been scaled to $^{17}K_c(T)$ and are plotted with the latter data as a function of temperature. The $T \rightarrow 0$ (orbital shift) limits of these shifts plots have been determined as discussed in the text, leading to an estimate for the zero of the spin components of the shifts and of the spin susceptibility. Scales for the various shift components are shown on the left. The scale for $\chi_s(T)$ has been estimated from the knowledge of HF constants (see text) and is shown on the right side of the plot. The solid line represents the estimate of $\chi_s(T)$ so determined. The inset shows an Arrhenius plot of ln($\chi_s(T)$) versus 1000/T(K). The linear behavior seen illustrates the gapped nature (i.e. pseudogap) of $\chi_s(T)$, where the gap energy is found to be $\Delta_{Gap} \simeq 104$ K

Fig. 4.15b to T = 0, then, yields $^{17}K_{ax}(0) = 0.013(2)\%$. One can then use (4.2.2) to obtain the other orbital shift values, which are $^{17}K_c(0) = -0.014(7)\%$, $^{17}K_{iso}(0) = 0.024(11)\%$, and $^{63}K_{ab}(0) = 0.30(2)\%$. These values are virtually unchanged from those determined earlier for YBCO7 (See Table 3.2).

Using the relation $\chi_s(T) =^{63} K_{ab}(T)/[f_{Cu}(A_{ab} + 4B)]$, where $A_{ab} + 4B \simeq 198\, kOe/\mu_B$ [192], and the estimate $f_{Cu} \simeq 0.7$ for the fraction of χ_s which resides on the Cu(2) site, we have determined the scale shown for $\chi_{spin}(T)$ in Fig. 4.16. The parameters derived from a Mila-Rice fit to the data in Sect. 3.2 (see Table 3.4) would give a slightly larger value for $(A_{ab} + 4B)$. Other fits to the shift and relaxation data in Chap. 5 will also give slightly different values for the scale of $\chi_{spin}(T)$. These scaling factors may have a slight dependence on composition, a point that will be discussed further in Chap. 5.

In any case, Fig. 4.16 represents a very precise characterization of the pseudogap effect in YBCO6.63. One can probe the energy gap origin of the decline of the the quasiparticle density of states in Fig. 4.16 with an Arrhenius plot of the $\chi_{spin}(T)$ data, i.e. by setting $\chi_{spin}(T) = \chi_0 \exp(-\Delta_{Gap}/T)$. Such a plot is shown in the inset to

Fig. 4.16, where it is found to yield a rather accurate representation of the data, with $\Delta_{Gap} \simeq 104$ K. Such a result was also implied by the fitting procedure of Uldry and Meier [195], to be discussed in Chap. 5. To our knowledge, this representation has not been predicted by any model calculation up to now and has not been emphasized in the literature. This result will be compared with INS data on a similar sample below.

We may complete the NMR shift picture for YBCO6.63 by adducing the ^{89}Y NMR shift and relaxation data which came two years later [191]. The corresponding shift data for both YBCO6.63 and YBCO7 cases are plotted versus T in Fig. 4.17a, where the "6.63" data are seen to exhibit a very similar temperature dependence as the scale plot Fig. 4.16. This idea is tested in Fig. 4.17b where the latter data for both field orientations are plotted versus $^{17}K_c(T)$ data from Fig. 4.15. The resulting curves are straight lines within narrow error limits and the extrapolation to $^{17}K_c(0) = -0.014(7)\%$ then gives the chemical shifts $^{89}K_{orb,c} = 160(15)$ and

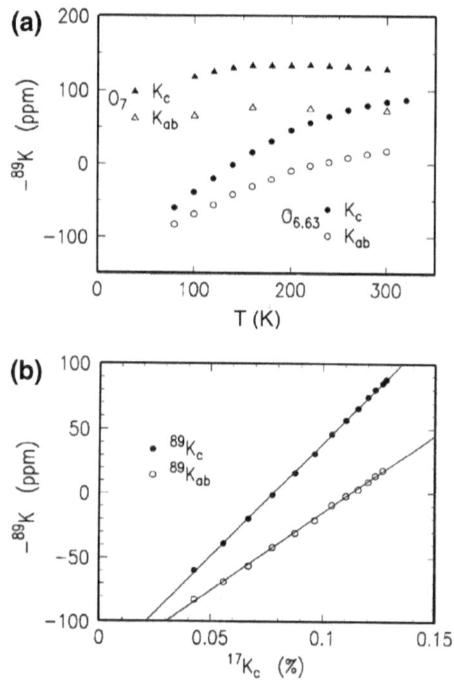

Fig. 4.17 **a** The negative of the ^{89}Y NMR shift in YBCO6.63 is plotted as a function of temperature for applied field both parallel and perpendicular to the c axis of an oriented powder sample [191], which is actually the same sample used in the foregoing Cu(2) and O(2, 3) NMR studies [190]. For comparison, previous data on YBCO7 [191] are also shown. **b** The YBCO6.63 shift data from part (**a**) are plotted versus the oxygen shift data $^{17}K_c(T)$ from [190]. Extrapolation of these curves to $^{17}K_c(0)$ yields the chemical shift values for ^{89}Y given in the text. These linear fits also demonstrate that all NMR shifts in YBCO6.63 follow the same temperature dependence, namely that of $\chi_s(T)$

$^{89}K_{orb,ab} = 152(10)$ ppm. Thus, the ^{89}Y shift data also follow the curve in Fig. 4.16 to high precision.

Scaling of HF Coefficients to Match YBCO7

We therefore have a complete characterization of NMR shifts for all significant sites and field orientations in YBCO6.63 in terms of a single universal curve [190]. The resulting shift values at $T = 300$ K are listed in Table 4.2 ("300 K Values"). Also listed there is the estimated Cu(2) susceptibility tensor at 300 K from [190]. We suggest, however, that the latter values are very approximate and that better values for both the susceptibility tensor and the HF coefficients can be assigned by scaling the HF tensor components so that they match the YBCO7 values from Tables 3.4 and 3.6 as closely as possible. For convenience, the latter set of values ("YBCO7 Values") is also listed in Table 4.2.

A "trial" set of HF coefficients can be defined from the data in Table 4.2 using the relation from (3.3.4)

$$\mathcal{A}_i = N_A \mu_B K_i / z_i \chi_{si}, \tag{4.2.3}$$

where K_i are the "300 K" values from the table, z_i are the numbers of Cu(2) neighbors, and χ_{si} is the appropriate susceptibility estimate from the same line of the table. The $\mathcal{A}_i$ are multiplied by a scaler S, which is then adjusted to minimize a weighted total fractional squared error defined by

$$\mathcal{E} = \sum_i \frac{w_i\,[S\mathcal{A}_i - A_i]^2}{A_i^2} = \sum_i w_i\,[1 - SR_i]^2, \tag{4.2.4}$$

where the YBCO7 coefficients are labelled A_i, and $R_i = \mathcal{A}_i / A_i$. This procedure will also yield a new estimate of the susceptibility tensor, from (4.2.3), of χ_{si}/S. The ratios R_i are listed in Table 4.2, and are seen to be grouped closely around a value $\simeq 1.36$.

It is easy to show that the errors are minimized when

$$S = \frac{\sum_i w_i R_i}{\sum_i w_i R_i^2} = 0.723. \tag{4.2.5}$$

The optimally scaled coefficients obtained are stated in Table 4.2 ("YBCO6.63"). Correspondence with the YBCO7 values is seen to be quite satisfactory. The resulting (scaled) susceptibility components are also given. They are somewhat larger than the values from Fig. 4.16, but are $\sim 80\%$ of the corresponding values in Table 3.4, in reasonable agreement with the relative values shown by experimental data given in [190].

One important consequence of this fitting procedure is that the reduction factor 0.70 used to estimate the HF coefficients in Table 3.4 is corroborated by the YBCO6.63 shift scaling (Fig. 4.16) results. These results strongly suggest that the composite Cu(2) shift coefficient, for which we find the value $A_{ab}^{(2)} \simeq 155 kOe/\mu_B$

Table 4.2 The NMR shift tensors for ^{17}O(2, 3) and ^{89}Y in YBCO6.63 at 300 K [190, 191] are listed in the top line, then the corresponding HF coefficients are stated for YBCO7 from Tables 3.4 and 3.6, below which the scaled coefficient values for YBCO6.63 derived from the stated shift values are listed for comparison. The scaling procedure is described in the text, wherein the weighted sum of fractional errors defined by (4.2.4) is minimized. Weighting factors w_i are designed to give each species the statistical weight it has in the unit cell. NMR shift values are given in %, the spin HF coefficients in kOe/μ_B, and susceptibilities in 10^{-6} emu/Cu(2)

Shift comp.	$^{63}K_{s,ab}$	$^{17}K_{s,a}$	$^{17}K_{s,b}$	$^{17}K_{s,c}$	$^{89}K_{s,ab}$	$^{89}K_{s,c}$	$\chi_{s\perp}$ (300 K)	$\chi_{s\parallel}$ (300 K)
300 K values	0.218	0.205	0.124	0.143	0.0173	0.0255	58	72
Spin HF Coef.	$A_{ab}^{(2)}$	C_a	C_b	C_c	D_{ab}	D_c		
YBCO7 (A_i)	155	72	44	38	1.71	1.86		
YBCO6.63	152	71.4	43.2	40.0	1.51	1.87	80	100
R_i	1.370	1.386	1.372	1.470	1.231	1.338		
w_i	4	8/3	8/3	8/3	1	1		

(Table 4.2), must, in fact, be rather smaller than the value $\simeq 198\ kOe/\mu_B$ given by [192].

Next, we consider the anisotropy of the ^{17}O(2, 3) and ^{89}Y HF coefficients C_α and D_α, respectively. Not surprisingly, the C_α have a strong axial component along the Cu(2)-Cu(2) bond axis owing to the $2p_\sigma$ hybridization. The observed c/a axis anisotropy is, however, mainly a g-factor effect. The constants C_α to be used in the main Hamiltonian (3.3.4) are proportional to $g_\alpha C_\alpha$, which differs by only a few percent between the b and c axes. This is within the expected errors. The D_α values, however, display a clear anisotropy if we consider the quantities $g_\parallel D_c$ versus $g_\perp D_{ab}$, where the former is ~30% larger. The origin of this effect is not clear, but the Y^{3+} HF response is clearly more than simply the isotropic core polarization HF field [190].

Dipolar Contribution to ^{89}Y HF Coupling

One candidate effect for the D_α anisotropy is direct dipolar coupling from the Cu^{2+} moments, which has been considered by Takigawa et al. [191]. Since the ^{89}Y HF couplings are rather small for YBCO, one must consider the possibility that the dipolar interaction between any Cu^{2+} moment $-\sum_\alpha g_\alpha \mu_B S_\alpha \hat{u}_\alpha$, where $\hat{u}_\alpha$ is a cartesian unit vector, and a ^{89}Y nuclear spin moment $^{89}\mathbf{I}$,

$$\mathcal{H}_{dipolar} = -\frac{\gamma_{89}\hbar\mu_B}{r^3}\left[\sum_\alpha g_\alpha S_\alpha {}^{89}I_\alpha - 3\frac{({}^{89}\mathbf{I}\cdot\mathbf{r})\sum_\alpha g_\alpha S_\alpha r_\alpha}{r^2}\right] \tag{4.2.6}$$

is a significant HF energy. The importance of this is that (4.2.6) does not have the simple diagonal form of (3.3.4), thereby changing the nature of the dynamics calculations if it is an important effect.

To consider the direct dipolar HF coupling with the ^{89}Y, we note that owing to hybridization, the Cu^{2+} moment at the origin (Fig. 2.1(b)) actually consists of a fraction $1-4\delta$, with fractions δ located on the four nn O(2, 3) sites. At the hybridization level $\delta \simeq 0.08$, the two O(2, 3) sites nearest the ^{89}Y are quite significant, since they are rather closer to the Y^{3+} than the Cu^{2+}. We ignore the two more distant O(2, 3) sites in this estimate. The positions of various ions in the crystal (Fig. 3.1) are such that the Cu(2)–O(2, 3) horizontal distance is $d = 1.95$ Å, and the vertical distances between ^{89}Y and Cu(2) and O(2, 3) are $\Delta c1 = 1.65$ Å and $\Delta c2 = 1.42$ Å, respectively. Using (4.2.6) we then find for the total contribution to the HF coefficients

$$\Delta D_{dip,ab} = -8\mu_B \left| \frac{1-4\delta}{r_{Cu}^3} \left[1 - 3\frac{d^2}{r_{Cu}^2} \right] + \frac{\delta}{r_O^3} \left[2 - 3\frac{d^2}{r_O^2} \right] \right| \quad (4.2.7a)$$

$$\Delta D_{dip,c} = -8\mu_B \left| \frac{1-4\delta}{r_{Cu}^3} \left[1 - 3\frac{\Delta c1^2}{r_{Cu}^2} \right] + \frac{2\delta}{r_O^3} \left[1 - 3\frac{\Delta c2^2}{r_O^2} \right] \right| , \quad (4.2.7b)$$

where $\mathbf{r}_{Cu} = \hat{\mathbf{i}}d + \hat{\mathbf{j}}d + \hat{\mathbf{k}}\Delta c1$ and $\mathbf{r}_O = \hat{\mathbf{i}}d + \hat{\mathbf{k}}\Delta c2$. Inserting the foregoing numbers into these equations, we find $\Delta D_{dip,ab} = 0.14\,\mathrm{kOe}/\mu_B$ and $\Delta D_{dip,c} = -0.28\,\mathrm{kOe}/\mu_B$. These are of the order of 20–30% of D_α and so they are quite appreciable as suggested in [190]. However, since they have the opposite sign of the observed anisotropy, it is clear that there is another source of anisotropy which is twice as large.

The other noteworthy point is that there are a number of dipolar terms that are effective for the T_1 process but which have an angular dependence that does not vanish at the "magic angle" and therefore may be considerably larger than the shift coefficients estimated above. Thus, one may not find that T_1 for ^{89}Y exhibits the expected anisotropy from the measured D_α coefficients. The dipolar effects for the ^{89}Y were also discussed in detail by Auler et al. [245]. We shall address this matter further in the next subsection.

4.2.2 *A Simple Model Theory of Spin Dynamics for YBCO6.63*

The Spin-Hamiltonian formulation of HF coupling at the Cu^{2+} sites combined with the Mila-Rice-Shastry [118, 146] picture of transferred HF couplings from the Cu^{2+} moments to the Y and O(2, 3) sites provides a definite basis for discussing electron spin-based static and dynamic NMR properties. In the last subsection we saw that one finds a consistent picture of HF couplings from NMR shift data, which apply to both YBCO7 and YBCO6.63. The coupling constants we refer to here are found in Tables 3.4, 3.6 and 4.2. They were used in the Mila-Rice discussion of T_1 in Chap. 3.

Meanwhile, here we consider the relaxation properties of YBCO6.63, where spin-spin correlations will be used to discuss comparative temperature variations of T_1 in terms of dynamical spin fluctuations. As was noted in Sect. 3.3.3, dynamical spin correlations are the apparent cause of the T-varying ^{63}Cu(2) T_1 anisotropy for YBCO7 [143]. Although a comparable data set is not available for YBCO6.63, we shall find application in this case for similar formulations of the ^{89}Y and ^{17}O(2, 3) relaxation properties as well as for those of the ^{63}Cu(2).

Let us anticipate some of the themes as well as stumbling blocks which arise in the course of trying to account for the relaxation behavior of the YBCO6.63 (60 K) phase. First, we note the pervasive characteristic temperature variation $1/T_1T \propto K_s(T)$, which arose in an empirical way, but then found a measure of theoretical justification [196]. Secondly, the case of ^{89}Y has two special problems, which have been discussed by Takigawa et al. [191]: (i) The direct dipolar interaction is large enough to make an important contribution to the overall HF coupling, as we saw in the previous subsection. Thus, the HF coupling Hamiltonian does not have the simple form of (3.3.4). This complicates the relaxation calculations. (ii) The ^{89}Y relaxation is driven by Cu^{2+} spin moments in neighboring CuO_2 planes (see Fig. 3.1), so the question of interplanar spin correlations is an important one for T_1. As there is neutron-scattering evidence for such correlations [199], the ^{89}Y T_1 phenomena are more complex on that account.

T_1 *Enhancement and a Model Dynamic Susceptibility for Underdoped Cuprates*

It was first noted empirically that the quantity $(T_1T\, K_s(T))^{-1}$ is *approximately constant* for ^{89}Y and ^{17}O(2, 3) in both the YBCO7 and YBCO6.63 phases [148]. Moreover, the same quantity for the ^{63}Cu(2) no longer exhibits a peak around 150 K, but executes a roughly power-law rise as T approaches T_c in the normal state. These behaviors are shown in Fig. 4.18 using data from various sources (see figure caption). Interestingly, the ^{63}Cu(2) data for YBCO7 show a similar temperature charactistic. The ^{89}Y and ^{17}O(2, 3) data shown indicate very little change with T, i.e. $1/T_1T \propto \chi_s(T)$. Later, more complete data, however, deviated from this simple picture.

The data plotted in Fig. 4.18 is scaled as $[T_1(Korr.)/T_1(T)]\,[K_s(300\,K)/K_s(T)]$, where

$$T_1(Korr.) = N_s\gamma_e^2\hbar/(4\pi k_B T\gamma_n^2 K_s^2), \tag{4.2.8}$$

which is the usual Korringa expression with a factor N_s to account for multiple sources of HF fluctuations. Here, $N_s = 4, 2,$ and 8 for Cu(2), O(2, 3) and Y, respectively. Note that $N_s = 4$ for the c-axis Cu(2) T_1 process, which is dominated by the four *nn* couplings. The shift has also been normalized arbitrarily to its room-temperature value, so that the absolute values plotted are only a relative indication of fluctuation enhancement levels. Clearly, for Cu(2) the YBCO6.63 case is smartly enhanced over the YBCO7, and both of these by a considerable margin over the ligands. These effects were attributed to AFM fluctuations, which barely touch the ligands, but could strongly enhance the Cu(2) relaxation and increase quickly with carrier dilution.

The proportionality between $1/T_1T$ and $K_s(T)$ is suggested by the model dynamic susceptibility form [148, 161]

$$\chi''(\mathbf{q}, \omega) = \frac{\pi\chi_0\omega}{\Gamma}G(\mathbf{q}) \tag{4.2.9}$$

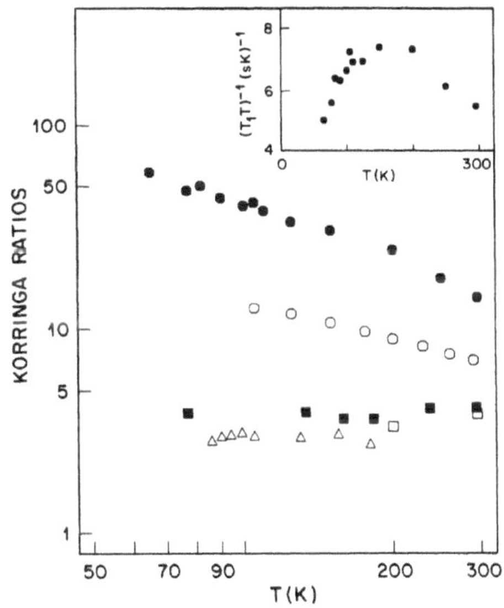

Fig. 4.18 The quantity $[T_1(Korr.)/T_1][K_s(300\,K)/K_s(T)]$ is plotted versus T in a log-log plot over the range $50\,\mathrm{K} \leq T \leq 300\,\mathrm{K}$ for a series of cases as follows: Closed and open circles are for ^{63}Cu(2) in YBCO6.63 [148] and YBCO7 [143], respectively. Closed squares are ^{89}Y data for YBCO6.63 from Alloul et al. [166] using, however, an independently determined value for the ^{89}Y chemical shift [148]. (A similar value was reported later by Alloul et al. [200]). The open squares are data for YBCO7 from [154] using the same chemical shift. Finally, the open triangles represent ^{17}O(2, 3) data on YBCO7 from [178]. The inset shows preliminary data for $^{63}Cu(2)$ from [148]. See text for a definition of $T_1(Korr.)$

for the low-frequency dynamics ($\omega \ll \Gamma$) of the quasiparticles, where χ_0 is interpreted as an estimate of the static susceptibility, and the factor $G(\mathbf{q})$ embodies any q-dependent enhancement effects which occur. Since $1/T_1T \propto \sum_{\mathrm{q}} G(\mathbf{q})$, this form suggests that $1/T_1T \propto \chi_0$. To use this form even when χ_0 has the gapped temperature characteristic of an underdoped cuprate is to go beyond the usual mean-field theory interpretation of (4.2.9). Nonetheless, this procedure was adopted in order to treat T_1 phenomena in spin-gapped superconductors [162].

However, not long after the latter work appeared, a theoretical model was developed which actually exhibited such behavior. This was the "attractive" (i.e. negative-U) Hubbard model of Randeria, Trivedi et al., which was studied using QMC techniques [196]. The authors only studied local, **q**-independent HF couplings, and emphasized that this was not a suitable model for cuprate physics in other ways. Nonetheless, their results for the gapped behavior of both susceptibility and $1/T_1T$, shown in Fig. 4.19, give a remarkably realistic version of the effects seen in Fig. 4.18 for, e.g. the ^{89}Y and ^{17}O(2, 3) sites. To account for the Cu(2) data would require inclusion of **q**-dependent enhancement effects in the negative-U Hubbard model, which has not, to our knowledge, been done. In any case, the results of Fig. 4.19 offer some support for the phenomenological treatment based on (4.2.9) with $\chi_0 = \chi_s(T)$.

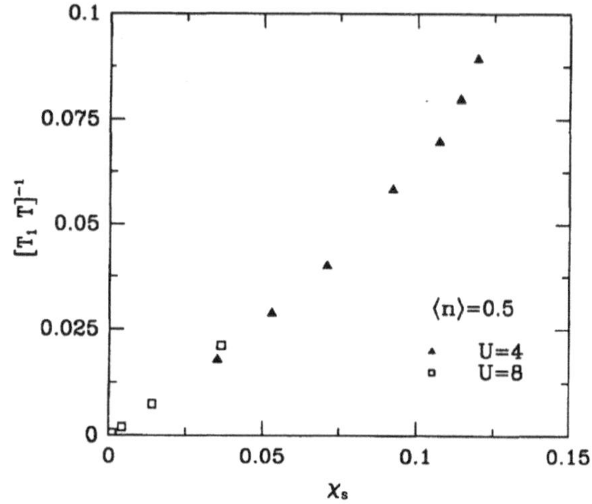

Fig. 4.19 The variation of $1/T_1T$ versus χ_s, with T as an implicit parameter, for the attractive-U Hubbard model calculations of [196]. Over the lower 40% of the variation of $1/T_1T$, this quantity shows a very clear linearity with the spin susceptibility, with a constant of proportionality which is essentially independent of the value of U

The Problem of ^{89}Y T_1 Behavior in YBCO6.63

In Fig. 4.18 ^{89}Y T_1 data are plotted so as to test the pseudogap scaling with $\chi_s(T)$. The result is a very flat curve, based on data from [166]. However, there are two more recent sets of T_1 measurements, by Takigawa et al. [191] and by Alloul et al. [200], neither of which agree with the previous conclusion. Data plots from [191] are shown in Fig. 4.20. In Fig. 4.20a are plotted $1/T_1T$ for ^{89}Y in YBCO6.63 with both field orientations as a companion to the NMR shift measurements in Fig. 4.17 on the same sample, as well as the ^{17}O(2, 3) relaxation data from Fig. 4.23 for comparison. It is clear that $1/T_1T$ for the oxygen declines rather less at low temperatures than either ^{89}Y measurement. This point is illustrated further in Fig. 4.20b, where $^{17}T_{1c}/^{89}T_{1c}$ is plotted for $100\,\mathrm{K} \le T \le 300\,\mathrm{K}$. At 100 K this ratio is seen to have increased by more than 20% over its value at 300 K. Similar data are given in [200] for both $K_s(T)$ and $1/T_1T$ for ^{89}Y in YBCO6.64. Although the latter data have somewhat larger error bars, they are consistent with the data shown in Fig. 4.20. Since these more recent data are much more complete than those shown in Fig. 4.18, we conclude that $(T_1TK_s(T))^{-1}$ for ^{89}Y is not constant in this composition, in contrast with the ^{17}O(2, 3) data in Fig. 4.23.

We discuss the data in Fig. 4.20 further by means of the general T_1 formulas (3.5.22) and (3.5.21). Taking the ratio of these equations, we find

$$\frac{^{89}T_{1ab}}{^{17}T_{1c}} = \frac{\gamma_{17}^2}{\gamma_{89}^2}\frac{[C_a^2 + C_b^2]}{[D_{ab}^2 + D_c^2]}\frac{[1 + \mathcal{K}_{01}]}{[1 + 2\mathcal{K}_{01} + \mathcal{K}_{12} + \mathcal{K}^{IC}]} \tag{4.2.10}$$

where we have added an interplanar correlation coefficient $\mathcal{K}^{IC}$ to take account roughly of that possibility for the ^{89}Y. In the high temperature limit where $\mathcal{K}_{ij} \ll 1$, one would expect $^{89}T_{1ab}/^{17}T_{1c} \to \gamma_{17}^2[C_a^2 + C_b^2]/\gamma_{89}^2[D_{ab}^2 + D_c^2]$. Inserting parameter values from Table 4.2, we find a limiting ratio of 2360, as compared with the

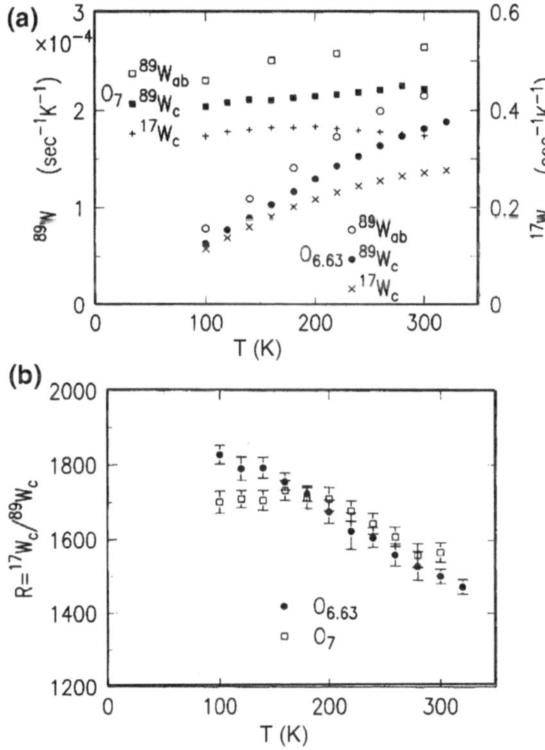

Fig. 4.20 **a** Data for $1/T_1T$ for ^{89}Y ($^{89}W_{ab,c}$) in both YBCO7 (O_7) and YBCO6.63 ($O_{6.63}$) and for both "ab" and "c" axis field orientations are plotted from 100 K to above 300 K [191]. For comparison, $1/T_1T$ for ^{17}O(2, 3) ($^{17}W_c$), measured on the same samples, is also plotted. **b** With the same notation as in (**a**) the ratios R $= ^{17}W_c/^{89}W_c$ are plotted for both compositions and the same temperature range [191]

experimental ratio of 1305 at T = 300 K (Fig. 4.20) [191]. It seems, then, that the estimated rate $1/^{89}T_1$ is too small by as much as a factor ~ 2 when analyzed on the same basis as the ^{17}O(2, 3).

On the other hand, this analysis has some positive points. First, the anisotropy of $^{89}T_1$ according to the model (3.5.22) is given by $^{89}T_{1c}/^{89}T_{1ab} = [D_{ab}^2 + D_c^2]/[2D_{ab}^2]$ $\simeq 1.17$ by the data in Table 4.2. With modest error bars on that value, the data in Fig. 4.20 are in reasonable agreement with the latter estimate for both YBCO7 and YBCO6.63 and at all temperatures shown. Furthermore, when correlations become important at low temperatures, one would certainly expect that

$$[1 + 2\mathcal{K}_{01} + \mathcal{K}_{12} + \mathcal{K}^{IC}] < [1 + \mathcal{K}_{01}]. \tag{4.2.11}$$

One would therefore expect $^{89}T_{1\alpha}/^{17}T_{1\alpha}$ to increase at low T, as it does in Fig. 4.20b.

If we examine the effects of dipolar terms, in the case of negligible correlations they add less than 10% to the estimated rates. The case of dipolar terms in the presence of intraplanar and interplanar correlations has been discussed in detail by Takigawa et al. [191]. While they could become significant in such a case, they have the opposite anistropy to the D_α's, which is a serious deterrent. One cannot see how a calculation based entirely on the model Hamiltonian (3.3.2) can heal the disparity in the estimated T_1 for ^{89}Y. Additional calculations with the full dipolar Hamiltonian have been performed by Auler et al. [245]. These are discussed in Sect. 5.3, where a substantial contribution to the ^{89}Y relaxation was reported.

Another possible answer for the ^{89}Y T_1 problem is that there may be enough Y(3d)-hole character hybridized into the Cu^{2+} spin states to generate a significant orbital T_1 effect. After all, the measured D_α coefficients in Table 4.2 are undoubtedly the result of such an hybridization, which has its main effect through the core-polarization term. The orbital HF effects are typically of the same order, but are probably more nearly isotropic. Since there is no Fermi-surface orbital shift effect related to the orbital T_1 term, the latter would just appear as a "background" relaxation effect unrelated to the spin HF effects. There could also be a small Van Vleck orbital NMR shift as part of the positive, temperature-independent chemical shifts recorded for the ^{89}Y.

4.2.3 INS Data for Dynamic Susceptibility of Underdoped YBCO

In this and the following subsection we discuss the T_1 processes in underdoped YBCO in terms of the treatment of nuclear relaxation given in Sect. 3.5, where the electron spin dynamics are represented by the dynamic susceptibility $\chi''(\mathbf{q}, \omega)$. For that purpose we review in this subsection available INS data for underdoped YBCO. In Sect. 4.1 of this chapter the same treatment was presented for LSCO, for which absolute, quantitative $\chi''(\mathbf{q}, \omega)$ data have been available for more than two decades.

INS data for underdoped YBCO appeared in the early 1990's [197, 199], along with evidence that $\chi_I''(\mathbf{q}, \omega)$ obeys an ω/T scaling law [202]. However, it was not until 2004 that high–quality, quantitative data taken on a single crystal specimen of YBCO6.5 were published, including calibration of $\chi_I''(\mathbf{q}, \omega)$ on an absolute scale [201]. With suitable preparation, the latter data will be used in Sect. 4.2.4 to calculate values of T_1 versus T with no adjustable parameters. This constituted another opportunity to test and validate the formulation of T_1 in terms of $\chi''(\mathbf{q}, \omega)$ spelled out in Sect. 3.5. Before [204], no such comprehensive test had been conducted since this result was derived in 1963 by Moriya [58]. A preliminary test of this result was performed on LSCO in 1994 [239] with some measure of success, but dynamic spin–spin correlations such as appear in (3.5.20) were not taken into account in the latter test, leading to apparent discrepancies. With correlation effects included, a different conclusion was reached [204], namely that an important term in $\chi''(\mathbf{q}, \omega)$ embodying the pseudogap had not yet been resolved for LSCO, YBCO, or any other cuprate. Summarizing the results from [201] briefly, the dynamic susceptibility was found to exhibit incommensurate peaks near $\mathbf{Q} = (\pi, \pi)$, with an unresolved splitting when scanned along [100] (see Fig. 4.21a), but with no such splitting along [010] (Fig. 4.21b). Thus, there appeared to be an anisotropic incommensurability. Moreover, as with the results given in [202], data for $\chi_I''(\mathbf{q}, \omega)$ were found to exhibit ω/T scaling at the low–frequency end of the scale. The ω/T scaling effect was compared with ^{63}Cu T_1 data in a somewhat odd fashion in Ref. [201], as shown in Fig. 4.22.

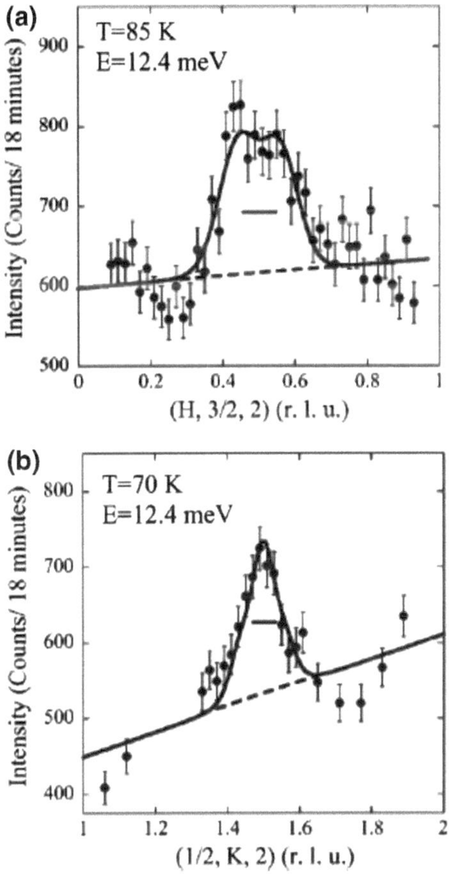

Fig. 4.21 **a** This is a [100] scan through an incommensurate peak centered at (π, π) at the energy and temperature shown, taken from the absolute $\chi''(\mathbf{q}, \omega)$ data in Ref. [201]. The solid line is a fit to the functional form given by Chou, et al. [203], as described in the text. These data are used to calibrate the integrated area $I(\omega, T)$ that follows ω/T scaling for YBCO6.5. See text for discussion. **b** A [110] scan transverse to that shown in part (**a**), showing diminished incommensuration along the y axis. The solid line is a fit to the functional form given by Chou, et al. [203], discussed in the text, in order to calibrate the area under the dynamical susceptibility data

There, $1/T_1T$ is plotted as a function of T, along with the peak amplitude of $\chi''(\mathbf{q}, \omega)$ recorded at an energy of 12.4 meV. These plots display an obvious discrepancy in that $1/T_1T$ exhibits a clear pseudogap feature in rolling over and declining at T well above $T_c \sim 60$ K, while $\chi''_I(\mathbf{q}, \omega)$ shows no such feature. This discrepancy will be a principal concern throughout this section. With ω/T scaling, the integral of $\chi''_I(\mathbf{q}, \omega)$ over the BZ varies simply as 1/T.

What is omitted in Fig. 4.22 is that it is the integral of $\chi''_I(\mathbf{q}, \omega)$ over the BZ, rather than $\chi''_I(\mathbf{q}, \omega)$ itself, that must be compared with $1/T_1T$. Even then, such a comparison neglects dynamic correlation effects between neighboring electronic moments that also affect T_1 behavior. As in the case of LSCO [278], the authors of [201] argue that the behavior of $\chi''_I(\mathbf{q}, \omega)$ for underdoped YBCO is largely caused by the occurrence of dynamical stripes in this cuprate as well as in others. The details of stripe effects go beyond the intended limits of our discussion here, so we refer the interested reader to [193, 194, 201] and references cited therein for a more in–depth look at the stripe phenomenon.

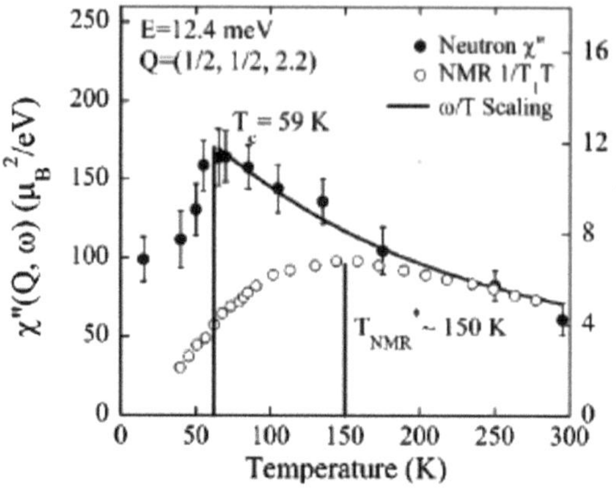

Fig. 4.22 The peak susceptibility at 12.4 meV as a function of temperature. For the normal phase, the solid curve is a fit to the analysis of Birgeneau, et al. [202] (see text for details). This fit confirms ω/T scaling at low frequencies for these data. A clear suppression of the scattering is found in the superconducting state. The low–frequency susceptibility sensed by the ^{63}Cu relaxation ($1/T_1T$) is suppressed below a temperature $T^* \sim 150$ K, while the peak susceptibility continues to rise down to $T_c \simeq 59$ K. This evident pseudogap effect is analyzed in the text. NMR data are for YBCO6.64 [201]

What we need to interpret the T_1 data is, first, a quantitative calibration of the quantity $\chi_I''(\mathbf{q}, \omega)/\omega$ at very low frequencies ω, and second, a realistic representation for the $\mathbf{q}$ dependence of $\chi_I''(\mathbf{q}, \omega)$, both of these over temperatures ranging from below T_c up to room temperature. These results have been derived from INS data with a fairly elaborate analysis, which we summarize here in the next few paragraphs. To begin, $\chi_I''(\mathbf{q}, \omega)$ is represented as

$$\chi_I''(\mathbf{q}, \omega) = B\hbar\omega f_p(\mathbf{q}), \tag{4.2.12}$$

where ω is the NMR frequency (i.e., $\hbar\omega \ll kT$), $f_p(\mathbf{q})$ describes the form taken by $\chi_I''(\mathbf{q}, \omega)$ in $\mathbf{q}$ space and is, along with B, an implicit function of T. $f_p(\mathbf{q})_{peak}$ is of order unity, so that B(T) is essentially the peak value of $\chi_I''(\mathbf{q}, \omega)/\hbar\omega$.

The incommensurate peaks in YBCO6.5 are well represented by the squared Lorentzian form

$$f_p(\mathbf{q}) = \frac{1}{[1 + \xi_0^2(\mathbf{q} + \mathbf{q}_{inc})^2]^2} + \frac{1}{[1 + \xi_0^2(\mathbf{q} - \mathbf{q}_{inc})^2]^2}, \tag{4.2.13}$$

where $\mathbf{q}_{inc} = (0, \pi\delta/a)$ is the incommensuration vector for peaks displaced from the origin at $(\pi/a, \pi/a)$. Fits to data such as the scans in Fig. 4.5S (S = Stock [201]) yield the parameter values $\delta = 0.06$ and $\xi_0 \simeq 20$ Å. Accordingly, peak values for $\chi_I''(\mathbf{q}, \omega)$ have been measured at $\hbar\omega = 12.4$ meV as a function of temperature. Such data are plotted in Fig. 4.9S, top panel, for T varying from well below T_c to $T = 300$ K.

Using the empirical form $\chi''_I(\mathbf{q},\omega)_{peak} = Ctan^{-1}(\hbar\omega/0.9kT)$, a fit has been made to the 12.4 meV data that leads to the $\hbar\omega \ll kT$ result wherein $\chi''_I(\mathbf{q},\omega)$ obeys ω/T scaling. This is represented by the solid line passed in Fig. 4.9S. The width parameter ξ_0 is said to be essentially constant up to 300 K. Adopting the above form, we determine the value of C using $\chi''_I(\mathbf{q},\omega)_{peak} = 171\,\mu_B^2/eV$ at T = 59 K, yielding $C = 140.5\,\mu_B^2/eV$. Then, expanding the expression $\chi''_I(\mathbf{q},\omega)_{peak} = Ctan^{-1}(\hbar\omega/0.9kT)$ for $\hbar\omega/0.9kT \ll 1$, we obtain the result

$$\frac{\chi''_I(\mathbf{q},\omega)_{peak}}{\hbar\omega} = (9.745 \times 10^{13}/kT)\frac{\mu_B^2}{erg^2 - f.u.}, \tag{4.2.14}$$

where $\hbar\omega$ and k_BT are expressed in ergs. Using this result and $f_P(\mathbf{q})_{peak} = f_P(\mathbf{q}_{inc}) = 1.045$ we find, with (4.2.12), B = $(9.523 \times 10^{13}/kT)\,\mu_B^2/erg^2 - f.u.$ Finally, (4.2.12) gives

$$\chi''_I(\mathbf{q},\omega)/\hbar\omega = [9.523 \times 10^{13} f_P(\mathbf{q})/kT]\frac{\mu_B^2}{erg^2 - f.u.} \tag{4.2.15}$$

Equations (4.2.13) and (4.2.15) are now ready to be used to calculate the T_1 parameters for the 60 K phase of YBCO.

4.2.4 T_1 Data and Dynamical Susceptibility–Based Analysis for YBCO6.63

As we saw in Sect. 4.2.1, and in particular in Fig. 4.4, YBCO6.3 presents a classic pseudogap case in its static susceptibility and NMR shift behavior. Here we will present and analyze this system's equally classic pseudogap behavior for the T_1 of planar ^{63}Cu and ^{17}O nuclei, where we employ INS–generated data for the dynamic susceptibility $\chi''(\mathbf{q},\omega)$ to add insight to our understanding of the spin–fluctuation physics of this system. The latter analysis, which appeared in 2011 [204], was the first of its kind in solid–state NMR. A principal conclusion from [204] was that an important piece of the dynamic susceptibility $\chi''(\mathbf{q},\omega)$ was *missing* from the available INS data. As of this writing, no further data have appeared on this system.

In this subsection a similar analysis of data for T_1 and $\chi''_I(\mathbf{q},\omega)$ is presented for the 60 K phase of YBCO as was for LSCO in Sect. 4.1. The result is a very similar outcome, with an even more predominant pseudogap term $\chi''_P(\mathbf{q},\omega)$ to be modeled and parameterized, and with a markedly higher pseudogap temperature. For a pair of systems having such contrasting structures and static magnetic properties, the sub–components of dynamic behavior for the two of them are remarkably similar.

Relaxation (T_1) data for ^{63}Cu(2) and ^{17}O(2, 3) in YBCO6.63 corresponding to the NMR shift and spin susceptibility results of Figs. 4.16 and 4.17 are shown in Fig. 4.23, along with similar data for the composition YBCO7 for comparison. These

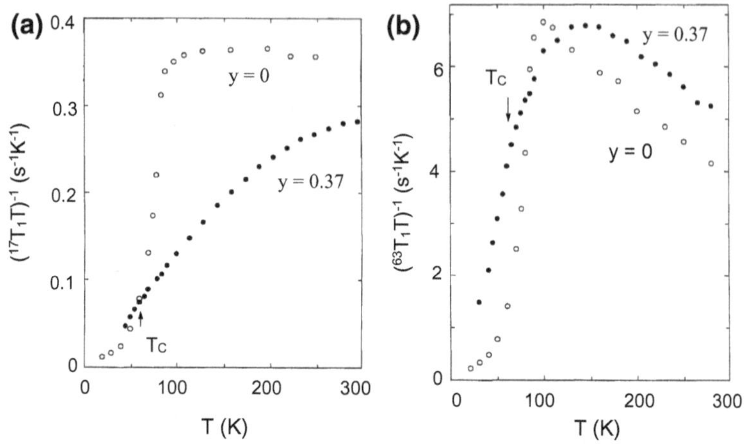

Fig. 4.23 **a** Data for $1/T_1T$ for $^{17}O(2, 3)$ in both YBCO7 ("y = 0", open circles) and YBCO6.63 ("y = 0.37", closed circles) are plotted from 300 K down to well below T_c [190]. **b** With the same notation and symbols as in (**a**), the corresponding data for $^{63}Cu(2)$ are shown [190]

are companion data to the shift results plotted in Fig. 4.4 and are of remarkably high quality [190]. The T_c points indicated refer to YBCO6.63, where $T_c \simeq 64$ K. This is substantially lower than the value $T_c \simeq 92$ K for YBCO7 that is evident in Fig. 4.23, as is characteristic of pseudogap behavior.

For YBCO6.63 (closed circles) the overall curves are very different from YBCO7, and neither data plot shows any visible feature at T_c (indicated by arrows), where both curves are smooth and continuous. As shown in the inset to Fig. 4.1a, however, the resistive superconducting transition remains as sharp as ever. As with $\chi_s(T)$ (Fig. 4.16), then, the low-frequency spin dynamics show only a smooth decline in the vicinity of T_c. *This behavior shows that the pseudogap merges smoothly into the superconducting gap which presumably forms below* T_c.

We present an analysis of $\chi''_P(\mathbf{q}, \omega)$ for YBCO6.63 that is closely analogous to that for LSCO in Sect. 4.1.4. Again, we employ (4.1.6–4.1.9), with $\chi''_P(\mathbf{q}, \omega)$ substituted for $\chi''_I(\mathbf{q}, \omega)$, to define the $\mathcal{K}_{nP}$'s, τ_{eP}, $1/^{63}T_{1cP}$, and $1/^{17}T_{1cP}$, as well as the INS–measured $\chi''_I(\mathbf{q}, \omega)$ and its associated parameters. As a preliminary step, we evaluate the $\mathcal{K}_{nI}$'s using 4.2.13 with the parameter values given. For the crystal sample of [201], which consists of 70% of one domain orientation and 30% of the other, one finds a slight anisotropy of $\mathcal{K}_{1I}$ and $\mathcal{K}_{3I}$, leading to

$$\begin{aligned} \mathcal{K}_{1Ia} &= -0.895 \\ \mathcal{K}_{1Ib} &= -0.962 \\ \mathcal{K}_{2I} &= 0.865 \\ \mathcal{K}_{3Ia} &= 0.639 \\ \mathcal{K}_{3Ib} &= 0.877. \end{aligned} \tag{4.2.16}$$

Those values only apply, of course, to the INS crystal. For the oriented powder NMR sample we can only assume equal occupation of both domains, so that we average the "a" and "b" values from (4.2.16) to give $\mathcal{K}_{1I} \simeq -0.929$; $\mathcal{K}_{2I} \simeq 0.865$; and $\mathcal{K}_{3I} \simeq 0.758$.

Using the latter numbers for T_1 interpretation, we now employ (4.1.8) and (4.1.9) with T_1 data from Fig. 4.23 and HF constants from Table 4.2 to obtain estimates of τ_e for both ^{63}Cu and ^{17}O. These are plotted in Fig. 4.24a as dash–dot lines labeled for the two nuclear species. Here we encounter a peculiar and unexpected problem, namely that the dash–dot lines are expected to converge in the vicinity of T_c, as they did for LSCO in Fig. 4.13. In this case, however, they remain rather widely separated near T_c. The source of this behavior was not hard to find, stemming from a large contrast in the width parameters ξ_0 for the INS and NMR samples. Thus, if the effective width parameter for the powder NMR samples is allowed to be ~ 2.5 times larger than that measured for the INS crystal, then the values of $\mathcal{K}_{nI}$ used to calculate the $1/T_{1Ic}$'s from (4.1.8) and (4.1.9) will be strongly affected and the curves of τ_e versus T will actually be brought into coincidence near T_c. This result is shown in the τ_e plots for ^{17}O (circles) and ^{63}Cu (squares) in Fig. 4.24a.

Combining (4.2.15) with (4.1.7), we also find a value for τ_{eI}, which is independent of T because of ω/T scaling. For consistency, the plot τ_{eI} = constant must also meet the two τ_e curves at $T \simeq T_c$ in Fig. 4.24a, because that is where the INS curve $\chi_I''(\mathbf{q},\omega)$ versus T breaks downward in Fig. 4.9S [201]. But here we find another inconsistency, namely that the calculated value of τ_{eI} based on the fit to INS data via (4.2.15), shown in Fig. 4.24a as a dashed line, is ~50% too large to meet the τ_e curves at T_c. We therefore suggest that the curve for τ_{eI} should be the horizontal solid line in Fig. 4.24a. Why the result for τ_{eI} is slightly at odds with the T_1 data is not known, but the earlier problem with the two τ_e curves from the T_1 data suggests that inhomogeneous broadening of the $f_p(\mathbf{q})$ profile may be the source of the discrepancy. One may also blame both of the problems here on the general difficulty of obtaining high–quality single crystals of YBCO, while such specimens of LSCO have been available from the earliest days of high–T_c.

The results so far with YBCO6.63 T_1 data bear a close resemblance to those for LSCO in Sect. 4.1. In Fig. 4.24a we see a tremendous increase in the values of τ_e over τ_{eI} as temperature rises toward 300 K. Thus, we suggest again that there is another susceptibility term $\chi_P''(\mathbf{q},\omega)$ that eventually becomes an order of magnitude larger than $\chi_I''(\mathbf{q},\omega)$. Again, there is no obvious clue from other physical data suggesting the appearance of $\chi_P''(\mathbf{q},\omega)$. As before, then, we proceed to model this effect with a squared Lorentzian centered on (π,π) and having a width parameter of the order of the incommensurate splittings observed in these systems. We return to this model below after discussing the analysis of τ_{eP} values that are derived from such a term.

The parameters that determine τ_{eP}'s for the two NMR nuclei of interest must merge smoothly with those from the INS data in the vicinity of $T = T_c$. Thus, we assume values of the $\mathcal{K}_{nP}$ essentially the same as the $\mathcal{K}_{nI}$ near T_c. That, plus the values of the τ_{eP}'s determines the amplitude and width parameters for $\chi_P''(\mathbf{q},\omega)$. As T is increased above T_c, the width parameter has to be adjusted so as to keep the τ_{eP}'s the same, as is mandated by the fluctuation–dissipation model of (3.5.19)–(3.5.21).

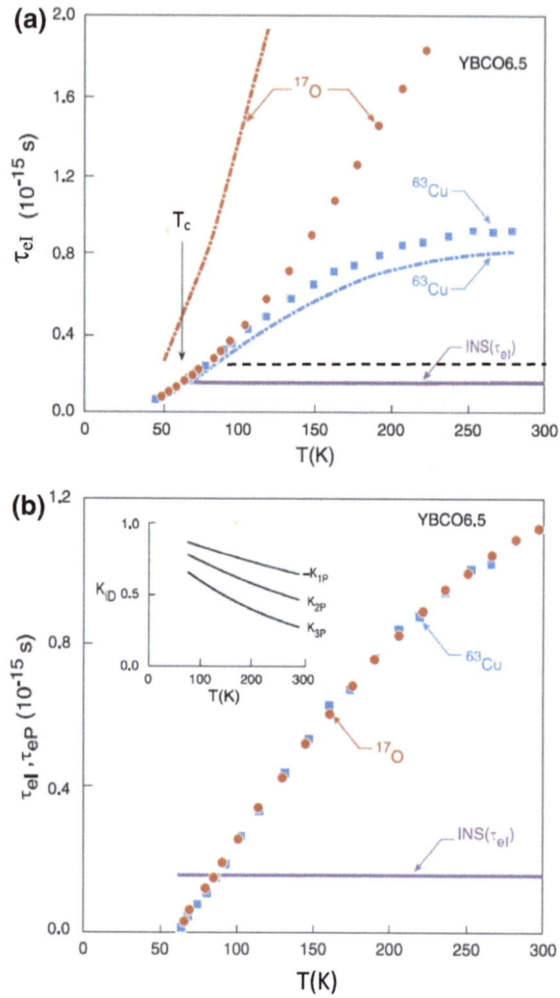

Fig. 4.24 **a** Plots of τ_e and τ_{eI}, as defined in the text, as a function of ambient temperature T. The dash-dot lines show values of $\tau_e(T)$ derived from T_1 data in Fig. 4.23 using $\mathcal{K}_{nI}$'s obtained with INS data using (4.1.7). Recalculating the τ_e's with a broader estimate for the profile of $\chi_I''(\mathbf{q}, \omega)$ leads to data points plotted as circles for ^{17}O and as squares for ^{63}Cu. See text for details. An estimate of T–independent τ_{eI} based on (4.1.6) and (4.2.15) is shown as a dashed line. This result was judged ~50% too high to be consistent with the τ_e data described. A more nearly valid estimate of $\tau_{eI}(T)$ is shown as a solid line. See text for discussion. **b** Values of τ_{eP} deduced from T_1 data and the squared Lorentzian model for $\chi_P''(\mathbf{q}, \omega)$ are plotted for both ^{17}O (circles) and for ^{63}Cu (squares). By adjusting the $\mathcal{K}_{nP}$'s as described in the text, the two curves for τ_{eP} versus T are made to coincide. This process yields the variation of the $\mathcal{K}_{nP}$'s with T (see inset) and estimates of the width parameter for $\chi_P''(\mathbf{q}, \omega)$. The resulting curve for $\mathcal{K}_{1P}$ is represented quite accurately by the temperature function given in the text. For comparison, the adjusted value of $\tau_{eI}(T)$ is shown as a solid line

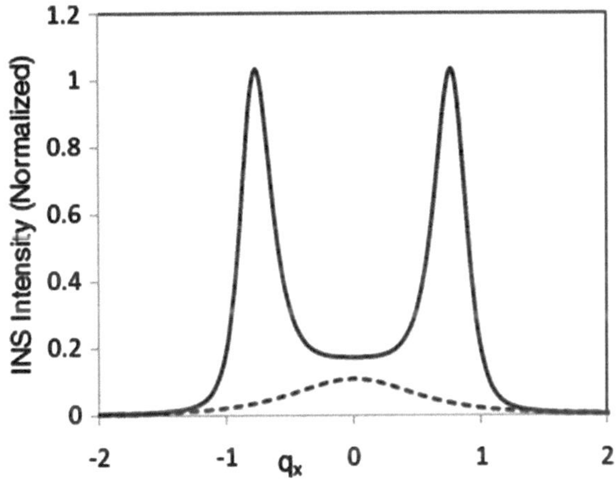

Fig. 4.25 Plot of low–frequency **q**–scan through incommensurate peaks and (π, π), based on discommensuration model for LSCO along with a squared Lorentizian form for $\chi''_P(\mathbf{q}, \omega)$, as described in the text. Relative magnitudes represent conditions at $T \sim 100\,\text{K}$. For the case of YBCO6.5 the pseudogap peak intensity is more difficult to resolve, because the incommensurate peaks are themselves not as well resolved

Not surprisingly, the principal adjustment is to values of $\mathcal{K}_{1P}$. The other correlations are not tightly constrained. For simplicity, then, we assume the correlations to vary exponentially with bond length. As for LSCO, the temperature variation was found to follow a simple exponential form $\mathcal{K}_{1P}(T) = -0.87\exp[-(T - 62)/725]$, with corresponding expressions for the other two coefficients. These results are plotted in an inset to Fig. 4.24b. The resulting plot of τ_{eP} values is shown in the main part of this figure, along with the 'adjusted' value $\tau_{eI} \simeq 0.18\times10^{-15}s$. The agreement between τ_{eP} values for the two nuclear species is well within the scatter of the data. A successful account of the data with this picture is therefore achieved for both LSCO and YBCO6.63. For YBCO6.63 the statistical weight of the pseudogap term is, by room temperature, an order of magnitude greater than that of the incommensurate peaks sensed by INS.

An important question arises as to why the INS doesn't 'see' the pseudogap term. To discuss this point, we return to the squared Lorentzian model for $\chi''_P(\mathbf{q}, \omega)$. If the latter model were superimposed on $\chi''_I(\mathbf{q}, \omega)$ for LSCO with parameters suitable for $\text{T} = 100\,\text{K}$, then a scan through the incommensurate peaks, intersecting the model for $\chi''_P(\mathbf{q}, \omega)$ at its maximum point $(\pi/a, \pi/a)$ would appear as shown in Fig. 4.25. The increase in intensity between the peaks as shown would be extraordinarily difficult to resolve. An equally interesting question is, of course, that of the physical origin of the pseudogap term in $\chi''(\mathbf{q}, \omega)$.

We turn now to pseudogap phenomenology as seen with other probes.

4.3 Pseudogap Phenomenology with Other Probes

In this section we address briefly several topics relating to pseudogap effects in the cuprates, but not directly related to NMR. Thus, we are discussing "background material", the point of which is to give a deeper context in which to present and discuss the NMR results. It is essential to point out that pseudogap effects have been observed and studied in great detail with several other techniques. In particular, it is interesting to note that the presence of the pseudogap is felt in both spin and charge-based properties. One example is the in-plane conductivity $\sigma_{ab}(T)$ of underdoped YBCO at 250 cm^{-1}, which according to Rotter et al. [207] exhibits a gap-like feature at low temperatures very similar to $1/T_1T$ for ^{89}Y and ^{17} O(2, 3) for that case (see Figs. 4.1b and 4.23a). A clearly presented review of pseudogap effects has been given by Randeria [208].

4.3.1 *Pseudogap Effects in the Electronic Specific Heat*

We offer brief descriptions of experimental results which detect and elucidate the pseudogap effect in cuprates, the first of which is the electronic specific heat. By using a very precise differential technique, J. W. Loram and co-workers have produced an extraordinarily clear and detailed picture of the behavior of low-energy electronic states in the cuprates [209, 210, 230]. We review here a selection of their results on YBCO7-δ.

A sampling of specific heat results is shown in Fig. 4.26a, b. In (a) one sees that, first, the peak in γ^{el}(T) ($\equiv C_p(T)/T$) at T_c gradually vanishes as δ is increased from 0, and simultaneously a pseudogap maximum appears at temperatures $T \gg T_c$, causing γ^{el} to decline as it approaches T_c from above. The peak at T_c makes a rather weak reappearance at $T_c \sim 60$ K, then declines further as superconductivity fades away at larger δ values.

In Fig. 4.26c the entropy $S^{el} = \int_0^T \gamma^{el}(T')dT'$ is plotted as S^{el}/T versus T. Such a plot closely resembles $\gamma^{el}(T)$. The superconducting gap feature is very pronounced for $\delta \ll 1$, persisting up to $\delta \sim 0.20$. Then entropy loss at $T > T_c$ becomes a major feature—i.e. the pseudogap—as δ increases, while the decline in S^{el}/T begins at ever higher temperatures. Note, however, that a vestige of the superconducting gap feature beginning at $T \sim T_c$ remains in place until superconductivity disappears altogether.

These effects on γ^{el} and S^{el}(T) seem to be entirely consistent with what one would expect based on the spin susceptibility and low-energy spin excitations reflected in the $1/T_1T$ results given in Sect. 4.2. Loram et al. present a companion set of spin susceptibility curves [230] to confirm this point. There is a clear conclusion of *non-separation of spin and charge behavior*, contrary to some of the early models of high-T_c properties [123, 231].

4.3.2 *ARPES Studies of Pseudogapped Cuprates*

Angle-resolved photoemission studies (ARPES) on cuprates exhibiting pseudogap effects have also given a great deal of insight into the microscopic physics of cuprate

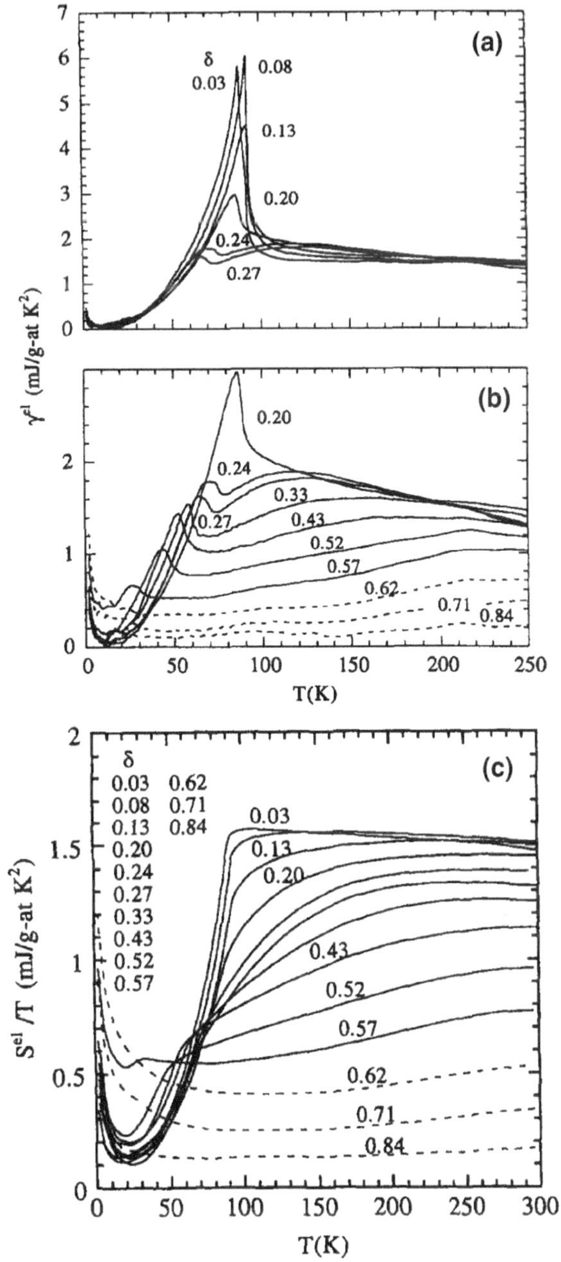

Fig. 4.26 **a** and **b** Data for the electronic specific heat coefficient γ^{el} of YBCO7-δ, obtained via differential measurements with YBCO6.0, are plotted for different doping levels δ as a function of temperature up to 250 K [230]. In **b** there is clear evidence for a pseudogap creating broad maxima in γ^{el} at progressively higher temperatures with increasing carrier dilution. **c** The quantity S^{el}/T, where S^{el} is the electronic entropy calculated from the data in **a** and **b**, is plotted as a function of temperature up to 300 K. The superconducting gaps occurring at T_c as well as the pseudogap effects in these data are discussed in the text

superconductors as well as the pseudogap phenomena per se. Since this method offers good resolution of energies as well as of $\boldsymbol{k}$ values, one is able to study both gap energies and their anisotropy over the Fermi surface. Extensive investigations of the cuprate properties have been conducted. We shall briefly review studies of pseudogap effects in the Bi2212 family of superconductors by Harris, Shen et al. [211–213].

What ARPES reveals is not only that gaps appear anisotropically around the Fermi surface, but also that quasiparticle peaks in the normal state are greatly diminished. They appear mainly at temperatures below T_c. Gap energies are measured, then, as the distance between E_F and the "leading edge midpoint" of the photoemission intensity. The pseudogap is found to behave in similar fashion to a d-wave superconducting gap, namely to be at its maximum along the cut from $(\pi, 0)$ to (π, π) and to vanish along the diagonal cut $(0, 0)$ to (π, π). With that in mind, we examine the temperature dependence of experimental gap data, which offers some insight into the relation between the pseudogap behavior and that of the superconducting gap which sets in below T_c.

Harris et al. reported a study of $Bi_2Sr_2Ca_{1-x}Dy_xCu_2O_{8+\delta}$ untwinned thin-film crystals with variable doping controlled by the parameters x and δ [211]. Their gap energy data scans extend over 1/8 of the (2D) Fermi surface.[6] Two separate doping levels yielded a sample with a "strong" pseudogap effect ($T_c = 42$ K) and a second one with a much "weaker" pseudogap ($T_c = 78$ K), where "strong" and "weak" are to be defined presently.

In Fig. 4.27a, b experimental gap energies are shown along an arc of the Fermi surface which extends from the diagonal $(0.4, 0.4)\pi$ where the abscissa is zero to $(1, 0.2)\pi$, where it is nearly one. The abscissa is defined as $0.5|cos\, k_x a\text{-}\cos k_y a|$, which would yield a linear variation of gap energy with a d-wave ($d_{x^2-y^2}$) superconducting gap. For the data in Fig. 4.27a, the superconducting state data (closed circles) show a nearly linear variation of gap energy along this path. In a conventional superconductor there would be no gap energy above T_c; thus, the curve would remain flat. However, this $T_c = 78$ K sample has a "weak" pseudogap, so that there are substantial gap energies at 100 K and 150 K which behave qualitatively like the superconducting gap.

On the other hand, the $T_c = 46$ K sample exhibits a "strong" pseudogap, which means that the gap energy profile shown is the same both above and below T_c, exhibiting a flat region near the null (abscissa $= 0$) and rising to a maximum which is the same near the point $(1,0.2)\pi$ both above and below T_c. At 150 K the gap energy shows a small decline. Thus, with a "strong" pseudogap, the gap energy profile in **k**–space is very nearly the same above and below T_c. For either a weak or strong pseudogap, the gap energy behaves in **k**–space in a very similar fashion to a superconducting gap, giving the impression that these effects are related.

[6] The region of **k**-space considered is triangular, extending from $(0, 0)$ to $(\pi, 0)$, then to (π, π) and back to $(0, 0)$. The Fermi surface is an arc extending from $\sim(0.4, 0.4)\pi$ on the diagonal around to the point $\sim(1,0.2)\pi$ on the $(0, \pi)$–(π, π) axis.

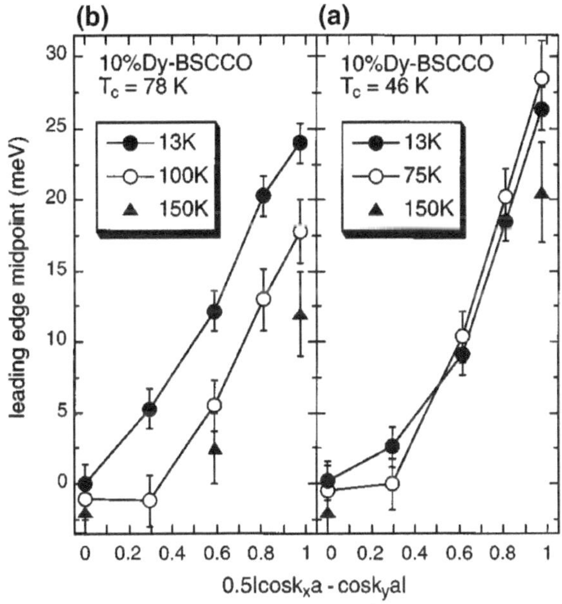

Fig. 4.27 **a** Measurements of anisotropic gap energies taken as described in the text on untwinned thin film crystals of $Bi_2Sr_2Ca_{1-x}Dy_xCu_2O_{8+\delta}$ are plotted at a series of points near the Fermi energy for two different doping levels and three temperatures for each [211]. The abscissa (see text) is such as to give a straight-line characteristic for a $d_{x^2-y^2}$ superconducting gap. **b** Gap energy data obtained for an underdoped sample of $Bi_2Sr_2CaCu_2O_{8+\delta}$ [211] in the same fashion as in (**a**) are plotted versus temperature for the square axis region (filled circles) where the gap is maximum and for the zone diagonal (open circles) where the gap is null. These data show the pseudogap for this sample vanishing at $T \sim 250\,\text{K}$

At what temperature, then, does the energy gap finally collapse in an underdoped sample? This question is answered in Fig. 4.27, where measured gap energies at the maximum point ($\sim(1, 0.2)\pi$) (solid circles) and at the null point ($\sim(0.4, 0.4)\pi$) (open circles) are plotted versus T for temperatures up to 250 K ($T_c = 85\,\text{K}$) [211]. At the uppermost temperature the pseudogap has fully collapsed, the decline beginning at some point below T_c. Note that these data were taken on a slightly different sample from those used in Fig. 4.27.

The anisotropic gap behavior found in this system and presumably occurring widely in the cuprates would have interesting consequences for the behavior of $\chi''(\mathbf{q}, \omega)$ and therefore for T_1 measurements of pseudogap systems. Although to our knowledge no such calculation has been reported, it seems very likely that the observed anisotropy of gap effects over **k** space would result in some distribution of gapping effects in **q** space as well. Speculating about the consequences of this, it seems likely that nuclei such as ^{89}Y and ^{17}O(2, 3) in YBCO, which depend for relaxation on **q** values over the entire zone except near the $(0, \pi)$ axes would therefore always be subject to pseudogap effects. On the other hand, the pseudogap anisotropy may allow situations where the fluctuations peak strongly at (π, π) and are driven by **k** vectors on or near the diagonals. Such behavior could tend to dominate the planar ^{63}Cu T_1 process, which would not then be strongly affected by the pseudogap behavior. Such a case is found for $La_{2-x}Sr_xCuO_4$, as was discussed in Sect. 4.1.

4.4 Pseudogap Behavior of $YBa_2Cu_4O_8$ (Y248)

Another pseudogap system which has been studied intensively is Y248, a superconductor which is unique among the cuprates in having a *fixed, stoichiometric carrier doping level*, which pegs it as slightly underdoped. It appears to have a pseudogap not unlike that of YBCO6.63 presented in Sect. 4.2. The structure of Y248 is closely related to YBCO7, except that there are two oxygenated chain layers between the pairs of CuO_2 planes instead of one. Cu(1) sites in the chain layers are staggered along the b (chain) axis, so that the b-axis oxygen sites O(4) for one chain are apical oxygens for Cu(1) sites in the neighboring chain. Thus, planar arrays separated by a double chain layer are also staggered, making the unit cell $\sim$ 27 Å high [214]. This structure of chains is unusually stable, as the chain oxygens remain bound in place up to temperatures beyond 700 K.

Thus, Y248 is a nearly ideal cuprate on which to conduct high-temperature studies. The equilibrium stoichiometric structure of Y248 also yields unusually narrow NMR lines, offering the possibility of high-precision NMR studies. We shall see some examples of this in Sect. 5.1.4.

It is very difficult to measure the ratio of mobile holes which occupy the Cu(1) and Cu(2) regions of the Y248 structure. It is, however, useful to estimate the doping level with the simple valence method. Assigning the usual valences, there is a single mobile hole per formula unit of $YBa_2Cu_4O_8$, which gives 0.25 carrier holes per Cu ion. With YBCO6.63, by comparison, the situation is more complicated in that 63% of the chain sites are oxygenated Cu^{2+} and 37% are two-fold coordinated Cu^+ ions. On that basis one calculates 0.24 mobile holes per Cu^{2+} ion, giving very nearly equal doping for these two systems. There are marked differences in T_c, however, which is $\sim$ 81 K for the "best" samples of Y248, but only 60–65 K for YBCO6.63. As we shall see, there are other important differences as well.

4.4.1 NMR Shift and Relaxation Behavior in Y248

It is both interesting and important to compare the pseudogap character of Y248 with that of YBCO6.63 and other systems. Owing to the simple and seemingly immutable character of this compound, a big responsibility seems to have been placed upon it, with a great deal expected of it. Here we examine the general behavior of the pseudogap effect as reflected in NMR shift (i.e. susceptibility) and relaxation data. Our point of reference will be the simple behavior found for YBCO6.63 as reflected

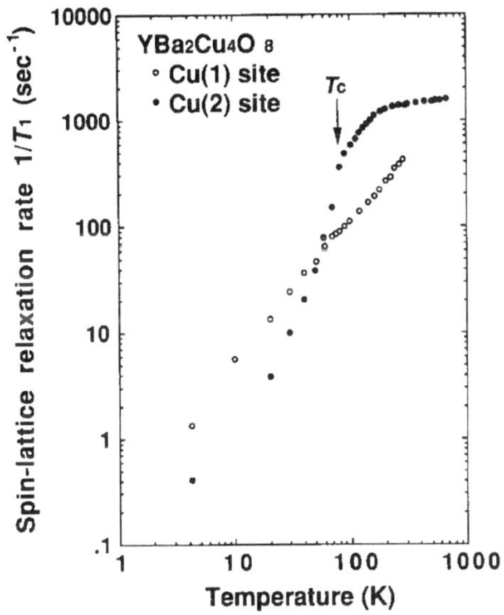

Fig. 4.28 NQR results for $1/T_1$ versus T are shown for the ^{63}Cu(1) and ^{63}Cu(2) sites in Y248 and for temperatures ranging from 4.2 K up to 700 K [215]. The log-log plot emphasizes major features in the data. See text for detailed discussion

by the data in Figs. 4.16 and 4.23. As concerns the immutability issue, we look at reproducibility of experimental measurements in data from several research groups.

Let us first look at the overall relaxation behavior in Fig. 4.28. The log scales give a broad overview of the Cu(1, 2) sites both above and below T_c [215]. The Cu(1) relaxation rate crosses through T_c looking Korringa-like, much as it did in YBCO7. For the Cu(2) there is a near discontinuity of slope at T_c which is much more pronounced than in the YBCO6.63 data of Fig. 4.23. Below T_c there is a steep drop into a very brief T^3 region, which then flattens into a presumably disorder-related (i.e. states in the gap) Korringa runout below $\sim 20\ K$. Above T_c there is a near-Korringa stretch which flattens into a very gradual increase above $\sim 200\ K$. Similar data have been given by Zimmerman et al. [229]. Apart from the pseudogap onset point around $\sim 200\ K$, the normal state behavior is quite featureless. Nonetheless, it is useful to have data extending as high as 700 K. For another view of ^{63}Cu(2) data for Y248 see the presentation by Yasuoka [217].

In Fig. 4.29 similar data for ^{63}Cu(2) are plotted as $1/T_1T$ versus T [216]. These data were also taken with NQR, but on a different sample, and are in good agreement with Fig. 4.28. In Fig. 4.29, data for the ^{17}O(2, 3) are compared with those for ^{63}Cu(2) up to 250 K. The ^{17}O(2, 3) data also show a very definite kink at T_c, below which they fall well below the corresponding curve for YBCO6.63. Recall that the latter ^{17}O(2, 3) T_1 data in Fig. 4.23a show no feature whatever at T_c. The pseudogap effect in Y248 is therefore somewhat like the $T_c = 78$ K BSCCO sample in Fig. 4.27a for which the superconducting gap has its own distinct character, while YBCO6.63 is like the $T_c = 46$ K (b) sample in that same figure, where the SC gap is completely merged into the pseudogap.

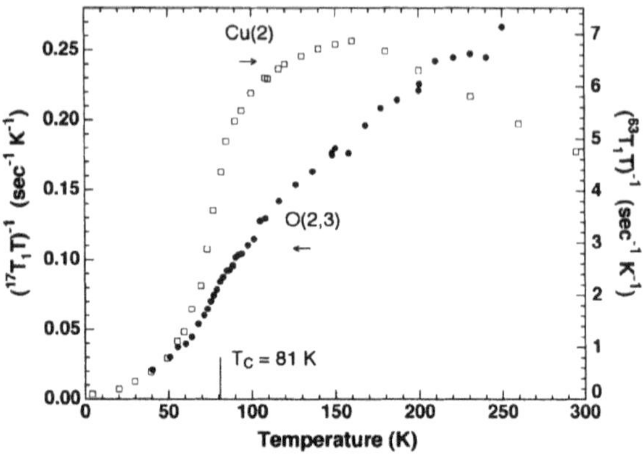

Fig. 4.29 Relaxation data for both ^{63}Cu(2) and ^{17}O(2, 3) in Y248 are plotted as $1/T_1T$ versus T, showing detailed behavior near T_c. Like the ^{63}Cu(2) plot in Fig. 4.28, the planar ^{17}O show a visible break in slope at T_c

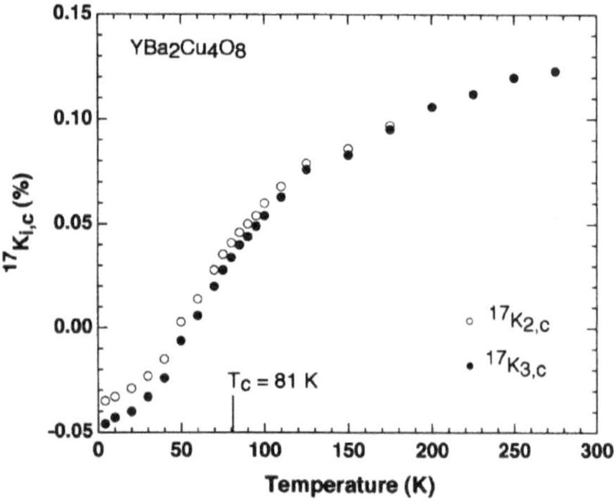

Fig. 4.30 ^{17}O(2, 3) NMR shift data for Y248 are plotted versus T for temperatures up to 300 K [215]. Note that the O(2) and O(3) NMR lines are resolved below $T \sim 150$ K, but merge above that point. See text for discussion and possible interpretation

In contrast with T_1 data, neither the Cu(2) NMR shift [215] nor that of the ^{17}O(2, 3) show any abrupt change at T_c. The latter shift data $^{17}K_{i,c}(T)$, i = 2 and 3, are shown in Fig. 4.30 where, unlike the case of YBCO6.63, they are resolved and separately measurable. The latter data presumably offer a reasonable measure of the pseudogap, but unfortunately these data have a very large, negative intercept at T = 0, and no value of the zero shift has been given for either site. We shall make some guesses about the T = 0 shifts and diamagnetic corrections in our analysis below.

Another peculiar feature of the latter ^{17}O(2, 3) NMR shift data is the fact that these two sites have distinct and well-resolved T = 0 shift values, but as the temperature rises above T_c the two lines gradually merge together. This could be entirely coincidental, resulting from slightly different HF coefficients. Alternatively, there could be chemical migration of O^{2-} ions in the CuO_2 planes at a rate τ_{cm}^{-1} which increases with temperature and eventually becomes fast enough

$$\gamma_{17} H \Delta K_{2,3} \tau_{cm} \ll 1 , \tag{4.4.1}$$

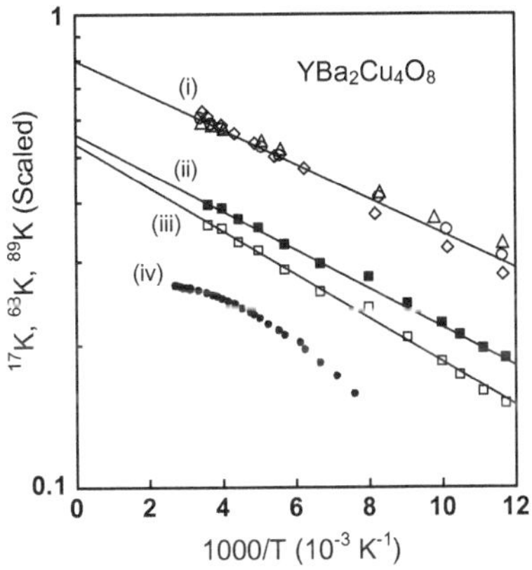

Fig. 4.31 An Arrhenius plot of several sets of NMR shift data for Y248 is presented to make estimates of the pseudogap energy for the density of states. The data are scaled to arbitrary positions along the vertical axis for display purposes. The data shown are, set (i): $^{17}K_{ab}(T)$ (triangles), $^{17}K_c(T)$ (diamonds), $^{65}K_{ab}(T)$ (circles) from [218]; set (ii) and (iii) (same data, see text) $^{17}K_c(T)$ (open and closed squares) from [216]; and set (iv) $^{89}K_c(T)$ (dots) from [219]. The lines drawn are approximate fits to the data for the purpose of extracting pseudogap energies from the slopes

so that the two NMR lines merge into one. The latter possibility seems a bit remarkable, but would avoid an even more remarkable coincidence with the HF coefficients. It would be easy to test the chemical migration hypothesis with selective saturation techniques. As far as we know this has not been done.

We now inquire as to whether the Y248 NMR shift data can provide a simple measure of the pseudogap energy, as they did for YBCO6.63 (see the inset to Fig. 4.16). Accordingly, several NMR shift data sets are plotted as ln(K(T)) versus 1000/T in Fig. 4.31 for temperatures down to T_c. The linear slope of such a plot gives one measure of the pseudogap energy. The data shown here have been scaled up and down for display purposes. At the top we see data set (i), $^{17}K_{c,ab}(T)$ and $^{65}K_{ab}(T)$ (triangles, diamonds and circles) from Bankay et al. [218]. The straight line drawn through these points defines a gap energy $T_G \simeq 84\,K$, which is $\sim$20% smaller than that of YBCO6.63. The open and filled squares are both plots of the data for ^{17}O(3) from Fig. 4.30, with two different assumptions regarding $^{17}K_{3,c}(0)$. In (ii) $^{17}K_{3,c}(0)$ is assumed to be -0.04%, with a diamagnetic correction of -0.005%, while in (iii) it is assumed to be -0.02% with a diamagnetic correction of -0.025%. Both assumptions lead to a reasonable straight line, with $T_G \simeq 96$ K and 106 K for (ii) and (iii), respectively. Option (ii) probably makes better physical sense and gives a gap energy closer to that of data set (i). Our main conclusion from plots (i)–(iii) is that Y248 NMR shifts show Arrhenius-type behavior with pseudogap energies 10-20% smaller than YBCO6.63.

Data plot (iv) in Fig. 4.31 presents the ^{89}Y NMR shift data by Williams et al. [219], which the authors presented in a proposed test of the pseudogap isotope effect. The undifferentiated data points shown are for both ^{16}O and ^{18}O-doped samples, which gave indistinguishable behavior over the range of temperatures (131–370 K) shown. The data are very precisely measured as one can see from the small scatter of the points, but the most striking thing is, first, how differently these ^{89}Y data points behave compared with the other two data sets, and secondly, that they do not appear to correspond to a well-defined gap energy in this kind of plot. Noting that in a stoichiometric sample of Y248 with low disorder one would expect NMR shifts of the ^{89}Y, ^{17}O(2, 3), and ^{63,65}Cu(2) to follow a common temperature dependence as they did in YBCO6.63 (Fig. 4.16), the results shown suggest that the sample material used for data set (iv) is somewhat different from that used for data sets (i)–(iii).

In the following subsection we review efforts which have been reported to measure the pseudogap isotope effect in Y248.

4.4.2 *Measuring the Pseudogap Isotope Effect in Y248 with NMR*

The ^{89}Y data set (iv) in Fig. 4.31 was reported as part of an effort to evaluate the isotope effect, if any, on the pseudogap and on T_c in Y248 [219]. The samples employed for that purpose were annealed with ^{16}O and and with ^{18}O, respectively, and were found to have $T_c(^{16}O) - T_c(^{18}O) \simeq 0.65$ K, in reasonable agreement with previous work [220]. The shift data, as noted in the previous subsection, show no isotope effect to the unaided eye for $T > 130\ K$, while undergoing a pseudogap-related decrease of ~40% from their peak values at 370 K. It is interesting at this point to inquire as to how large a change in the NMR shift curve is expected if the isotope effect parameter $\alpha_g = \alpha_c$?

First, let us define $\alpha_{c,g} = \mathrm{d}ln(T_{g,c})/\mathrm{d}ln(M)$, where M is the oxygen atomic weight. Note that if $\alpha_c = \alpha_g$, then $\Delta T_g/T_g = \Delta T_c/T_c$. We shall now use this in a very simple way to estimate the change which might be expected in the ^{89}Y NMR shift curve. If we consider the gap function K(T) = $K_0 \exp(-T_g/T)$ (or any function purely of T_g/T), then the temperature interval ΔT between equal values of shift for the two curves is $|\Delta T/T| = |\Delta T_g/T_g|$. The fits to shift data in Fig. 4.31 gave $T_g \sim$ 90 K; at T = 130 K, then, we expect curves shifted by $\Delta T \sim 1.1$ K.[7] At the lower end of the data curve the slope is ~ 1 ppm/K, leading one to expect deviations with the foregoing temperature shift of ~ 1 ppm. As the inset shows, there are deviations of that order, but the fluctuations are large. To resolve such an effect would require a very precise fitting curve and a large number of data points. In the light of the analysis given below for a different experimental effort to achieve the same end, we suggest that the ^{89}Y NMR shift data given were excellent, but the foregoing analysis in the present case does not quite rule out the possibility of a pseudogap isotope effect.

A second effort to measure the pseudogap isotope effect was made via NQR studies of the ^{63}Cu(2) relaxation curve for Y248, with the basic data plot of $1/T_1T$ versus T shown in Fig. 4.33b [221]. The data shown are for two samples, one with

[7] This estimate is somewhat smaller than that of the authors [219], which is not stated explicitly.

^{16}O and one with $\sim 90\%\,^{18}O$, as in the previous case. The peaks in the two curves are seen to occur at $T \sim 150$ K, where there is a razor-thin margin between the curves fitted to the two sets of data. At 0.47 K, the measured difference in T_c values for the ^{16}O and ^{18}O samples is slightly smaller than in the previous case. The key to analyzing the isotope effect here is to find a functional fit to the data curves. For this purpose the authors have used

$$\frac{1}{^{63}T_1 T} = \frac{C}{T^a}\left[1 - tanh^2\left[\frac{\Delta}{2k_B T}\right]\right], \tag{4.4.2}$$

where the factor $T^{-a}, a \sim 1$, fits the Curie-like decline of $1/T_1T$ at high temperatures, and $\Delta = k_B T^*$ acts as the spin pseudogap energy—see discussion below. The data are a good fit to (4.4.2), but the deviations were not quite random. The authors have therefore tinkered with the temperature scale a bit [221] and thus achieved the high-quality fit to the data shown in Fig. 4.33. This fit gives a shift in T^* for the two samples test of 0.96 K. It is interesting that this temperature shift is less than was predicted for the ^{89}Y NMR shift curves in Fig. 4.32, but was resolved using a precisely fitting function and a large number of data points.

The pseudogap energy value turned out to be $T^* = 246.9$ K for the $1/T_1T$ curves [221]. This is more than twice the values found with the Arrhenius plots in Fig. 4.31. Further, although (4.4.2) is just a fitting function, T^* really serves as an energy gap, since the asymptotic form of (4.4.2) is $1/T_1T \propto T^{-a}\exp(-T^*/T)$ for $T^*/T \gg 1$. Note, however, that the NMR shifts represent the residual density of states, whereas T^* represents the DOS squared. The appropriate comparison is therefore between $T^*/2$ and $T_g \sim 90$ K from the NMR shift results. Whether you consider $T^*/2$ or T_g to be the pseudogap energy is a matter of preference.

It seems, then, at least for the dynamics there is a pseudogap isotope effect which is similar to that of T_c. So far as we are aware, there is no theoretical prediction available for the variation of the pseudogap with the atomic weight of the oxygen.

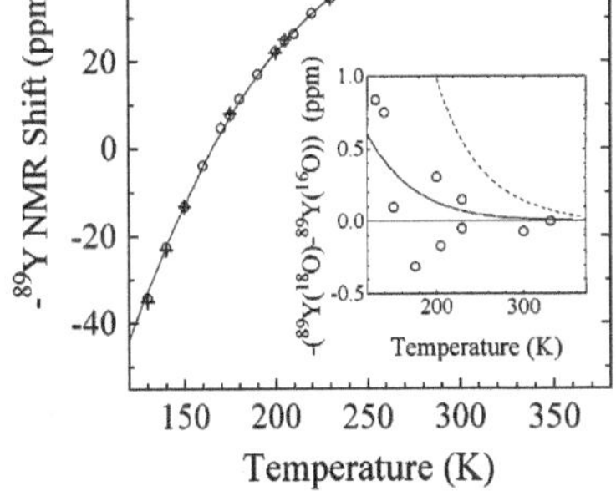

Fig. 4.32 ynmrshift@^{89}Y NMR shift data measured on two samples of Y248, one with nearly 100% ^{16}O (+ symbols) and the other with $\simeq 94\%\ ^{18}O$ (circles) [219]. The difference between these data plots is clearly smaller than the plotting symbols. Inset: Plot of the differences along with calculated difference curves. See text for discussion

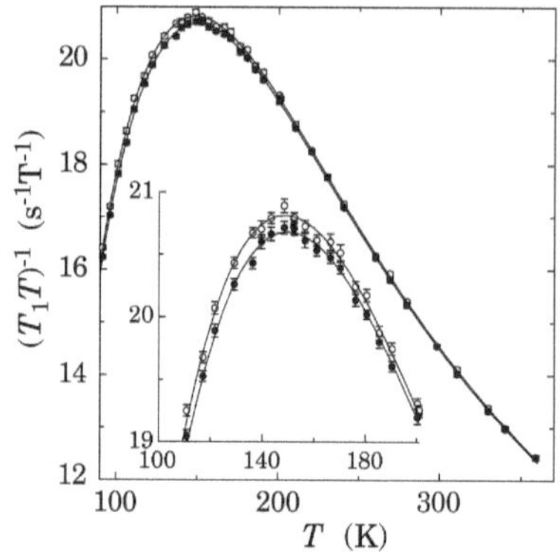

Fig. 4.33 T_1 data are plotted as $1/T_1T$ versus T for two samples of Y248, one with ^{16}O and the other with $\sim 88\%$ ^{18}O [221]. Using the functional fit shown (see text) the difference in these two relaxation curves near their peak at $T \sim 150\,K$ is analyzed to find a difference in the pseudogaps roughly equal to that found for T_c. See text for detailed discussion. Inset: Detailed view of the peak vicinity

4.4.3 *Theory of the Pseudogap by Varma*

The only successful theory of the pseudogap which we are aware of is that by Varma [222]. Varma has constructed a theory of cuprate superconductivity and ancillary properties, based on the existence of a *quantum critical point* (QCP) at the center of the *superconducting dome* of the phase diagram shown in Fig. 4.34, where the abscissa is the doping level or carrier concentration in the CuO_2 planes. Region I is the marginal Fermi liquid regime proposed and expounded in [159]. The pseudogap effects occur in Region II below the line $T_p(x)$ [222]. In Region II there is a broken symmetry based on time-reversal violating (TRV) states in which fluctuating currents flow in the CuO_2 plane. The existence of such currents has been verified experimentally via neutron-scattering experiments [223].

The aforesaid theory also predicts anisotropic gapping in **k**-space near the chemical potential in Region II of the phase diagram, as was depicted by the gapping curves in Sect. 4.3.2. Corresponding calculations of the density of states in the pseudogap regime have been presented for comparison with experimental data on Y248. In Fig. 4.35 the ^{89}Y NMR shift data from [219] (see Fig. 4.32) are compared with a calculated pseudogapped density of states from [222], where an excellent correspondence is seen to occur. It would, of course, be most interesting if more detailed calculations of, e.g. $\chi''(\mathbf{q}, \omega)$ would be forthcoming, for comparison with the **q**-dependent relaxation effects in these systems.

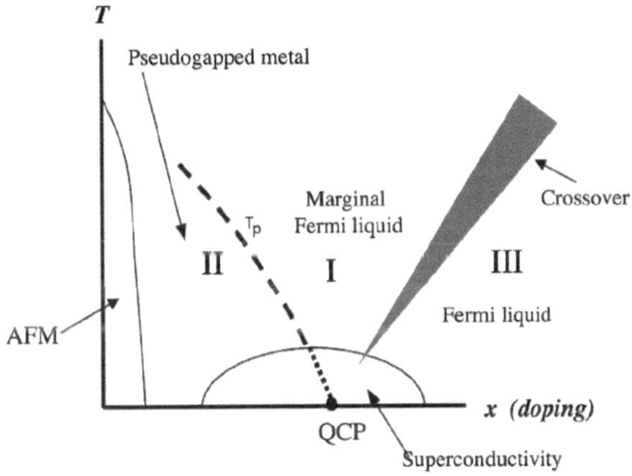

Fig. 4.34 Phase diagram of doped cuprates showing the putative quantum critical point at the center of the superconducting dome [222]. From normal metal behavior at high doping levels (III), there is a crossover to the marginal Fermi liquid behavior in (I) followed by pseudogap behavior below the line $T_p(x)$ emanating from the QCP. The calculations presented in [222] are for Region II

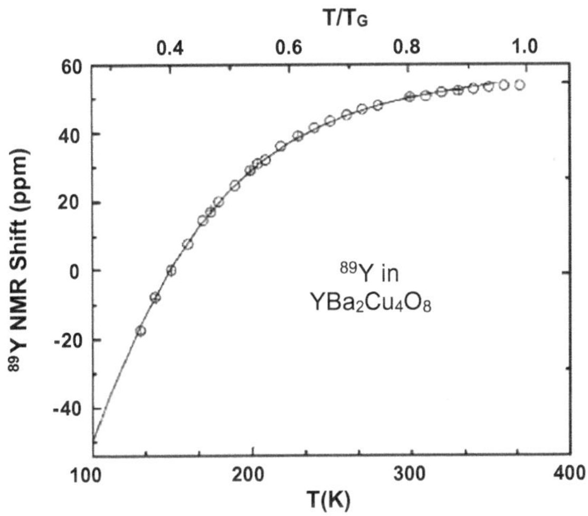

Fig. 4.35 The ^{89}Y NMR shift data from Y248 (see Fig. 4.32) is shown here fitted by the pseudogap theory given by Varma [222] (solid line), yielding an excellent fit

4.4.4 Effect of Zn Impurities on the Pseudogap in Y248

From early on it has been known that Zn impurities—thought to substitute for the Cu(2)—cause a dramatic reduction of T_c in high-T_c cuprates [224]. Even though Zn^{2+} is a nonmagnetic ion core, its presence in the CuO_2 plane (as well as Li^+) is thought to cause potential scattering which destroys d-wave pairs, and it also causes the occurrence of a small localized magnetic moment in the same vicinity [225]. The latter is a very interesting effect in its own right, even exhibiting a Kondo effect [226, 227].

The strategy of Zn-doping to eliminate the effect of the SC transition on the pseudogap was adopted by Zheng et al. [228]. However, instead of the collapse of AFM spin fluctuations as found with YBCO7-x, the Zn-doped Y248 undergoes an

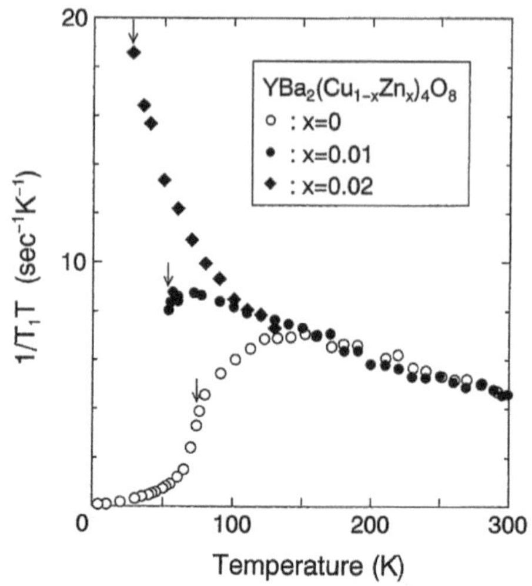

Fig. 4.36 A plot of data for $1/T_1T$ for $^{63}Cu(2)$ in Y248 made with samples which are undoped (open circles) and doped with 1% Zn (closed circles) and 2% Zn (closed diamonds), showing suppression of T_c from 74 K to 53 K and $\sim 27\,K$, respectively, combined with large progressive increases in the value of $1/T_1T$ at $T = T_c$ [228]. The corresponding change in the NMR shift curve is slight and barely resolvable

enhancement of $1/T_1$ for $^{63,65}Cu$ similar to the Curie-Weiss Behavior of T_1T in LSCO:x [279], as shown in Fig. 4.36. With 2% Zn substitution, the peak value of $1/T_1T$ is almost three times the peak value of that quantity in the undoped case. Meanwhile, the NMR shift profile remains spin-gapped with effectively no change [228], in spite of a reduction of T_c by a factor ~ 3. The latter result suggests that the SC transition was already having very little effect on the shift profile in the undoped case.

The Zn impurities, while suppressing T_c, have the very specific effect of ungapping the Fermi surface regions which contribute to $\chi''(\mathbf{q}, \omega)$ near $\mathbf{q} = (\pi, \pi)$, or so it would appear. In connection with these remarkable data, the results of two additional relaxation measurements would be of great interest. First, a comparative plot of T_{2g} for the $^{63}Cu(2)$ over the same temperature range would give a direct measure of the doping effect on $\chi'(\mathbf{Q})$ (see Sect. 6.1). Second, the T_1 behavior of either ^{89}Y or of $^{17}O(2, 3)$ would test the effect of Zn doping on low–frequency fluctuations away from $\mathbf{Q}$ as a corroboration of the unchanged NMR shift results. In other words, these measurements would further test the scenario where the Zn doping effect appears to be confined to the region of the AFM peak.

Chapter 5
Relaxation Models for Cuprate NMR

From the very beginning of cuprate NMR studies, it was clear that understanding the relaxation properties for all sites with nuclear spin species would be one of the major challenges for the theorist and experimentalist alike. In Chap. 3 the Mila-Rice-Shastry model [118, 146] was presented, including the quantum-chemical basis for transferred HF couplings and the spin Hamiltonian treatment of local spin and orbital HF effects on the Cu^{2+} sites. With the latter formulation a complete characterization of the NMR shift tensors for YBCO7 and YBCO6.63 has been presented in Chaps. 3 and 4. Then, using general expressions for $1/T_1$, Mila and Rice discussed T_1 anisotropy for ^{63}Cu(2) in YBCO7 (Sect. 3.3.4). A simplified model given in Sect. 4.1.2 offered an interpretation of T_1 anisotropies and spin correlation effects for the ^{89}Y and ^{17}O(2,3) sites. These developments set the stage for the most serious challenge, which is the locally-varying, overall temperature dependences of the T_1 processes in these compounds. Such a panorama of complex relaxation behavior was never encountered before the cuprates were discovered.

A variety of models treating the relaxation processes have been put forward in the literature. Here, we will review the results of several of these models in an effort to gauge the present status of our understanding of these matters. There have been phenomenological models put forward by Monien and Pines [160] and Millis et al. [161, 162], where the AFM fluctuations are taken into account by means of an RPA treatment of the AFM exchange interactions. Moreover, it is assumed and is consistent with experiment that the Cu^{2+} spin variables are itinerant and undergo pairing in the superconducting state. Similar modeling along these lines has also been put forward by Horvatić et al. [250].

These models were presented early on and gave a reasonable account of the data which were available at the time. In particular, they met the challenge of explaining the behavior of all measured relaxation processes with a single dynamical variable per Cu site, and did so using only HF couplings generated by the Cu^{2+} spin moments, in

© Springer-Verlag GmbH Germany 2018

R.E. Walstedt, *The NMR Probe of High-T_c Materials and Correlated Electron Systems*, Springer Tracts in Modern Physics 276,
https://doi.org/10.1007/978-3-662-55582-8_5

a picture consistent with the Mila-Rice model [118] of on-site spin Hamiltonian and transferred HF couplings of the general form $\sum_\alpha A_\alpha S_\alpha I_\alpha$. A possible deviation from this form may occur in the case of ^{89}Y in YBCO compounds, where the transferred coupling is very weak and the dipolar interactions, which do not have the latter form, come into play.

In Chap. 3 a relaxation formalism was reviewed in which, using the fluctuation–dissipation theorem, $1/T_1$ may be expressed as a summation over the Brillouin zone (BZ) of the dynamic susceptibility $\chi''(\boldsymbol{q}, \omega)$ multiplied by Fourier components of HF coupling energies having the form given in the previous paragraph. Thus, T_1 may be estimated from absolute and quantitative data for $\chi''(\boldsymbol{q}, \omega)$, combined with HF coupling data derived from NMR shift and susceptibility measurements. In Chap. 4 such data derived from INS studies and the NMR literature are presented, and quantitative analyses of $T_1(T)$ are reviewed for optimally doped LSCO and for oxygen-deficient YBCO6.5. Over limited ranges of temperature and composition quantitative agreement between INS and NMR data are found to occur, with no adjustable parameters. That is the strength of this scheme. The weakness is that the INS data, in particular, are extraordinarily difficult to obtain, and it was found for both systems studied that the term in $\chi''(\boldsymbol{q}, \omega)$ that exhibits the pseudogap effect is inaccessible to conventional INS methods, while such a term completely dominates the T_1 process at temperatures well above T_c.

It is clear from the foregoing, however, that a "theory" of the T_1 process is essentially a theory of $\chi''(\boldsymbol{q}, \omega)$. A slightly different approach to modeling the dynamical susceptibility has been pursued by Auler et al. [245], where the basic form of the dynamic susceptibility was taken directly from INS data. This scenario had the unusual feature that the range of spin-spin correlations was fixed, although their strength was allowed to increase as temperature was lowered. These authors also made the unusual step of using spin-echo decay time (T_{2g}) measurements to determine the strength of the AFM susceptibility peak. In reviewing the results of their analysis, we shall find that a satisfactory account of the data can be achieved in this fashion.

Beyond phenomenology, there have also been first-principles model calculations by Bulut and Scalapino [158] and by Si et al. [254]. The latter are, of course, more difficult to formulate and execute, but these have yielded impressive insights. Finally, there has been a proposal by Varma which departs from the usual spin HF-only picture wherein an intersite orbital matrix element is suggested to make an important contribution to the Cu(2)-site relaxation process [244]. The essence of these various results will be summarized below.

This chapter will begin, however, with discussion of a simple, yet ingenious scheme to characterize the cuprate relaxation processes developed by Uldry and Meier [195] (UM). This is a scheme of a fairly general nature for characterizing the spin dynamics and spin-spin correlations which determine the relaxation behavior in the CuO_2 planes. It serves as a useful framework for evaluating almost any experimental data or theoretical picture.

5.1 The Uldry-Meier Parameterization Model

The Uldry-Meier (UM) approach to the analysis of cuprate relaxation data relies only on a general formulation of spin-fluctuation T_1 processes, which is characterized by just three elements, (i) the HF constants, (ii) real-space, dynamical spin-spin correlation coefficients, and (iii) an effective correlation time variable in which the temperature variation of the spin dynamics is also implicit in a very general way. The HF parameters used are those derived from the Mila-Rice-Shastry model (3.3.4) for each site in and near the CuO_2 planes.[1] The spin-spin correlations and correlation time which characterize the relaxation processes are assumed to apply to all nuclear species considered, e.g. the ^{63}Cu(2), ^{17}O(2,3), and the ^{89}Y in YBCO7 and YBCO6.63, etc. The spin-spin correlations are allowed to be anisotropic, but the correlation time variable is assumed to be isotropic and common to all spin fluctuation processes. After describing the UM analysis, the broader significance of this methodology will be discussed at the end of this section.

5.1.1 *Basic Formulation of the UM Analysis*

In the first step, UM define single-axis T_1 components

$$\frac{1}{T_{1\alpha}} = U_\beta + U_\gamma\,, \tag{5.1.1}$$

where the subscripts are a permutation of the axes (a, b, c). Thus, U_α is an axis-specific fluctuation term given by

$$U_\alpha = \frac{1}{4}\sum_{i,j}[A_{i\alpha}A_{j\alpha}\mathcal{K}^\alpha_{ij}]\tau_{\mathit{eff}}\,, \tag{5.1.2}$$

where $A_{i\alpha}$ is the HF coefficient for the i_{th} neighbor spin $\boldsymbol{S}_{i\alpha}$ (see (3.3.2)), and the $\mathcal{K}^\alpha_{ij}$ are now defined as

$$\mathcal{K}^\alpha_{ij} = 4\langle S_{i\alpha}S_{j\alpha}\rangle\,, \tag{5.1.3}$$

where the index α allows the spin-spin correlations to be anisotropic. The only $\mathcal{K}^\alpha_{ij}$'s which will occur in the analysis are $\mathcal{K}^\alpha_{ii} = 1$ and $\mathcal{K}^\alpha_{01}$, $\mathcal{K}^\alpha_{12}$, and $\mathcal{K}^\alpha_{13}$, where the subscripts refer to the sites in Fig. 3.6. However, as was done in Chap. 4, we shall relabel these coefficients $\mathcal{K}^\alpha_1$, $\mathcal{K}^\alpha_2$ and $\mathcal{K}^\alpha_3$, respectively, for the first three neighbor pairs of Cu(2) sites in the CuO_2 lattice. This notation is used for the remainder of Chap. 5.

[1] The Cu(1) sites and environs in YBCO are set aside as a separate problem.

The correlation factors, originally introduced by Mila and Rice [118], were defined and discussed in Sects. 3.2 and 3.5. In particular, they are easily expressed in terms of $\chi''(\boldsymbol{q},\omega)$ (see (3.5.17) and (3.5.18)). The parameter τ_{eff} plays the role of a correlation time in (5.1.2), but also contains the effects of quasiparticle dynamics in the calculation. At low temperatures, for example, $\tau_{eff} \propto T$. In Sect. 3.5 a similar quantity $\mathcal{X}_{at}$ was introduced, where we can make the identification $\tau_{eff} = (k_B T/\hbar)\mathcal{X}_{at}$. In fitting (5.1.2) to the experimental data, the principal assumption is the form of the HF coupling Hamiltonian (3.3.4), where it should be noted that deviations from this form (e.g. dipolar coupling) are not unusual.

In order to fully characterize the relaxation process of a given nuclear site, it is necessary to have T_1 data from all three axes, since from (5.1.1) we have $U_\alpha = \frac{1}{2}[1/T_{1\beta} + 1/T_{1\gamma} - 1/T_{1\alpha}]$. It simplifies, of course, if there is *ab* plane symmetry. Thus, it is necessary to have a full data set for all nuclear sites of interest in order to apply this scheme of analysis to any given cuprate. In Fig. 5.1a–c, a 100–300 K data set is shown for ^{63}Cu(2) [143, 156], ^{17}O(2,3) [205], and ^{89}Y [191] in YBCO7, plotted as $1/T_1T$ versus T. Using (5.1.1), the corresponding (interpolated) plots for $^{63}U_\alpha$, $^{17}U_\alpha$, and $^{89}U_\alpha$ are shown in Fig. 5.1d–f. It is remarkable how similar the latter curves look when plotted as $1/T_1$ rather than as $1/T_1T$.

The next step is to make a simplifying approximation regarding the $\mathcal{K}_n$'s. Since the three coefficients described above correspond to spin sites separated by distances $r_1 = a/2$, $r_2 = a/\sqrt{2}$, and $r_3 = a$, respectively, UM have assumed that $\mathcal{K}_2^\alpha = (\mathcal{K}_1^\alpha)^{\sqrt{2}}$, and $\mathcal{K}_3^\alpha = (\mathcal{K}_1^\alpha)^2$, which is equivalent to the assumption of the form $\mathcal{K}_n^\alpha = \exp(-2r_n/a\lambda_\alpha)$. At each temperature, then, the fit procedure only has to determine the two values $\lambda_{ab,c}$, rather than six $\mathcal{K}_n^\alpha$'s. If this approach should prove unsatisfactory it could easily be modified.

It is also convenient to define, along with the $^kU_\alpha$'s, a parallel set of coefficients $^kV_\alpha$, given by $^kU_\alpha = ^kV_\alpha\,\tau_{eff}$. Then from (3.5.20), (3.5.21), and (3.5.22) we have

$$^{63}V_\alpha(T) = \frac{1}{4}[A_\alpha^2 + 4B^2 + 8A_\alpha B\mathcal{K}_1^\alpha(T) + 8B^2\mathcal{K}_2^\alpha(T) + 4B^2\mathcal{K}_3^\alpha(T)] \quad (5.1.4a)$$

$$^{17}V_\alpha(T) = \frac{1}{2}C_\alpha^2[1 + \mathcal{K}_1^\alpha(T)] \quad (5.1.4b)$$

$$^{89}V_\alpha(T) = 2D_\alpha^2[1 + 2\mathcal{K}_1^\alpha(T) + \mathcal{K}_2^\alpha(T)], \quad (5.1.4c)$$

where superscripts are used to label the isotope concerned. Since the HF coefficients used here relate directly to (3.3.2), their anisotropies as well as those of the λ_α's should stem directly from the data in a natural way.

Finally, it is important to note that the Eq. (5.1.4) derived by UM with a very simple equation of motion, are exactly equivalent to (4.1.8) and (4.1.9) for ^{63}Cu and ^{17}O, respectively, in the case of isotropic correlations, derived originally in Chap. 3.

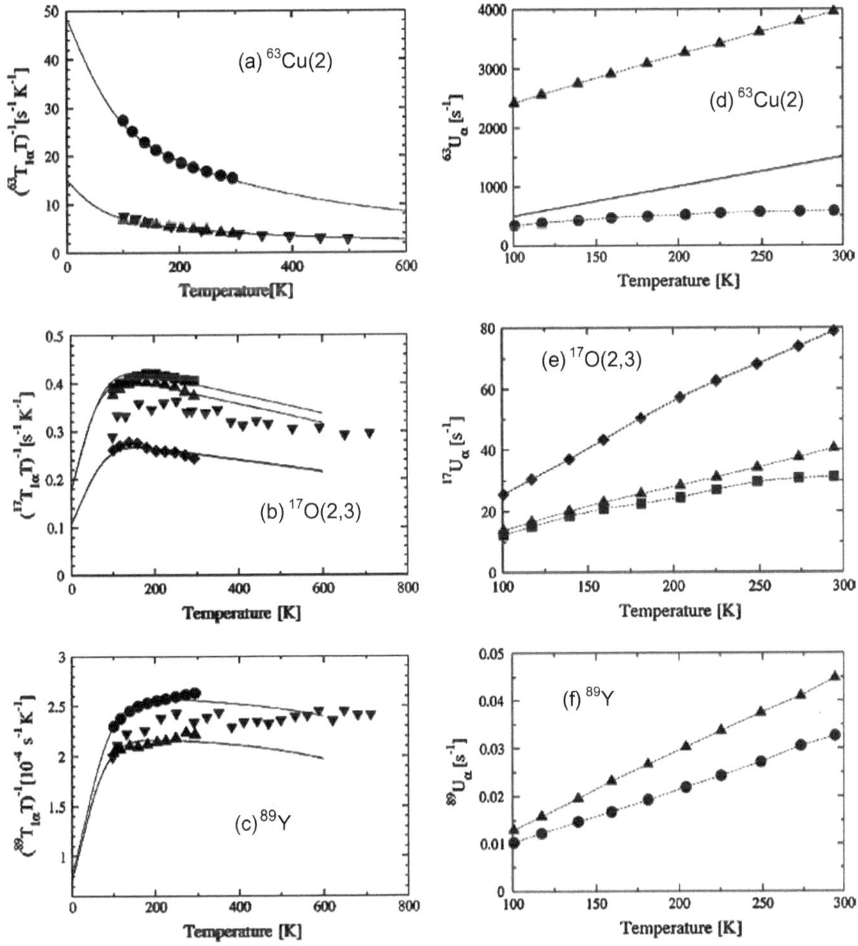

Fig. 5.1 **a–c** Relaxation data for YBCO7 plotted as circles, upward triangles, and diamonds from [156] ($^{63}T_{1c}$), [143] ($^{63}T_{1ab}$), [205] ($^{17}T_{1\alpha}$), and [191] ($^{89}T_{1\gamma}$) are plotted as $1/T_1T$ versus T. Solid lines drawn are fitted curves derived from the UM model calculations (see text). The downward triangles are supplemental data included to illustrate high-T behavior for the ^{17}O(2,3) [147] and the ^{89}Y [206], as discussed in the text. **d–f** Interpolated values of $^kU_\alpha$(T) derived from the data shown and described in (**a**)–(**c**), with which the actual fitting process was conducted. The solid lines drawn are guides for the eye. The straight line drawn in panel (**d**) shows $1/^{63}T_1$ versus T for Cu metal for comparison

5.1.2 The Uldry-Meier Analysis of YBCO7

To apply this formulation to the data in Fig. 5.1d–f, the following procedure was employed. Defining the six ratios,

$$R_1 = \frac{^{63}U_c}{^{17}U_c};\ R_2 = \frac{^{63}U_{ab}}{^{17}U_a};\ R_3 = \frac{^{63}U_{ab}}{^{17}U_b};\ R_4 = \frac{^{17}U_b}{^{17}U_c};\ R_5 = \frac{^{89}U_c}{^{89}U_{ab}};\ R_6 = \frac{^{17}U_a}{^{89}U_c}, \tag{5.1.5}$$

the HF constants in (5.1.4) and the λ_α's are then determined by minimizing the quantity

$$\chi^2 = \frac{1}{n_R}\sum_i^{n_R}\sum_j^{n_p}\frac{[R_i^{exp}(T_j) - R_i^{mod}(T_j)]^2}{R_i^{exp}(T_j)^2}, \tag{5.1.6}$$

where n_R and n_p are the numbers of ratios (= 6) and the numbers of points (= 10), respectively, and where the R_i^{mod} for the "model" are actually calculated from (5.1.4) in terms of the HF constants and the λ_α's. The HF parameters so obtained are listed in Table 5.1, where for comparison a "standard" set of parameter values from UM are given, as well as the YBCO parameters derived from NMR shift data in Chaps. 3 and 4. We consider these comparisons in detail below.

The λ_α's and the results for τ_{eff} from the fitting procedure are displayed in Fig. 5.2a and b, respectively. The correlation lengths, which are of the order of the lattice constant, are seen to execute a smooth progression toward longer values as T decreases, a process which accelerates below 150 K. They also find consistent anisotropy of ~20%, the stronger c-axis correlations suggesting the possibility of a corresponding exchange anisotropy between neighboring Cu^{2+} spins. Although the fitting scheme with the ratios r_i promotes consistency among τ_{eff} values it is nonetheless remarkable how nicely they each describe a very similar arc with temperature. As we shall see below, some of the most subtle physics of the cuprate spin fluctuations is embodied in the behavior of τ_{eff}, where even at this point a great deal remains to be understood.

Table 5.1 HF coefficients defined in (3.3.4) and derived from the UM fit procedure with the T = 100 K relaxation data shown in Fig. 5.1, as described in the text, are listed and compared with data presented by Nandor et al. [206] as well as with data derived from NMR shift measurements as described and presented in Tables 3.4 and 3.6. HF coefficient values are given in units of 10^{-6} eV. See text for discussion of these results

HF constant	$^{63}A_c$	$^{63}A_{ab}$	B	$^{17}C_a$	$^{17}C_b$	$^{17}C_c$	$-^{89}D_{ab}$	$-^{89}D_c$
UM fit	−1.68	0.168	0.438	0.259	0.173	0.196	0.00280	0.00349
Nandor	−1.6	0.29	0.4	0.25	0.13	0.156	0.0048	0.0048
Shift Coef.	−1.67	0.05	0.409	0.355	0.217	0.209	0.00286	0.00369

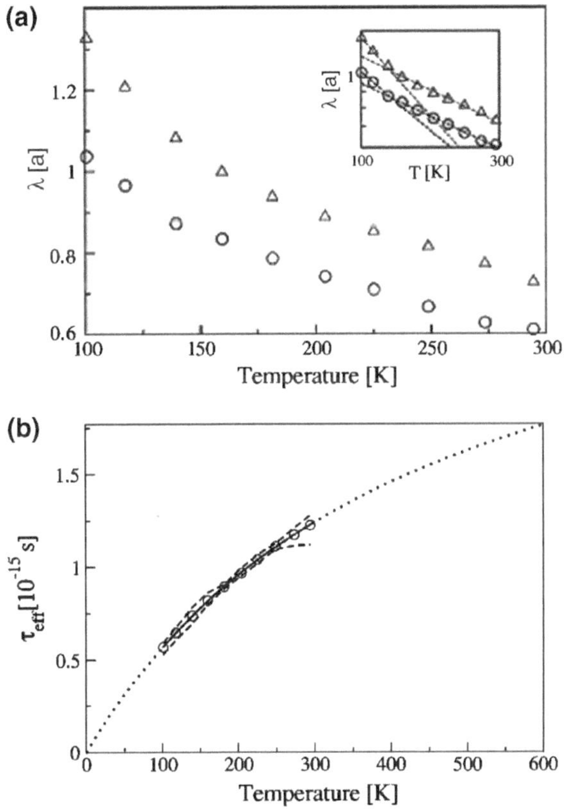

Fig. 5.2 **a** Correlation lengths λ_{ab} (circles) and λ_c (triangles) derived from the UM fit to YBCO7 are plotted versus T over the range of the fit. Inset: Low and high-T asymptotic behavior used to extrapolate the λ_α's is shown with dotted lines. **b** The average (circles) and extremal (dashed lines) values for the seven sets of τ_{eff} parameters obtained from the T_1 data fit for YBCO7 are plotted versus T. The dotted line is a fit to these data which is extrapolated down to $T = 0$ and upward to $T = 600$ K. This fit and its interpretation are described in the text

The HF coefficients from the fit are compared with values from other sources in Table 5.1. The HF constants labelled "Nandor" [206] are actually mostly derived from Barzykin and Pines [240], and are seen to be in rough accord with the fit. The "Shift Coef." values shown have been converted from the primed coefficients in Tables 3.4 and 3.6 using (3.3.4) so that g-factor anisotropy (3.3.7) comes into play. The value of B given, for example, comes from c-axis shift measurements. The value of $^{63}A_{ab} = 0.05 \times 10^{-6}$ eV is obtained by subtraction of large values and is correspondingly unreliable. The agreement of the $^{63}A_c$ and of the $^{89}D_\alpha$ tensor is very good, indeed.

The fitted values of the $^{17}C_\alpha$ are 10–20% below those derived from the NMR shifts. This is an interesting point, because the "T_1 problem" for the ^{89}Y discussed in Sect. 4.2.2 has been managed in the UM fit process not by enhancing the $^{89}D_\alpha$ tensor coefficients, but by reducing the $^{17}C_\alpha$ ones. Either way, there will be a discrepancy between shift coefficients and those derived from relaxation. We continue to believe that there is an additional ^{89}Y relaxation process (orbital or dipolar), which is beyond the effects of the MR Hamiltonian (3.3.2). In spite of this effect, there is quite a good correspondence between HF constants derived from the UM fit and those from NMR shifts.

It is also noteworthy that the fitting procedure based on (5.1.5) is independent of scale. That is, all HF constants (i.e. the ${}^kV_\alpha$'s in (4.1.4)) can be multiplied by a uniform scale factor and the results will not be affected. Thus, any set of results for the HF constants needs to be keyed to some particularly reliable value. In Table 5.1 the results are, in effect, keyed to ${}^{63}V_c$.

The UM fitting procedure leads to temperature dependences for the correlation lengths λ_α shown in Fig. 5.2a and to seven estimates (from the ${}^kV_\alpha$'s in (5.1.4)) for the correlation time curve $\tau_{eff}(T)$, which are summarized in a plot in Fig. 5.2b. In the latter case the extrema of the bundle of seven curves are shown as dashed lines, where the circles are the average values. The dotted line is a functional fit described below. Considering the ^{89}Y relaxation rate anomaly noted above and in Sect. 4.2.2, the degree of consistency among τ_{eff} values is surprisingly good.

Extrapolation of the Fits to Higher Temperatures

In Fig. 5.1a–c we see that the fits are generally very good over the fitting temperature range. However, UM went on to extrapolate their fits to higher temperatures with a fitting function *ansatz* as follows. The general behavior of $\tau_{eff}(T)$ in Fig. 5.2b is one of downward curvature, but linear in T at the low-T end. The latter feature is expected in metallic hosts, but the curving over is much greater than could be attributed to loss of Fermi-statistical degeneracy. They propose a fitting function wherein a Fermi-statistical term $\tau_1 = aT$ is combined in an inverse fashion with a constant term τ_2 which dominates at high temperatures. Thus [195],

$$\frac{1}{\tau_{eff}} = \frac{1}{\tau_1} + \frac{1}{\tau_2} = \frac{1}{aT} + \frac{1}{\tau_2}, \tag{5.1.7}$$

where $a = 7 \times 10^{-18}$ s/K and $\tau_2 = 3 \times 10^{-15}$ s. The line representing this function is plotted in Fig. 5.2b as a dotted line up to 600 K. In Sects. 5.4 and 5.5 we shall look for the possible origin of this flattening effect in the Hubbard models. However, it does not appear to have been successfully identified yet.

The physics of the ansatz (5.1.7) is a bit unfamiliar, because normally when two rates combine inversely, it is the "faster" of the two which dominates. Here, however, at high temperatures when the Korringa T_1 is very short, it is undercut by a weaker relaxation process. As for the form of τ_1, Pines and Slichter [241] suggest a picture where $\tau_1 = \tau_c T/T_F$ is made up of a correlation time factor $\tau_c \sim \hbar/E_F$ representing the dwell time of a carrier in the unit cell, and T/T_F represents a Fermi-statistical scattering probability factor. Thus, $a = \hbar/k_B T_F^2$, from which UM estimated $T_F \sim$ 1000 K. Meanwhile, τ_2 appears to represent the behavior of the gas of quasiparticles when the degeneracy is completely lifted.

This extrapolation of the curve for τ_{eff}(T) is also used to extrapolate the fit curves in Fig. 5.1a–c. It is interesting to compare these extrapolations with high-temperature data by Nandor et al. [206] for ^{17}O(2,3) and for ^{89}Y, shown in panels (b) and (c) as downward triangles. The data for ^{17}O(2,3) fall somewhat below the fitted data, but show a temperature dependence which is in general accord with the extrapolated model prediction. For the ^{89}Y, however, contrary to the model the data are flat if not slightly rising with temperature. This behavior is consistent with the finding in Sect. 4.2.2 that there is an additional relaxation term for the ^{89}Y beyond what is derived from (3.3.2). A possible orbital term was mentioned in Sect. 4.2.2. There are also dipolar terms which do not follow the spin-spin correlations [191]. The dipolar terms have also been calculated by Auler et al. [245] and are discussed further in Sect. 5.3.

The Spin Correlations and the "Basic" Relaxation Mechanism

The fit to YBCO7 data illustrates rather dramatically the role played by antiferromagnetic (AFM) spin-spin correlations in the relaxation processes of all nuclei. It is interesting in this connection to examine the behavior of ${}^{k}V_\alpha(T)$. According to the definitions (5.1.4), the ${}^{k}V_\alpha(T)$ can vary from a high-temperature limit where $\mathcal{K}_n \to 0$, to a fully correlated limit at low T where (in principle) $\mathcal{K}_1 \to -1$ and $\mathcal{K}_2, \mathcal{K}_3 \to 1$. The $T \to \infty$ limits for the ${}^{k}V_\alpha(T)$ are defined to be ${}^{k}V_\alpha^0$, and are given by

$$^{63}V_\alpha^0 = \frac{1}{4}(A_\alpha^2 + 4B^2) \quad (5.1.8a)$$

$$^{17}V_\alpha^0 = \frac{1}{2}\,C_\alpha^2 \quad (5.1.8b)$$

$$^{89}V_\alpha^0 = 2\,D_\alpha^2\,. \quad (5.1.8c)$$

The "basic" relaxation mechanism is what it would be in the absence of AFM correlations and is given by $1/{}^{k}T_{1\alpha} = {}^{k}V_\alpha^0 \tau_{\mathit{eff}}$. Thus, apart from a scale factor, it is the same for all nuclei in the system.

The AFM correlation factors are the only thing which distinguishes the behavior of different nuclei. Figure 5.3a–c illustrate this point by plotting the ${}^{k}V_\alpha(T)$ versus T for the YBCO7 fit. (Figure 5.3d–f plot the same quantities for YBCO6.63—see discussion below.) The high-temperature ($\mathcal{K}_n = 0$) limits are shown as solid bars on the right, and the low-temperature limits by solid bars on the left side of the figures. The $T \to 0$ behaviors are estimated by extrapolating the correlation lengths in Fig. 5.2a by means of the dotted lines shown in the inset. The behavior seen in these plots shows (i) that the correlation causes the relaxation rates of the ^{63}Cu(2) to increase and those of the ^{17}O(2,3) and the ^{89}Y to decrease, and (ii) that the behavior of all three nuclear species is changed by similar amounts as a result of the correlations. In this picture it is something of an accident that the ^{17}O(2,3) and the ^{89}Y relaxation processes appear to be Korringa-like in the normal state.

According to Fig. 5.3a–c the AFM correlations make sharp increases in the superconducting state. It must be emphasized that this is an extrapolation of normal-state behavior and is entirely hypothetical. As we shall see in Sect. 5.7, the actual behavior below T_c is quite different from what is shown here.

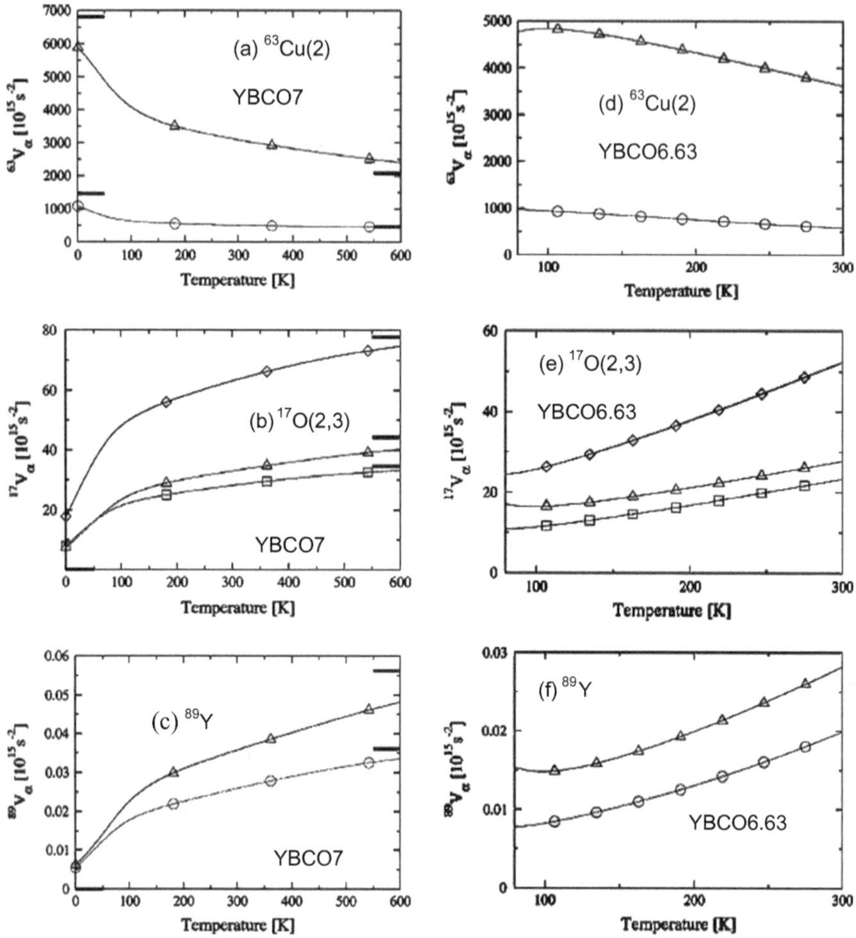

Fig. 5.3 **a–c** Curves are plotted representing $^{63}V_\alpha(T)$, $^{17}V_\alpha(T)$, and $^{89}V_\alpha(T)$ for the fit to YBCO7 in (**a**), (**b**), and (**c**), respectively. The solid bars on the right represent the values of $^kV_\alpha^0$ corresponding to zero spin correlations, and the bars on the left represent maximum correlations as explained in the text. The data plotted are seen to run between these two limits. The zero-temperature points for these curves were obtained by extrapolating the variation of the λ_α's near 100 K to zero temperature. Such an extrapolation is entirely hypothetical. **d–f** Curves are plotted representing $^{63}V_\alpha(T)$, $^{17}V_\alpha(T)$, and $^{89}V_\alpha(T)$ for the fit to YBCO6.63 in (**d**), (**e**), and (**f**), respectively. These results are similar to those on the left, but AFM correlations show a wider range and reversed curvature as compared with YBCO7

5.1.3 The Uldry-Meier Analysis of YBCO6.63

A very similar analysis to that described above was carried out by UM for YBCO6.63. The quality of the input data was perhaps not quite as good, and there were no data at all for ${}^{63}T_{1ab}$. The analysis was simplified a bit by adopting the HF tensor values shown in Table 5.1 for the YBCO6.63 analysis (recall that the values derived from NMR shifts in Chap. 4 were found to be nearly the same). The fitted curves (not shown) were only slightly less good, where, not surprisingly, the ^{89}Y fits were noticeably worse than the others.

The results for ${}^{k}V_{\alpha}(T)$ curves are shown in Fig. 5.3d–f, where they are seen to be quite different from those for YBCO7. The correlation lengths for YBCO6.63 were longer than for YBCO7, reaching $\sim$2.5 a or so. As a result the range and magnitude of the correlation factors driving the ${}^{k}V_{\alpha}(T)$ curves are considerably greater than for YBCO7. Another interesting feature is that the curvature of the ${}^{k}V_{\alpha}(T)$ curves is in all cases the opposite in these two systems.

There was also slightly more scatter in the $\tau_{eff}(T)$ curves for YBCO6.63, shown in Fig. 5.4a, where they execute a considerably greater range than for YBCO7 (Fig. 5.2b). This, of course, results from the gapped nature of the static and dynamical properties of YBCO6.63. The average curve for τ_{eff} (circles) was again fitted with the ansatz (5.1.7), except that a gap feature is introduced into the density of states with

$$\tau_1(T) = a\,T\exp(-T_g/T). \tag{5.1.9}$$

With that modification, (5.1.7) was fitted to the data in Fig. 5.4a and is shown as a dotted line extending from 0 to 600 K. The fitting parameters are listed below in Table 5.2, where a and τ_2 are seen to be quite similar to those for YBCO7. The gap energy $T_g = 94$ K is very close to that (104 K) derived from the susceptibility in Fig. 4.16, *rather than being twice that value* as we found for Y248 in Sect. 4.4. The slightly lower value for a suggests a somewhat lower degeneracy temperature for YBCO6.63 as shown in the table.

5.1.4 The Uldry-Meier Analysis of $YBa_2Cu_4O_8$ (Y248)

The foregoing analytical scheme was also applied, in a somewhat restricted fashion, to the case of Y248, analyzing the same body of NQR data for ^{63}Cu(2) as were discussed in Sect. 4.4. The method was restricted owing to a total lack of T_1 data for other sites and orientations. As a consequence, the ${}^{63}V_{ab}(T)$ curve for YBCO7 (see Fig. 5.3a) was adopted for the Y248 calculation, where

$$\frac{1}{{}^{63}T_{1c}} = 2\,{}^{63}V_{ab}\tau_{eff}. \tag{5.1.10}$$

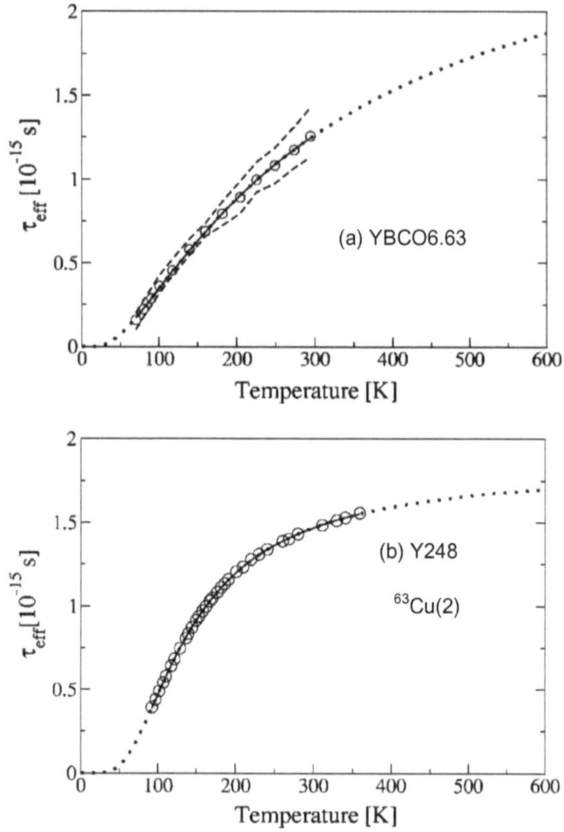

Fig. 5.4 **a** The data for $\tau_{eff}(T)$ derived from the fit to several relaxation curves from YBCO6.63, as described in Sect. 5.1.3. The circles plot the average values of τ_{eff}, and the dashed lines show the most extreme values. The dotted line is a fit of (5.1.7) and (5.1.9) to the average values, extrapolated to 0 and 600 K. **b** Values for $\tau_{eff}(T)$ derived from the fit to the relaxation curve for ^{63}Cu(2) in Y248, shown in Fig. 5.5, as described in Sect. 5.1.4. The dotted line is a fit of (5.1.7) and (5.1.9) to the data (circles), extrapolated to 0 and 600 K. Parameters from the fits shown are presented in Table 5.2) and discussed in the text

This is a fairly "neutral" assumption, since the temperature profile of $^{63}V_{ab}(T)$ is seen to be very flat for both YBCO7 and YBCO6.63 in Fig. 5.3, but particularly in the former case. In any case, the fit to the data for $1/^{63}T_{1c}T$, shown in Fig. 5.5a and b, is seen to be excellent, and furthermore, extends up beyond 700 K. Note that the 100–350 K data set from Williams et al. [219] (crosses) has been augmented with high-temperature data from Curro et al. [242] (triangles pointing right) and from Tomeno et al. [216] (triangles pointing down).

Although the profile of $\tau_{eff}(T)$ may be slightly different from what one would find with a full-blown analysis such as that for YBCO7, it behaves in a very interesting fashion as shown in Fig. 5.4b. Here there is only one data set from the fit (circles), which has very little scatter. The dotted line is a fitting of the ansatz (5.1.7) with the gap modification (5.1.9) used for YBCO6.63. The parameters from this fit are given in Table 5.2, where they may be compared with those for YCBO7 and YBCO6.63.

The parameters in Table 5.2 shown an interesting progression for the three YBCO-related systems. Although the forms of $\tau_{eff}(T)$ in (5.1.7) and (5.1.9) are really just educated guesses, they match the temperature dependence from the fits better than one

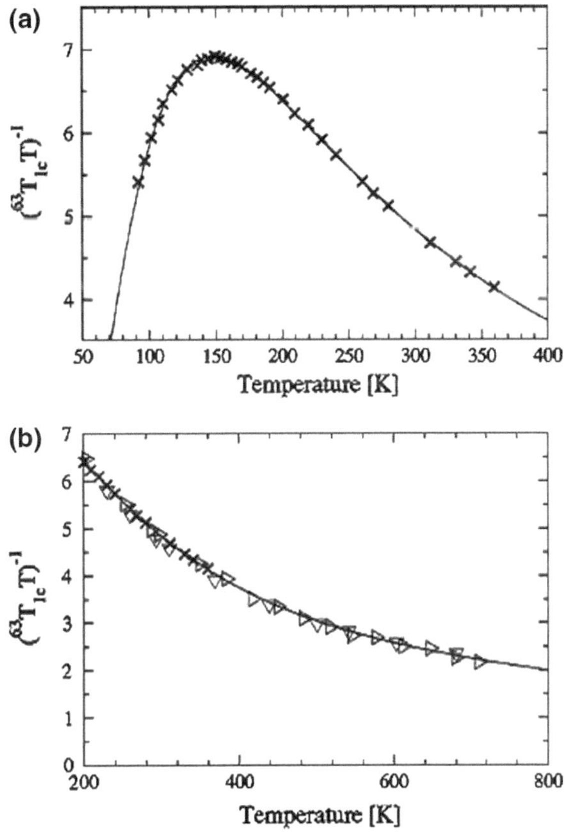

Fig. 5.5 **a** ^{63}Cu(2) NQR data from [221] for Y248 (crosses) are plotted as $1/T_1T$ versus T. The solid line shows the Uldry-Meier fit to these data described in Sect. 5.1.4. **b** The fit curve shown in part (**a**) is extended to above 700 K. Also shown there are high-temperature data by Curro et al. [242] (triangles pointing right) and by Tomeno et al. [216] (triangles pointing down). The data and the fit are seen to be extremely precise

Table 5.2 Parameters from fits to the ansatz (5.1.7) for YBCO7 and with the gap modification (5.1.9) for YBCO6.63 and Y248 are listed, showing a, τ_2, and the gap, crossover, and Fermi temperatures derived from their respective (average) data curves for $\tau_{eff}(T)$. The crossover temperature is where $\tau_1 = \tau_2$, and the estimated Fermi temperature is obtained from $T_F \sim (\hbar/k_B a)^{1/2}$. These results are discussed in the text

System	$a\,(s/K)$	$\tau_2\,(s)$	$T_g(K)$	$T_{cr}(K)$	$T_F(K)$
YBCO7	7×10^{-18}	3.0×10^{-15}	0	420	1,050
YBCO6.63	10×10^{-18}	2.9×10^{-15}	97	375	880
Y248	44×10^{-18}	1.9×10^{-15}	195	154	420

might have expected, giving very convincing extrapolations to higher and lower temperatures. After modest changes between YBCO7 and YBCO6.63, with the appearance of a ~100 K gap in the latter case, the jump to Y248 shows a doubling of the density of states ($\propto \sqrt{a}$) and a corresponding reduction by half of the estimated degeneracy temperature, as well as a doubling of the gap energy, which now exceeds the cross-over temperature.

A somewhat different view of the dynamic magnetism of Y248 via dynamical scaling theory is presented in Sect. 6.4.

The Wider Role of the Uldry-Meier Model Analysis

The UM analysis has offered a highly revealing and greatly unifying look into the microscopic phenomena associated with the cuprate T_1 processes. From this perspective, the difference between different sites and nuclear species essentially ceases to be an issue. This analysis is in no sense a theory, though it incorporates a widely accepted and time-tested formulation of the T_1 process, and it puts the Mila-Rice HF Hamiltonian to a severe test, which it mostly passes in good order. Note that this Hamiltonian is essentially unique to cuprates among metallic environments. It is worth emphasizing that the UM picture is also implicit in the general $\chi''(\boldsymbol{q}, \omega)$ formulation of the T_1 process, as demonstrated in Sect. 3.5.

Perhaps the most significant advantage of the UM analysis is that it partitions the T_1 problem into three independent phenomenological pieces. These are (i) HF constants, which are essentially inert, (ii) AFM correlation factors, which are much larger and universally more important than hitherto realized, and (iii) a universal fluctuation rate, which embodies all of the low-frequency spin fluctuation physics of the cuprates, including, but not limited to, Fermi liquid, spin pseudogap, and exchange-coupled local moments. These three elements are also very prominent in the INS-based analysis worked out in detail in Chap. 4, but that, of course, requires external data for $\chi''(\boldsymbol{q}, \omega)$. The ability to separate elements (ii) and (iii) in either an experimental or theoretical evaluation is a considerable simplification and an enormous advantage over previous practice. We emphasize that this analysis offers a useful way to evaluate model theories of $\chi''(\boldsymbol{q}, \omega)$. One can simply evaluate (ii) and (iii) as defined above and judge these components separately on their own merits. This procedure offers a new perspective on theory building for the cuprates and quite possibly for other materials as well.

5.2 The Millis, Monien, and Pines Model for T_1 in Cuprates

This straightforward phenomenological model [161, 162] served the important role of discussing and testing certain key ideas in the early days when both NMR shift and relaxation data were still unfolding. This model emerged at a time when there was still a great deal of uncertainty about what the ultimate theoretical basis for cuprate T_1 processes was going to be. Of the several lines of theoretical approach to cuprate spin dynamics which we discuss in this chapter, the conceptually simple MMP model is a good place to begin.

One of the main objectives of MMP was to show that the relaxation of ^{63}Cu(2), ^{17}O(2,3), and ^{89}Y can all be accounted for with a single dynamical variable associated with each Cu^{2+} site. In this matter they largely succeeded, although as noted in Sects. 4.4 and 5.1 there are other apparent contributions to the ^{89}Y relaxation process.

However, there is no evidence for another set of dynamical variables besides the Cu^{2+} spin moments.

In addition to the foregoing, these two papers [161, 162] give an in-depth discussion of the physics issues surrounding the NMR T_1 problem in cuprates. Although the actual model they expound is a phenomenological treatment of the Mila-Rice Hamiltonian [118], they present a derivation [162] showing that such treatment is equivalent to a tight-binding, band-theory model of an interacting Fermi liquid.

The MMP model is based on a mean-field treatment of AFM exchange interaction effects in a system of itinerant quasiparticles, i.e. a metal, which would be a simple Fermi liquid without the exchange interactions. We begin by reviewing their derivation of the mean-field dynamic susceptibility, in the course of which the relevant parameters are introduced and discussed. The following is just a brief summary. We refer the interested reader to the original papers for more details [161, 162].

The mean-field complex susceptibility $\chi(\boldsymbol{q},\omega) = \chi'(\boldsymbol{q},\omega) + i\,\chi''(\boldsymbol{q},\omega)$ of an interacting spin system is given by

$$\chi(\boldsymbol{q},\omega) = \frac{\overline{\chi}(\boldsymbol{q},\omega)}{1 - J_q\overline{\chi}(\boldsymbol{q},\omega)}\,, \tag{5.2.1}$$

where $\overline{\chi}(\boldsymbol{q})$ is the noninteracting susceptibility, and J_q is the interaction potential. Since NMR frequencies are generally much smaller that the electronic frequencies in such a system, we are concerned here with the region near $\omega \to 0$. In this limit it is assumed that

$$\overline{\chi}''(\boldsymbol{q},\omega)|_{\omega\to 0} \to \pi\frac{\omega}{\Gamma_q}\overline{\chi}'(\boldsymbol{q},\omega = 0) \equiv \pi\frac{\omega}{\Gamma_q}\overline{\chi}_q, \tag{5.2.2}$$

where Γ_q is a characteristic energy for spin fluctuations at wave vector $\boldsymbol{q}$. Correspondingly, in the same limit (5.2.1) gives

$$\chi''(\boldsymbol{q},\omega) \longrightarrow \pi\frac{\omega}{\Gamma_q}\overline{\chi}\frac{1}{(1-G_q)^2 + (\pi\omega/\Gamma_q)^2 G_q^2}\,, \tag{5.2.3}$$

where $J_q\overline{\chi}_q \equiv G_q$. As we are treating AFM fluctuations which appear to be diverging as the system is approaching a Neél point, G_q is assumed to be peaked at the AFM point $\boldsymbol{Q} = (\pi/a, \pi/a)$, at which point it has a value of slightly less than 1. For (5.2.3) we expand G_q about the AFM point: $G_q = G_Q - \boldsymbol{q}^2\beta_0^2$, where $\boldsymbol{q}$ is measured from $\boldsymbol{Q}$. The correlation length is defined by $G_Q = 1 - \xi_0^2/\xi^2$. Setting $G_q = 1$ in the second term of the denominator of (5.2.3), we obtain

$$\frac{\chi''(\boldsymbol{q},\omega)}{\omega} = \frac{\pi\overline{\chi}}{\overline{\Gamma}}\frac{(\xi/\xi_0)^4}{(1+q^2\xi^2)^2 + (\pi\omega/\overline{\Gamma})^2(\xi/\xi_0)^4} \simeq \frac{\pi\overline{\chi}}{\overline{\Gamma}}\frac{(\xi/\xi_0)^4}{(1+q^2\xi^2)^2}\,. \tag{5.2.4}$$

Because of the peaked nature of this function, the $\boldsymbol{q}$-dependence of the parameters $\overline{\chi}_q$ and $\overline{\Gamma}_q$ has been dropped. They argue further that when $\boldsymbol{q}$ is far away from $\boldsymbol{Q}$,

$\chi''(\boldsymbol{q},\omega)$ should be comparable with the uncorrelated value, so a $\boldsymbol{q}$-independent term of that magnitude is added in, giving

$$\left.\frac{\chi''(\boldsymbol{q},\omega)}{\omega}\right|_{\omega\to 0} = \frac{\pi\overline{\chi}}{\overline{\Gamma}}\left[1+\beta\frac{(\xi/a)^4}{(1+q^2\xi^2)^2}\right], \tag{5.2.5}$$

where $\beta = (a/\xi_0)^4(\overline{\chi}/\chi_0)(\Gamma/\overline{\Gamma})$, χ_0 is the uniform susceptibility, and Γ is the spin fluctuation frequency of the noninteracting system. Equation (5.2.5) may then be used to model quasiparticle dynamics for cuprate systems not too far from an AFM instability, i.e. when $\xi/a \gg 1$. From (5.2.4) we also see that the effective spin fluctuation frequency is

$$\omega_{SF} \sim \frac{\overline{\Gamma}\xi_0^2}{\xi^2}, \tag{5.2.6}$$

which can be greatly diminished if $\xi \gg \xi_0$.

The $\boldsymbol{q}$-dependent HF coupling Hamiltonian terms are denoted ${}^kF_\alpha(\boldsymbol{q})$ here (3.5.6) for nuclear spin species k and axis α. Thus, the relaxation rate ${}^kW_\alpha \equiv 1/{}^kT_{1\alpha}$ is given by [162]

$${}^kW_\alpha = \frac{k_BT}{4\hbar\mu_B^2}\sum_{\boldsymbol{q}}\left[{}^kF_\beta(\boldsymbol{q})^2 + {}^kF_\gamma(\boldsymbol{q})^2\right]\left.\frac{\chi''(\boldsymbol{q},\omega)}{\omega}\right]_{\omega\to 0}, \tag{5.2.7}$$

where (α, β, γ) are any permutation of the (a, b, c) axes. In (5.2.7) the ${}^kF_\alpha(\boldsymbol{q})$ and $\chi''(\boldsymbol{q},\omega)$ have the dimensions of s^{-1} and s, respectively (see (3.5.9)). For YBCO the ${}^kF_\alpha(\boldsymbol{q})$ are

$${}^{63}F_\alpha(\boldsymbol{q}) = {}^{63}A_\alpha - 2B(\cos q_xa + \cos q_ya), \tag{5.2.8a}$$

$${}^{17}F_\alpha(\boldsymbol{q}) = 2\,C_\alpha \cos^2(q_{x,y}a/2), \tag{5.2.8b}$$

$${}^{89}F_\alpha(\boldsymbol{q}) = 8D_\alpha \cos(q_xa/2)\cos(q_ya/2)\cos(q_za/2) \tag{5.2.8c}$$

(see (3.5.10)–(3.5.14)).

The fitting procedure, which is described briefly below, thus consists of estimating the HF constants (A, B, C, D's) from NMR shift and susceptibility data, and then adjusting the parameters Γ, β_0, and ξ from (5.2.5) to match the calculated rates ${}^kW_\alpha(T)$ as well as possible with experimental data.

The MMP Fitting Process for YBCO7

Here we summarize the issues regarding this fitting procedure as well as a brief description of the procedure itself. We first note that the squared Lorentzian form for $\chi''(\boldsymbol{q},\omega)$ in (5.2.5) is valid when $\xi/a \gg 1$, whereas the deduced YBCO7 values of ξ/a are of order unity. For this reason, MM also discussed the case of a Gaussian form for the peak at $\boldsymbol{Q}$ [162]. The latter is more nearly consistent with peaks in $\chi''(\boldsymbol{q},\omega)$ which have been reported from neutron scattering studies [199]. The latter results also showed some sign of incommensurability, which is also a major effect in LSCO. The question is whether the incommensurability observed by neutron scattering at energies of more than 1 meV is still present at NMR frequencies, which are roughly four orders of magnitude smaller. In Chap. 6 it will be shown that incommensurability in $\chi''(\boldsymbol{q},\omega)$ is incompatible with NMR data on LSCO. We assume this is also true here and adopt with MM [162] a picture with commensurate AFM fluctuations.

The HF parameters used by MM are compared with those from the UM fit as well as those from NMR shift values in Chap. 3 (see Table 5.1) in Table 5.3. In MM the HF values are given in units of μ_B^2/χ_0. Here we convert them to units of 10^{-6} eV. Since the susceptibility is assumed to be isotropic, we have used the powder average susceptibility *per Cu(2) ion* from Table 3.4 for this purpose. It was also assumed that $A_c + 4B = 0$, essentially consistent with earlier practice. The other MM values show some small differences with the other values listed. The ratio C_a to C_c is ~1.5, in rough agreement with the others. However, the ratio $C_a/D \simeq 69$ is rather smaller than the "Shift Coef." value ≃113, so that the "problem" with the T_1 of ^{89}Y (see Sect. 4.1.2) will not arise here. We firmly believe this problem to exist.

In the MM fit there was no value available for A_{ab}, so this was left as a parameter to be "fitted". From the Cu(2) T_1 anisotropy $^{63}W_{ab}/^{63}W_c$ it was decided that $0.1 \leq A_{ab}/4B \leq 0.2$, which is not unreasonable. This estimate is compared with other values in Table 5.3. The susceptibility parameters β and ξ from (5.2.5)—and their temperature dependences—were to be extracted from the data fits. This was not a simple matter, since these parameters have qualitatively similar effects on the results. There were two main questions to be decided: (i) Can the relaxation of all three nuclei be quantitatively accounted for with (5.2.5) and (5.2.8); and (ii) does a picture of AFM correlations emerge here, where $\xi(T)$ increases when T descends toward T_c?

The first answer is definitely "yes", but the second one is a bit more murky. It depended on the relative behavior of $^{17}W_\alpha(T)$ and $^{89}W_\alpha(T)$, for which the available data for the input parameters were not particularly accurate. Calculated ratios of these relaxation rates from MM are shown in Fig. 5.6 for the cases of Gaussian and Lorentzian models for $\chi''(\boldsymbol{q},\omega)$ and several sets of parameters. The calculations are based on the assumption that $^{63}W_c/^{17}W_c = 19$ at 100 K and $^{63}W_c/^{17}W_c = 10$ at 300 K, these values being taken from experiment. Shown is the variation of $^{17}W_c(T)/^{89}W_c(T)$ between 100 and 300 K for several sets of parameters, which are

Table 5.3 The HF parameters for YBCO7 and YBCO6.63 used by MM [162] are compared with values from the UM fit and those derived from NMR shift coefficients in Chap. 3 (both given in Table 5.1). The coefficients in MM were given in units of μ_B^2/χ_0, which have been converted to eV using the powder average susceptibility based on values given in Table 3.4, where χ_0 is interpreted to be the value for a single Cu(2) site. The values for C_a and C_c have been reversed, since the larger value corresponds to the Cu-O-Cu (a) axis and the smaller one to the c axis. Parameters are given in units of 10^{-6} eV

Source	$A_{ab} + 4B$	$A_{ab}/4B$	C_a	C_c	D
MM [162]	1.45	0.1–0.2	0.33	0.22	0.0048
UM fit	1.92	0.096	0.259	0.196	0.00303
Shift Coef.	1.69	0.03	0.355	0.209	0.00314

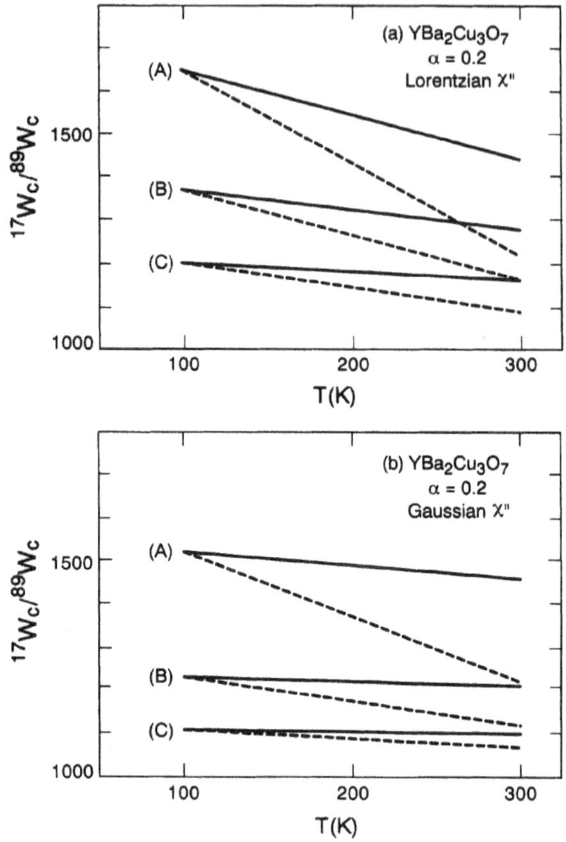

Fig. 5.6 Plots of the ratio $^{17}W_c(T)/^{89}W_c(T)$ versus T for $100\,\text{K} \leq T \leq 300\,\text{K}$ calculated with the model susceptibility shown in (5.2.5) (Lorentzian case) in panel (**a**) and a Gaussian version in panel (**b**). The Gaussian is written as $\beta\,\exp(-q^2\xi^2)$ and is centered at $\boldsymbol{Q}$. The calculations are based on the ratios $^{63}W_c/^{17}W_c = 19$ at 100 K and 10 at 300 K. Since the variation of $^{17}W_c(T)/^{89}W_c(T)$ is roughly linear, straight lines are drawn between values calculated for the two extremal temperatures. Calculations are carried out for constant β and T-varying ξ (solid lines) and for constant ξ with β varying (dashed lines). The parameter sets used are listed in Table 5.4. The same parameters are used at 100 K, but then different ones for 300 K. These calculated results are simply connected with a straight line

shown in Table 5.4. There are two assumed cases here: (i) β is constant and ξ varies with temperature, and (ii) ξ is constant and β varies with temperature. The experimental ratio is around 1,300 and is very nearly flat. These circumstances favor the Gaussian form for χ'' and the assumption of a constant β with temperature-varying ξ, and thus the solid line in Fig. 5.6b, case (B).

Thus, it was concluded that the constant β picture was more the satisfactory, because if β varies, while ξ is constant, then Γ must have a compensating temperature

Table 5.4 Parameter values for the calculated curves in Fig. 5.6 are listed for cases (A), (B) and (C), plotted both for a Lorentzian susceptibility peak (a) and for a Gaussian peak (b). Two curves are plotted for each case, both with the same parameters at T = 100 K and at 300 K with a constant value of β with ξ varied (first values listed) and with both ξ and β varied (values in parentheses)

Case	T(K)	ξ (Lor.)	β (Lor.)	ξ (Gauss.)	β (Gauss.)
A	100	1.9	30	1.4	30
A	300	1.3(1.9)	30(16)	1.1(1.4)	30(11)
B	100	2.4	10	2.0	10
B	300	1.7(2.4)	10(3.4)	1.4(2.0)	10(4)
C	100	3.7	3.0	3.3	3.0
C	300	2.5(3.7)	3.0(1.1)	2.2(3.3)	3.0(1.2)

dependence to keep the pre-factor in (5.2.5) approximately constant. This may be difficult to account for. MM also concluded that $\xi/a \simeq 1.0$–1.5 for the Gaussian case and 1.5–2.0 for the Lorentzian case, while $\beta \sim 15$ or greater.

It is interesting to view the results of the MMP analysis in the context of the results of the UM analysis of Sect. 5.1. The parameter values obtained in these cases are similar with some important differences. The MMP scheme assumed HF parameters derived from NMR shift measurements which are at least a rough facsimile of those obtained in Chap. 3 and of those obtained from the UM fit (Table 5.3). The coherence lengths appear to be similar, though it is not clear whether ξ from MMP can be directly compared with the λ_α's in Fig. 5.2a. The trend with temperature is, of course, the same.

One trouble spot in this comparison is the variation of the integrated area $\sum_q \chi''(\boldsymbol{q}, \omega)/\omega$ with temperature. For the UM picture, this simply goes as τ_{eff}/T, while in the MMP case (i.e. from (5.2.5)) this area is very nearly equal to

$$\sum_q \chi''(\boldsymbol{q}, \omega)/\omega \simeq \frac{\pi\overline{\chi}}{\overline{T}}[1 + \beta\xi^2]. \tag{5.2.9}$$

If the prefactor is independent of T as one might suppose for YBCO7, then the known behavior of $\tau_{eff}(T)$ from the UM fits tells us that $\beta\xi^2$ from the MMP model should be roughly constant. Thus, if ξ varies appreciably with T, β should vary inversely and twice as fast. This does not square well with the MM result.

On the positive side of things, one could say that the results of the UM analysis would be an excellent starting point for someone setting out to model the mean-field dynamic susceptibility as MMP did. As regards the possibility of accounting for cuprate T_1 behavior with an RPA type of model where the q-dependence of $\chi''(\boldsymbol{q}, \omega)$ (and thus ξ) has no temperature dependence, the UM analysis also gives a definite negation of any such possibility.

Millis and Monien: Discussion of YBCO6.63 and of Neutron Scattering Characterization of $\chi''(\boldsymbol{q}, \omega)$

MM analyze YBCO6.63 in a fashion similar to YBCO7. They note that rather larger values of ${}^{63}W_c/{}^{17}W_c$ indicate stronger AFM correlations in YBCO6.63, in accord with the UM result. As a result, they note that the constant-ξ option implies a somewhat larger increase in the ratio ${}^{17}W_c/{}^{89}W_c$ at lower temperatures, inconsistent with preliminary data available at the time. However, the subsequent T_1 results of Takigawa et al. [191] showed an increase in the latter ratio of ~20% below 300 K, which therefore cannot quite rule out the constant-ξ hypothesis. Again, the UM analysis of YBCO6.63, showing longer and more temperature-dependent correlation lengths in YBCO6.63, would preclude a fixed-ξ scenario.

Another troubling point about the MM discussion of YBCO6.63 is the assumption without justification that χ_0 in (5.2.5) be simply replaced by the gapped $\chi(T)$ shown in Fig. 4.16 (or its equivalent). Some kind of assumption along these lines is evidently necessary for this model, but then leads to difficulties in accounting for ${}^{89}W_\alpha(T)$ and ${}^{17}W_\alpha(T)$, as was noted in Sect. 4.2.2. This played, unfortunately, into the weakness of employing the MMP model to deduce the temperature dependences of both ξ and β from NMR data alone.

MM also offer a deep and useful discussion [162] of the behavior of $\chi''(\boldsymbol{q}, \omega)$ at higher frequencies, with an eye to gleaning further insight about its characteristics from inelastic neutron scattering (INS) data. Again, specifically, would it be possible to determine the behavior of $\xi(T)$ from the width of the peak at $\boldsymbol{Q}$? They note, with a detailed model example, that there is a tendency at typical INS frequencies for the measured peak width to appear independent of temperature while ξ is undergoing a significant variation with T_1 and that measurements must be conducted at very low energies to avoid this.

To our knowledge there has never been any successful study of the T-dependence of AFM peak widths in cuprates or of pseudogap effects on $\chi''(\boldsymbol{q}, \omega)$. If we take the INS results of Tranquada et al. [199] as an example, the desired effects are often masked by incommensurability. However, unknown at the time of MMP and MM, the most penetrating insights about pseudogap behavior in cuprates would come not from INS, but from ARPES (see a brief review of this in Sect. 4.3). This point will be considered further in Chap. 6, where extensive INS studies of LSCO will be discussed.

5.3 T_1 Estimate for YBCO Using INS Data to Model $\chi''(\boldsymbol{q}, \omega)$

In the first two subsections of this chapter we saw how both UM and MMP interpreted cuprate relaxation data in terms of AFM fluctuations with a temperature-dependent correlation length. In the present subsection we review a fit procedure performed by the Berthier group, wherein T_1 data from a YBCO6.9 sample are interpreted in terms

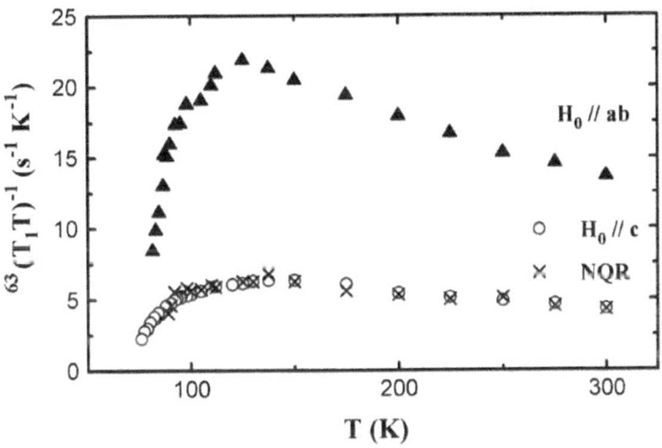

Fig. 5.7 **a** Relaxation rate $1/T_1T$ for ^{63}Cu(2) in YBCO6.9 [245], showing typical anisotropy for YBCO7, but pseudogap-like rolling off of the data below 120–130 K. Note that there is no break or other feature in the data at $T_c \sim 90$ K

of an AFM fluctuation peak characterized by INS measurements [245]. INS data are quoted which show the AFM peak to have a constant width, i.e. *the fluctuations have a temperature-independent correlation length.*[2] Here we review the data and analysis for this interesting case, which also includes extracting dynamic susceptibility parameters from data for T_{2g}, the Gaussian spin-echo decay from indirect spin-spin couplings (see Chap. 6). We compare this case with the UM fits in Sect. 5.1 in some detail. In the end, it is found that they are strikingly similar, and that the contrast between them may reflect a difference in samples rather than one of methods.

When large single crystals of YBCO7 became available, extensive neutron scattering measurements were reported, e.g. by Regnault et al. [246], Rossat-Mignod et al. [247], and by Tranquada, Sternleib et al. [199, 248]. NMR measurements on ^{63}Cu(2) and ^{89}Y in single crystals so characterized have been reported by Auler et al. [245], where they noted that the AFM peaks from INS data on their sample and related samples are well-fitted by a Gaussian, with a *peak width which is essentially independent of temperature*. The sample of YBCO6.9 in question gave some indication of disorder in spite of being nearly fully oxygenated. This was attributed to the presence of Sr^{2+} as a dilute substitutional impurity for Ba^{2+}, leading to fairly broad NMR lines with field $\perp$ c axis [249]. However, the NMR measurements were conducted normally. The other apparent consequence of this was that the sample exhibited a somewhat pseudogapped character compared with other YBCO samples with $T_c = 90 \pm 1$ K.

The pseudogap effect is noticeable in relaxation curves for ^{63}Cu(2) shown in Fig. 5.7, where the data are seen to roll off below $T \sim 125$ K for both field orientations. Furthermore, there is no visible break in the relaxation curve at T_c, which is a characteristic of pseudogapped material. On the other hand, the ratio $T_{1c}/T_{1ab} \simeq 3.7$ at $T \sim 100$ K and declines slightly at higher temperatures, which is typical behavior for YBCO7. This YBCO sample is therefore rather unique in having pseudogap features with $T_c \sim 90$ K.

[2]The MMP picture did not rule this possibility out, but in the absence of reliable INS data preferred the scenario with temperature-variable correlation lengths [162].

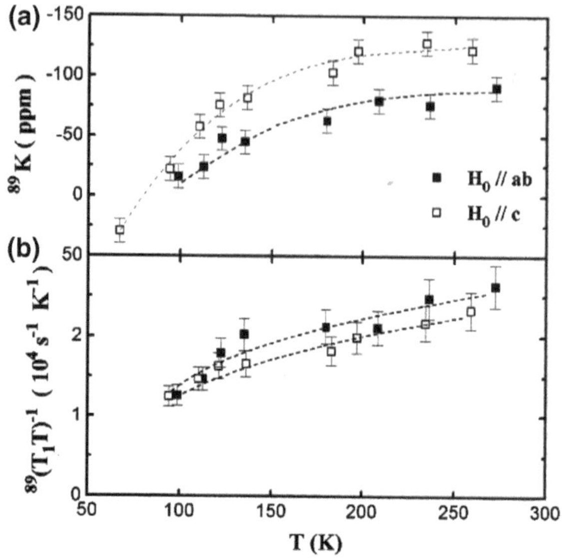

Fig. 5.8 **a** NMR shift data for ^{89}Y in YBCO6.9 from [245] for field oriented both $\parallel$ and $\perp$ c axis. $^{89}K_{\alpha}(T)$ shows pseudogap-like decline below $T \sim 200$ K. **b** Data for the ^{89}Y relaxation rate $1/T_1T$ for the same sample and conditions as in part (**a**)

Both NMR shift and relaxation data for the ^{89}Y nuclei are shown in Fig. 5.8. The NMR shift is seen to roll off at lower temperatures, much like one of the slightly deoxygenated samples in Fig. 4.2. Correspondingly, $1/T_1T$ in Fig. 5.8b is also seen to decline at lower temperatures, though not quite so much as for YBCO6.6 (see, e.g. Fig. 4.1b). Again, like the ^{63}Cu(2), the general character is that of a pseudogapped system.

5.3.1 *The Gaussian Model Susceptibility*

The authors introduce their Gaussian susceptibility model as follows [245]. Beginning with the standard form (cf. (4.2.2))

$$\left.\frac{\chi''(\boldsymbol{q},\omega)}{\omega}\right]_{\omega\to 0} = \frac{\chi'(\boldsymbol{q},0)}{\Gamma_{\boldsymbol{q}}}, \tag{5.3.1}$$

it is supposed that on account of AFM exchange forces $\chi'(\boldsymbol{q},0)$ is peaked at $\boldsymbol{q} \sim \boldsymbol{Q}$, and the spin fluctuation frequency $\Gamma_{\boldsymbol{q}}$ has a minimum at the same point. Further, it is assumed that $\chi'(\boldsymbol{q}) = \chi_{AF}(\boldsymbol{Q})f(\boldsymbol{q}-\boldsymbol{Q})$ and $\Gamma_{\boldsymbol{q}}^{-1} = \Gamma_{AF}^{-1}f(\boldsymbol{q}-\boldsymbol{Q})$ have the same functional form in the vicinity of $\boldsymbol{Q}$, which is taken to be a Gaussian

$$f(\Delta\boldsymbol{q}) = exp(-\mid \Delta\boldsymbol{q}\mid^2 \xi^2/4). \tag{5.3.2}$$

The value $\xi = 1.3a$ is taken from INS measurements on the same sample material [245]. Then, if we put the origin in $\boldsymbol{q}$-space at $\boldsymbol{Q}$, we may write $f(\Delta\boldsymbol{q}) = \exp[-0.4225\,(qa)^2]$. As in the MMP model, a noninteracting, $\boldsymbol{q}$-independent term χ_0/Γ_0 is added to give the relations

$$\left.\frac{\chi''(\boldsymbol{q},\omega)}{\omega}\right]_{\omega\to 0} = \frac{\chi_0}{\Gamma_0} + \frac{\chi'_{AF}(\boldsymbol{Q})}{\Gamma_{AF}(\boldsymbol{Q})}\exp[-0.845\,(qa)^2] \tag{5.3.3a}$$

$$\chi'(\boldsymbol{q}) = \chi_0 + \chi'_{AF}(\boldsymbol{Q})\exp[-0.4225\,(qa)^2]\,. \tag{5.3.3b}$$

The parameters in this model ((5.3.2) and (5.3.3)) were then extracted from experimental data. We give here just a brief summary of the procedure. $\chi_0(T)$ was defined to be the uniform susceptibility, the temperature-dependence of which is derived from NMR shift data in Fig. 5.8a, where the orbital shift value $^{89}K_{c,orb} = 160$ ppm is taken from [191]. The scale factor for $\chi_0(T)$ was obtained by comparing shift coefficients for ^{89}Y and ^{17}O with values from the literature [245]. The resulting curve of $\chi_0(T)$ versus T is shown in Fig. 5.9a. In that same figure is shown the curve for $\chi'_{AF}(T)$. This was extracted from data for $^{63}T_{2g}(T)$, which is shown in Fig. 5.9b, using the relations

$$\frac{1}{^{63}T_{2g}^2} = \frac{c_{63}}{8}\left[\sum_q (a^z_{ind}(\boldsymbol{q}))^2 - \left[\sum_q a^z_{ind}(\boldsymbol{q})\right]^2\right] \tag{5.3.4}$$

with

$$a^z_{ind}(\boldsymbol{q}) = \frac{\gamma_{63}\hbar\chi'(\boldsymbol{q})}{g_c^2\mu_B^2}[A_c + 4B\cos q_x a]^2\,. \tag{5.3.5}$$

These formulas are derived and discussed in Chap. 6. Here, we simply note that the T_{2g} data of Fig. 5.9b were analyzed with the foregoing equations to obtain the plot of $\chi'_{AF}(T)$ shown in Fig. 5.9a.

The final step, which is to evaluate the spin fluctuation parameters Γ_0 and Γ_{AF} as functions of temperature, requires a bit more explanation. Auler et al. [245] note that the dipolar interaction makes an appreciable contribution to $1/^{89}T_1$, which is not taken account of by the "D" terms of the HF tensor (3.3.2). This point has been noted by Takigawa et al. [191] and was discussed briefly in Sect. 4.2.2. Auler et al. provide a separate calculation to estimate the extra dipolar relaxation terms, which behave a bit like the ^{63}Cu(2) terms in that they do not cancel at $\boldsymbol{q} = \boldsymbol{Q}$. In evaluating Γ_0 and Γ_{AF} as functions of temperature, these authors present several "options" regarding the dipolar terms. We first describe their procedure for obtaining the Γ's and then describe the options.

The values of Γ_0 and Γ_{AF} are obtained by fitting the ^{63}Cu(2) and ^{89}Y T_1 data with the model susceptibility (5.3.3) with (5.3.2), using (3.3.11)–(3.3.13), i.e. the standard Mila-Rice HF Hamiltonian. As an example, these authors note that if the AFM susceptibility peak were narrow enough it would not make a significant contribution to $1/^{89}T_1T$, which would then be given by [245]

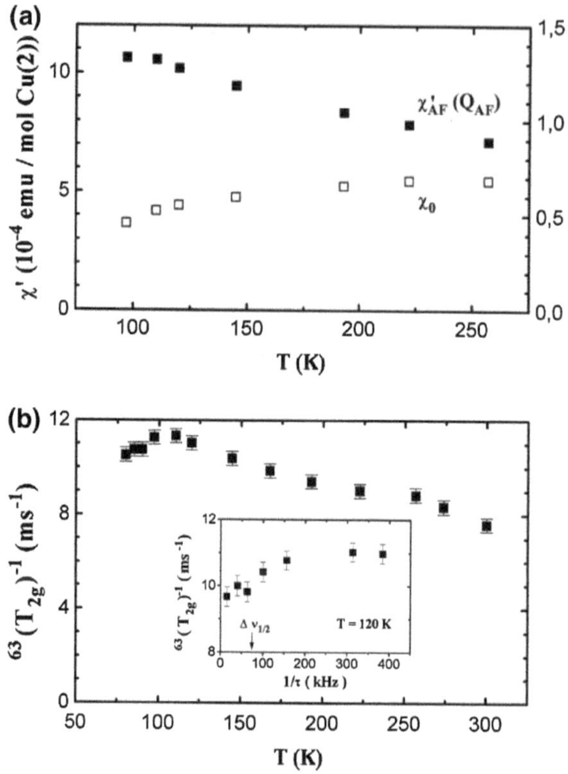

Fig. 5.9 **a** The susceptibilities χ_0 and χ'_{AF} are plotted as a function of temperature. χ_0 is the uniform susceptibility, with temperature dependence and amplitude scale derived principally from ^{89}Y NMR shift measurements as discussed in the text. χ'_{AF} is the amplitude of the AFM fluctuation peak at $\boldsymbol{Q}$. Its value as a function of temperature was derived from measurements of T_{2g} for the ^{63}Cu(2) NMR line. Equations (5.3.4) and (5.3.5) were used to interpret T_{2g} data shown in part (**b**) of this figure. See text and Chap. 6 for more details [245]. **b** Data for T_{2g} on ^{63}Cu(2) taken in the present study [245]. These data were interpreted to yield the values of χ'_{AF} shown in part (**a**) of this figure. See text and Chap. 6 for more details of interpretation. The inset shows how data vary with the bandwidth of the refocussing π-pulse of the echo sequence, where a large bandwidth is essential to obtain accurate data

$$\frac{1}{^{89}T_1 T} = \frac{2\hbar\gamma_{89}^2 k_B D_{ab}}{g_{ab}\mu_B \Gamma_0(T)}\, {}^{89}K_{ab}(T)\,, \tag{5.3.6}$$

so that $\Gamma_0(T)$ could be extracted directly from the data for $^{89}K_{ab}(T)$ and $^{89}T_1(T)$. In practice, (5.3.6) is not a good approximation, but serves to illustrate the method. In general, one has to use the data mentioned plus $^{63}T_1(T)$ to evaluate both Γ_0 and Γ_{AF} simultaneously. For this purpose they have used values of the ^{63}Cu(2) and ^{89}Y HF tensors, in units of 10^{-6} eV, as follows: $A_c = -1.56\,(-1.64)$, $A_{ab} = 0.35\,(0.16)$;

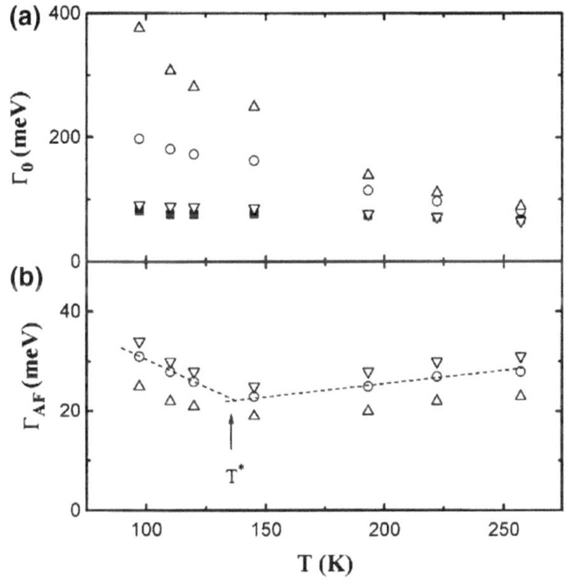

Fig. 5.10 **a** Values of the spin-fluctuation frequency Γ_0 deduced from fitting the relaxation data for both ^{63}Cu(2) and ^{89}Y to the form for $\chi''(\boldsymbol{q},\omega)$ (5.3.3) discussed in the text. Results from the oversimplified approximation in (5.3.6) are shown as solid squares. Open circles show values obtained when the dipolar terms for the ^{89}Y relaxation are taken into account, and down triangles show values obtained when they are not. The up triangles result when the dipolar terms are kept, and the value of A_{ab} is increased by 25 % over that given in the text. See text for discussion. **b** Values of Γ_{AF} (see (5.3.3)) obtained in same way as in (**a**), with symbols indicating the same cases listed there

$B = 0.39\,(0.41)$; $D_{ab} = -0.0035\,(-0.0028)$; and $D_c = -0.0043\,(-0.0035)$, where the values in parentheses are the fitted values from the UM analysis in Table 4.1 for comparison. Apart from A_{ab}, which is small and difficult to evaluate, these are all quite reasonable.

The values derived for Γ_0 and Γ_{AF} are plotted in Fig. 5.10a and b, respectively. The down triangles are obtained ignoring the dipolar terms altogether, while the open circles correspond to taking the dipolar terms fully into account. For the up triangles the value of A_{ab} was also increased by 25%. Our main concern here will be the open circles, which correspond to apportioning a sizeable fraction of the ^{89}Y relaxation to dipolar terms.

Since the susceptibility model has an AFM peak with a temperature-independent width, the temperature dependence of the T_1 data has to be accounted for by assigning the temperature variation (of $1/T_1T$) to the susceptibilities χ_0 and χ'_{AF} and to the spin-fluctuation parameters Γ_0 and Γ_{AF}. The value of Γ_0 is seen to approximately double from 250 K down to 100 K and Γ_{AF} to change by less than that. It is, of course, very difficult to know whether this is a reasonable scenario from the standpoint of microscopic physics, but it describes the data well enough. What is interesting is to compare this outcome with the results of the UM analysis, for which a temperature-dependent correlation length $\xi(T)$ was employed (Sect. 5.1).

5.3.2 Comparison with the UM Analysis Results for YBCO7

To effect a comparison with the UM analysis, we use (3.5.17) and (3.5.18) to calculate τ_{eff} and the $\mathcal{K}_{ij}$'s, since $\chi''(\boldsymbol{q},\omega)$ is fully specified by (5.3.3) and (5.3.2) with the parameter values in Figs. 5.9a and 5.10a and b (circles). For these evaluations we use the other parameter values from [245], which were stated above. We first note the contrast in parameterization of the $\mathcal{K}_{ij}$'s between UM and the present work. We adopt a general form

$$\mathcal{K}_{ij} = \mathcal{K}_0\, g(r_{ij}/\lambda)\,, \tag{5.3.7}$$

where $\mathcal{K}_0$ and λ are both possible functions of temperature. In the UM analysis, $\mathcal{K}_0$ was taken to be unity and $g(r_{ij}) = \exp(-2\,r_{ij}/a\lambda)$, with all temperature dependence in the correlation length λ. For the Gaussian $\chi''(\boldsymbol{q},\omega)$ model of Auler et al. [245], however, the width parameter ξ is fixed. The temperature variation has to reside in $\mathcal{K}_0(T)$. Thus, it is possible to have temperature varying correlations even when there is a fixed correlation length. Calculations of the $\mathcal{K}_n$'s (see below) reveal that, for example, $\mathcal{K}_1(T)$ reaches a value of -0.47 at 100 K and is about half that value at 250 K. These are typical numbers from the UM analysis.

For comparison purposes the quantities τ_{eff}(T) and the $^{63}V_\alpha$'s have been evaluated using the model susceptibility parameters given by the analysis of Auler et al. described above. First, we evaluate τ_{eff}(T) using (3.5.17) and (5.3.3), finding

$$\tau_{eff}(T) = \frac{\hbar k_B T}{4\mu_B^2 N_A}\left[\frac{\chi_0}{\Gamma_0} + 0.0936\,\frac{\chi'_{AF}}{\Gamma_{AF}}\right], \tag{5.3.8}$$

where the prefactor has been adjusted for susceptibilities in (emu/mole). Using the parameters from Figs. 5.9a and 5.10a and b (circles), we find the values plotted in Fig. 5.11a versus T (open squares). For comparison, The curves of τ_{eff}(T) for YBCO7 (open circles) and YBCO6.63 (closed circles) from the UM analyses (Figs. 5.2b and 5.4a, respectively) are plotted. Recalling that the YBCO6.63 curve was "gapped", we see that there is a remarkably close correspondence in absolute magnitudes for τ_{eff} and that the values from the present study are definitely "gapped" as well. As we noted earlier, the NMR shift and relaxation curves showed definite signs of a spin pseudogap.

Corresponding to the foregoing expression for τ_{eff}, we find for the $\mathcal{K}_n$'s,

$$\mathcal{K}_n = \frac{0.0936\,\chi'_{AF}\Gamma_0}{\chi_0\Gamma_{AF} + 0.0936\chi'_{AF}\Gamma_0}\,[-0.745;\ 0.555;\ 0.309]\,, \tag{5.3.9}$$

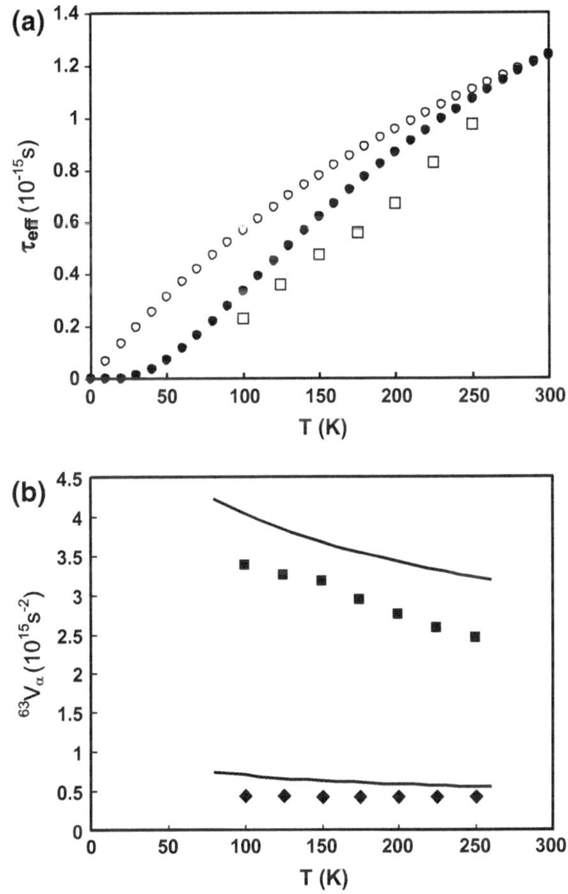

Fig. 5.11 **a** Data for τ_{eff}(T) have been calculated using (5.3.8) with interpolated data from Figs. 5.9a and 5.10a and b (open squares). These are compared with plots of τ_{eff}(T) for YBCO7 (open circles) and for YBCO6.63 (dots), which have been calculated using (5.1.7) and (5.1.9) with parameter values from Table 5.2. Comparison shows a marked similarity with a clearly gapped character for the present YBCO6.9 sample. The main difference between the present case and the UM YBCO7 fit is the marked pseudogap character of the results. **b** Plots of $^{63}V_c$ (filled squares) and $^{63}V_{ab}$ (filled diamonds) for the same series of temperatures shown for τ_{eff} in part (**a**). These quantities have been calculated with the HF parameter values from [245] given in the text using (5.3.9). Shown for comparison are the same quantities from the YBCO7 fit by UM (Fig. 5.3a) (solid lines), where the discrepancy in both cases is a scale factor of order unity

where the numbers in the brackets represent $\mathcal{K}_1$, $\mathcal{K}_2$, and $\mathcal{K}_3$, respectively. To plot $^{63}V_c$ and $^{63}V_{ab}$, this expression was used with interpolated data from Figs. 5.9a and 5.10a and b to calculate the $^{63}V_\alpha$'s shown in Fig. 5.11b (filled squares for $^{63}V_c$ and filled diamonds for $^{63}V_{ab}$). For comparison the same quantities for YBCO7 from the UM analysis (Fig. 5.3a) are plotted as solid lines. Again, considering that these materials are quite different in detail, the correspondence is remarkably close. Notice that the three $\mathcal{K}_n$'s are in a constant ratio of −0.745:0.555:0.309 at all temperatures, but their coefficient changes by about a factor two over $100\,\mathrm{K} \leq T \leq 250\,\mathrm{K}$.

It seems, then, that there may be a variety of parameterizations which will give the same form for τ_{eff}(T) and ${}^kV_\alpha(T)$, which is all that is needed to fit the relaxation data. If follows that it will be very difficult to work backward from such data to try to determine the details of some particular parameterization, as was the case with MMP. In first-principles model theories such as those in the next two subsections, however, we suggest that it would be very useful and informative to calculate both τ_{eff} and the $\mathcal{K}_n$'s as diagnostics.

5.4 A Small-U Hubbard Model of Cuprate Spin Dynamics

Extensive model calculations of NMR shift and relaxation behavior have been performed by Bulut et al. [158], using, among others, the 2D Hubbard model to represent a cuprate-like system with enhanced magnetic fluctuations near the AFM wavevector $\boldsymbol{Q}$. These calculations with the Hubbard model bear some similarity to the mean-field phenomenological model of MMP, the difference being that one is actually doing a first-principles calculation of quasiparticle dynamics using a simplified Hamiltonian and approximate methods. The up side of this is that the results of such calculations have been adapted to realistic cuprate (i.e. Mila-Rice-Shastry) HF couplings and form factors to obtain shift and relaxation behavior which is not unlike typical experimental data. If there is a down side, it is that the 2D Hubbard model, as solved in this way, presents some rather quirky behavior with certain parameters which have no clear relation to microscopic physics. There is also some "implausible fine tuning" of parameters to obtain desired behavior. The authors argue that this is simply a necessary aspect of using such a Hubbard model.

In this section we present a brief summary of Hubbard model behavior and some calculated results for the NMR shift and relaxation of ^{63}Cu(2) and ^{17}O(2,3) sites of YBCO7. Behavior at $T < T_c$ has also been treated in detail with these methods. The latter results will be presented in Sects. 5.7 and 6.2, along with other model results for the superconducting state.

5.4.1 *Basic Formulation of the Dynamic Susceptibility*

The following is a summary and discussion of results for the square-lattice Hubbard model of a doped system with metallic behavior. For derivations the reader is referred to the original papers [158] and references therein. The basic Hamiltonian is

$$\mathcal{H} = -t \sum_{\langle ij \rangle \alpha} [c^+_{i\sigma} c_{j\sigma} + c^+_{j\sigma} c_{i\sigma}] + U \sum_i n_{i\uparrow} n_{i\downarrow} \,, \tag{5.4.1}$$

consisting of nearest-neighbor hopping terms and on-site coulomb repulsion, where U is actually a renormalized interaction parameter. The basic (U = 0) band profile has width $W = 8t$ with a logarithmic singularity at the center. The band filling is $0 \leq n \leq 2$ particles per site. Near half filling there are commensurate and incommensurate spin-density wave phases, which must be avoided. For the assumed ratio $U/W = 0.25$, the critical value is $n_c = 0.865$, above which there is an "unphysical phase transition" for the Hubbard model where $U\chi(\boldsymbol{q}^*, 0) = 1$. The amount by which n falls below n_c determines the temperature T^* at which the amplitude of AFM fluctuations saturates on lowering the temperature. Thus, n is only qualitatively related to the doping level parameter of the cuprate being modeled.

The basic solutions to the Hubbard model (5.4.1) are as follows: The band energy is $\mathcal{E}(\boldsymbol{k}) = -2t(\cos k_x a + \cos k_y a)$ and the single-particle density of states is

$$N(\mathcal{E}) = \frac{1}{2\pi^2} K[(1 - (\mathcal{E}/4)^2)^{1/2}], \tag{5.4.2}$$

where K[ ] is the complete elliptic integral. The U = 0 (noninteracting) susceptibility is then given by

$$\chi_0(\boldsymbol{q}, \omega) = \frac{1}{N} \sum_{\boldsymbol{p}} \frac{f(\mathcal{E}_{\boldsymbol{p}+\boldsymbol{q}}) - f(\mathcal{E}_{\boldsymbol{p}})}{\omega - (\mathcal{E}_{\boldsymbol{p}+\boldsymbol{q}} - \mathcal{E}_{\boldsymbol{p}}) + i0^+}, \tag{5.4.3}$$

where $f(\mathcal{E}_{\boldsymbol{p}}) = 1/[\exp(\beta(\mathcal{E}_{\boldsymbol{p}} - \mu)) + 1]$ is the Fermi occupation function and μ is the chemical potential.

For the interacting system the authors use the RPA to approximate $\chi(\boldsymbol{q}, \omega)$, giving

$$\chi(\boldsymbol{q}, \omega) = \frac{\chi_0(\boldsymbol{q}, \omega)}{1 - U\chi_0(\boldsymbol{q}, \omega)}, \tag{5.4.4}$$

closely analogous to (5.2.1). However, while (5.2.1) gave a peak near $\boldsymbol{Q}$ which dominated the behavior of interest or could be modeled with something else, here it was pointed out that the RPA approximation (5.4.4) does not accurately render the detailed dependences of $K(T)$ and $1/T_1$ on U and on the band filling n [158]. The best one can hope for is to choose U and n appropriately to extract the essential features of large-$\boldsymbol{q}$ spin fluctuations.

If we choose a band filling n near to but less than n_c, then AFM fluctuations will grow with declining temperature down to some point T^* where their amplitude saturates. T^* is controlled by the value of n. By choosing $U/W = 0.25$ and $n = n_c$, the authors presented results for a case with large AFM enhancement effects and a peak in $\chi''(\boldsymbol{q}, \omega)$ near (but not precisely at) $\boldsymbol{Q}$. The calculated NMR properties then give for the shift at low temperatures

$$K = \frac{2\gamma_e}{\gamma_n} A(0) \frac{N(\mu)}{1 - UN(\mu)}, \tag{5.4.5}$$

for a HF interaction parameter $A(\boldsymbol{q})$, and for the relaxation rate

$$\frac{1}{T_1} = \frac{k_B T}{\hbar N} \sum_{q} A(\boldsymbol{q})^2 \frac{\chi''(\boldsymbol{q}, \omega_0)}{\omega_0}, \tag{5.4.6}$$

where $A(\boldsymbol{q})$ has the dimensions of frequency and $\chi''(\boldsymbol{q}, \omega)$ of time (see (3.5.8)), and where the actual form of $A(\boldsymbol{q})$ varies from site to site as usual. With the RPA susceptibility (5.4.4), (5.4.6) can be rewritten

$$\frac{1}{T_1} = \frac{k_B T}{\hbar N} \left\langle \frac{|A(\boldsymbol{p} - \boldsymbol{p}')|^2}{[1 - U\chi_0(\boldsymbol{p} - \boldsymbol{p}')]^2} \right\rangle, \tag{5.4.7}$$

where the brackets indicate that $\boldsymbol{p}$ and $\boldsymbol{p}'$ are to be summed over the Fermi surface.

The foregoing formalism has been used to calculate approximations to the NMR shift and relaxation behavior for ^{63}Cu(2), ^{17}O(2,3), and ^{89}Y in YBCO7. For this purpose calculations on the simple square-lattice Hubbard model have been generalized to include the effects of Cu-O hybridization in a Hubbard model of CuO_2 [158]. With a 3-center unit cell, the CuO_2 lattice is modeled using "partial susceptibilities", with coulomb repulsion only on the Cu site [158]. T_1 calculations were presented with and without this modification and, remarkably, the differences in T_1 behavior were found to be fairly subtle. Here we present some of the results for each model.

The NMR shifts at all sites undergo some Stoner enhancement according to (5.4.5), with very little temperature dependence. In Figs. 5.12 and 5.13, some of the T_1 results are displayed, to be compared with experimental data for YBCO7. The bandwidth W is set at 1.2 eV, with $U/W = 0.70$, and $n = 0.865$, very near to the critical value. The field is placed along the c axis. An on-site HF coupling form has been used for both the Cu(2) and O(2,3) for the first two results shown. In Fig. 5.12a, the relaxation curves for Cu(2) and O(2,3) are compared, where each is plotted as a fraction of its value at $T/W = 0.025$, which is approximately room temperature. The Cu(2) relaxation rate is seen to level off completely by room temperature, with rather more downward curvature than exhibited by the data (see Fig. 3.7), while the O(2,3) data also exhibit a slight downward curvature which is not unlike the experimental data (Fig. 4.23a). It seems that this extreme rollover of the Cu(2) relaxation is necessary in order to achieve the desired level of enhancement. The enhancement is shown for both Cu(2) and O(2,3) in Fig. 5.12b, where the Korringa products $(T_1TK^2)^{-1}$ are plotted versus T as fractions of the noninteracting values $(T_1TK^2)_0^{-1} = [(4\pi\hbar/k_B)(\gamma_n/\gamma_e)^2]$. This gives an enhancement of $\simeq$4 for the Cu(2) and slightly greater than 1 for the O(2,3). The authors quote experimental values of 5.5 for Cu(2) and 1.4 for O(2,3) [156], in reasonable accord with their calculations. The Cu(2) enhancement, however, has declined to near the noninteracting value by just above room temperature, which is rather unlike the experimental results.

Finally, we show in Fig. 5.13 a comparison of Hubbard model calculations with experimental data for ^{63}Cu(2) and ^{17}O(2,3) in YBCO7. For these calculations, the simple Hubbard model results were used [158] with realistic Mila-Rice-Shastry HF

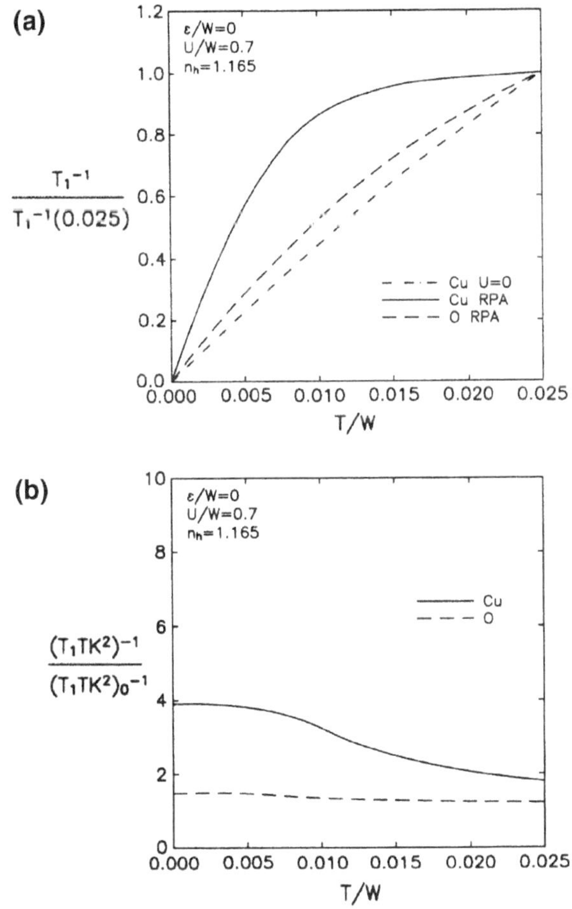

Fig. 5.12 **a** Relaxation rates $1/T_1$ calculated for the Cu(2) and O(2,3) sites in YBCO7 and scaled to their values at $T/W = 0.025$ (approx. 300 K) are plotted versus T/W. Calculations are made with the CuO_2 model described in the text and references, using on-site HF couplings. The U = 0 (dashed) curves are also shown for reference. Conditions shown are $\mathcal{E} = 0$, $U/W = 0.7$, and $n_h = 2 - n = 1.165$, where $\mathcal{E}$ is the energy of the O(2pσ) orbitals with $\mathcal{E}_d = 0$, and n_h is somewhat above the critical value $n_h = 1.135$. **b** The Korringa ratios for both Cu(2) and O(2,3) are plotted as a function of T/W in units of their noninteracting fermion values $(4\pi\hbar/k_B)(\gamma_n/\gamma_e)^2$. The low-T enhancement ~ 4 for the Cu(2) has decayed most of the way to its noninteracting, high-T limit by 300 K

coupling form factors (i.e. $| A_{ab}/A_c |= 0.21$ and $| B/A_c |= 0.22$), where we note that A_{ab} is a bit larger here than usual estimates. In this case the Hubbard model parameters were also slightly different from the foregoing, with $U/W = 0.25$ and $n = 0.86$ slightly below the critical value. Taken with the others, these calculations show that a fair range of T_1 behaviors can be obtained with this model.

5.4.2 Remarks on the Hubbard Model Calculations

One must, of course, evaluate the merits of this model in terms of the time frame of its development, in which case it was obviously highly successful. It was a first-principles model for the spin dynamics of superconducting cuprates which gave a remarkable facsimile of experimental behavior with a single dynamical variable. The model was also an important test for the use of suitable form factors to shield the O(2,3) sites from enhanced AFM fluctuations while the Cu(2) sites are very directly

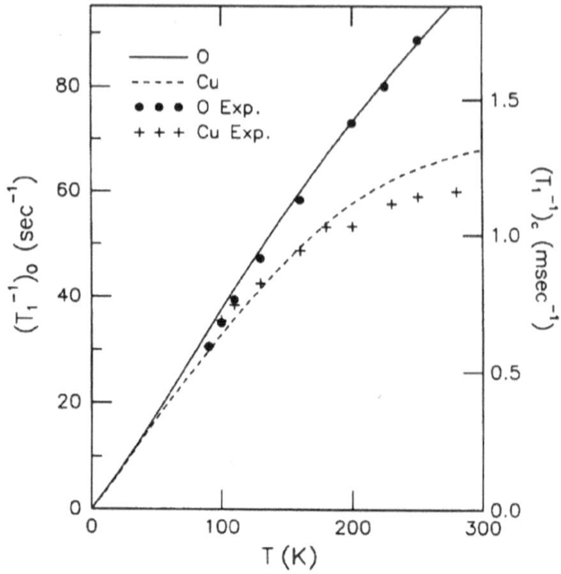

Fig. 5.13 Small-U Hubbard model calculations, similar to those for Fig. 5.12, have been applied to fitting T_1 data, where the Mila-Rice HF tensor given in the text has been used. Calculated T_1 curves have been fitted to ^{63}Cu(2) (+) and ^{17}O(2,3) (•) data curves from [156] and [243], respectively. The conditions are slightly different from Fig. 5.12, with $U/W = 0.25$ and $n = 0.86$

affected by them. That said, the authors note that this model probably cannot be made to give a truly quantitative account of the NMR data in the normal state. The physics of the high-temperature region is quite different from an actual cuprate, which typically approaches the degeneracy temperature not far above room temperature. The Hubbard model spin dynamics will clearly be Korringa-like up to any experimentally attainable temperature, while it seems that the cuprates are making a transition to insulator-like dynamics at high temperatures. The latter result is illustrated by the behavior of τ_{eff}(T) for several systems in Sect. 5.1.

In spite of the apparent drawbacks, however, the authors went on to treat relaxation in the superconducting state with the Hubbard model, with very interesting results. The latter results are discussed in Sects. 5.7 and 6.2.

5.5 The Large-U Hubbard Model of Si, Levin et al.

An RPA model of cuprate quasiparticle dynamics, which is of the same class of models as the MMP mean-field model [161, 162] and the small-U Hubbard model of Bulut et al. [158], has been developed and discussed extensively by Si, Levin, and co-workers [253–256]. This is a 3-band Hubbard model with U essentially infinite, wherein the nearest-neighbor AFM exchange from combined O^{2-}-mediated superexchange plus RKKY sources is incorporated via the RPA technique. Thus it is similar to the foregoing models, except that the authors report that they have microscopically derived the Lindhard (noninteracting) susceptibility. As we shall see, this is not only a first-principles calculation, but one with entirely realistic physical parameters. Here we will comment very briefly on their results, and discuss a comparison with INS data and with T_1 curves for ^{63}Cu and ^{17}O nuclei in the CuO_2 planes. We suggest below, however, that further calculations with this theory would be very interesting.

The interesting difference between this model and those mentioned in previous subsections is that on a decrease in doping level, the system nears a *localization transition* rather than a magnetic (AFM) one. Moreover, owing to strong coulomb correlation effects, the renormalized antibonding band narrows as the insulating limit is approached. Thus, the principal bandwidth parameter $t_{pd} \rightarrow t_{pd}(1 - n_d)^{1/2}$, where Mott localization occurs as the electron concentration $n_d \rightarrow 1$. The latter effect, then, lowers the *coherence temperature* at which Fermi liquid behavior gives way to temperature-independent relaxation. In cases under consideration, this has been estimated to occur at a few hundred Kelvin [253–255]. Evidence for such effects is clearly present in the Uldry-Meier analysis of cuprate T_1 data in Sect. 5.1. For example, in Table 4.2 the effective degeneracy temperature for $YBa_2Cu_4O_8$ (Y248) is seen to be as low as 420 K. It would be very interesting if the large-U, tight-binding, Hubbard model calculation could account for this effect.

Extensive calculations of dynamical susceptibilities have been made with the method described [256]. These results may then be compared with INS data and with T_1 data for both the ^{63}Cu and ^{17}O nuclei in the CuO_2 plane. Plots of the structure factor $S(\boldsymbol{q}, \omega)$ (i.e. $\chi''(\boldsymbol{q}, \omega)$) for two different scans in ***q****-space* (see figure caption) are shown in Fig. 5.14a for YBCO7 and b for YBCO6.7.[3] These results were calculated with the expression

$$\chi_{rr'}(\boldsymbol{q}, \omega) = \chi^0_{rr'}(\boldsymbol{q}, \omega) + \frac{\chi^0_{rd}(\boldsymbol{q}, \omega)[-J_H(\boldsymbol{q}]\chi^0_{dr'}(\boldsymbol{q}, \omega)}{1 + J_H(\boldsymbol{q})\chi^0_{dd}(\boldsymbol{q}, \omega)}, \tag{5.5.1}$$

where the $\chi_{rr'}$'s are "partial susceptibilities" [257], with $r = p$ or d, $\chi^0_{rr'}(\boldsymbol{q}, \omega)$ is the microscopically-derived Lindhard function [257], and $J_H(\boldsymbol{q})$ is the composite Cu-Cu exchange coupling, which is modeled here with

$$J_H(\boldsymbol{q}) = J_0[cos\, q_x a + cos\, q_y a]. \tag{5.5.2}$$

There is a critical value $J_0 = J_c$, above which $\chi_{rr'}(\boldsymbol{q}, \omega)$ is unstable. For YBCO7, J_0 was taken to be $0.5J_c$ and for YBCO6.7 $J_0 = 0.7J_c$, while the doping for these two was $x = 0.18$ and 0.36, respectively. For other details we refer the reader to [256] and [257].

Referring to the theoretical scan results in Fig. 5.14, plots of $\chi''(\boldsymbol{q}, \omega)$ are shown for $T \sim 12$ K and $T \sim 120$ K. The AFM peak is commensurate and noticeably narrower for YBCO6.7, as expected. Although the authors emphasize the constancy of these results with temperature [256], the peaks in all cases seem somewhat lower and wider at 120 K, this effect being incrementally greater for YBCO6.7. One can only imagine that at 300 K the peaks would be even wider, giving a palpable temperature dependence to the correlation lengths. Comparing these results with the INS data for reduced oxygen YBCO given by Stock et al. in Fig. 4.21 [201], there is some

[3]Similar results were also presented for LSCO. These are invoked in a discussion of that system in Chap. 6.

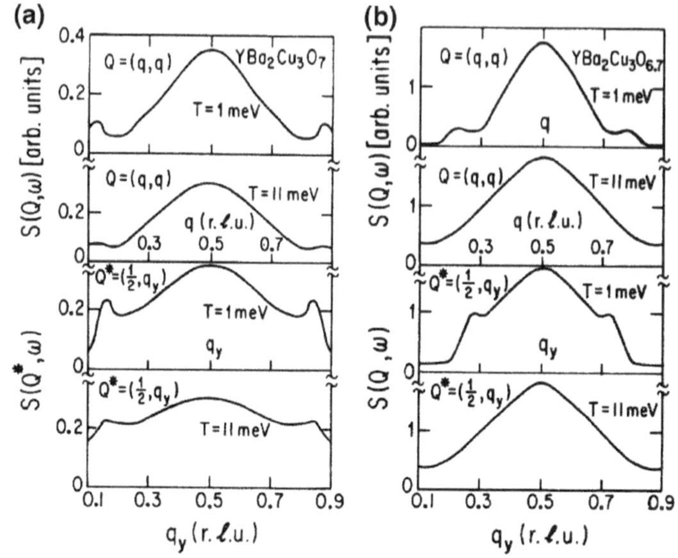

Fig. 5.14 **a** Cross sections of S($\boldsymbol{q}$, ω) for YBCO7 are shown for $\omega = 15$ meV, first along the diagonal at $T = 1$ meV and 11 meV, and then along the q_y axis for $\boldsymbol{q} = (\frac{1}{2}, q_y)$ for the same two temperatures. **b** Cross sections of S($\boldsymbol{q}$, ω) for YBCO6.7 are shown for the identical set of conditions as in part (**a**). All results were calculated as described in [256]

difference in peak shape, with the INS scan distinctly narrower. The frequency of the measurement was $\omega = 12.4$ meV, while that of the calculation is 10 meV. It was reported that the INS $\chi''(\boldsymbol{q}, \omega)$ peak width was independent of temperature [201]. However, there was some evidence for inhomogeneous broadening of the latter peak, casting some doubt on that result. See Sect. 4.2 for further discussion.

The authors also use their results for $\chi''(\boldsymbol{q}, \omega)$ to simulate relaxation rate curves for ^{63}Cu and ^{17}O as a function of temperature, using the same HF tensors as those employed by Bulut et al. in [158]. The results are shown for both compositions in Fig. 5.15 as temperature plots up to 360 K [256]. We make the following remarks about the nature and significance of these results.

- Comparing these plots with experimental data (see Fig. 4.23), certain trends are seen to be qualitatively correct. The deviation from Korringa behavior is greater for the ^{63}Cu than for the ^{17}O, and greater for the YBCO6.7 than for the YBCO7.
- The amplitude trends are dramatically different from the experimental data, where for ^{63}Cu, the experimental curves are intertwined so that YBCO6.7 rises above YBCO7 at $T \sim 150$ K, but falls below YBCO7 at room temperature. For ^{17}O the curves are reversed, where experimentally (Fig. 4.23a), $1/T_1T$ for YBCO7 is relatively flat and lying above the curve for YBCO6.7, which descends to a small fraction of its room temperature value at $T = T_c \sim 60$–65 K.
- The latter circumstances arise, in part, because YBCO6.7 exhibits a spin pseudo-gap effect in which the susceptibility, as well as $1/T_1T$ for the ^{17}O, decline at

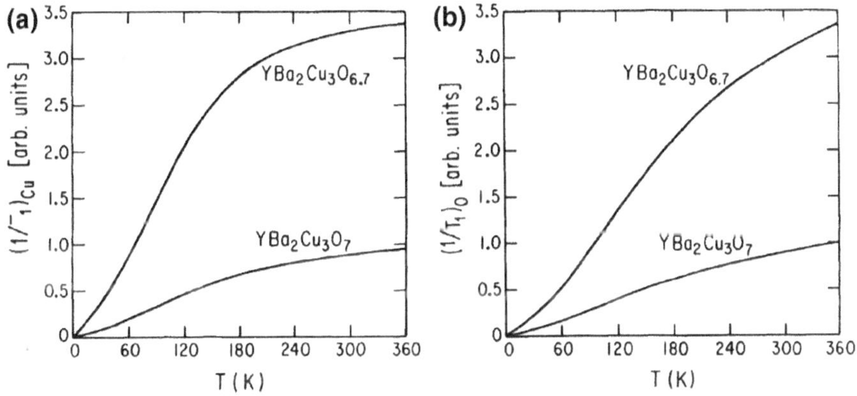

Fig. 5.15 **a** Relaxation rates $1/T_1$ for ^{63}Cu, calculated for both YBCO7 and YBCO6.7 using the results for $\chi''(\boldsymbol{q}, \omega)$ shown in Fig. 5.14 and the HF tensor given by Bulut et al. [158], are plotted as a function of temperature [256]. **b** Similar calculated results are plotted versus T for the case of planar ^{17}O, using the same resources

low temperatures $\propto \exp(-\Delta/k_BT)$, where $\Delta/k_B = 104\,\text{K}$ (see Sect. 4.1.2). The Hubbard model discussed here [256, 257] does not have this feature, so that comparison with YBCO6.7 data, especially at low temperatures, is not meaningful. The only successful model of low-frequency spin fluctuations for the case of a pseudogap is that of Varma [222].

- The authors have some concern that the ^{17}O results are not as Korringa-like as the experimental data for YBCO7 and/or are not compatible with their calculation for ^{63}Cu. The straightforward answer to this is found in the Uldry-Meier analysis of T_1 data, presented in Sect. 5.1. This type of analysis was not, of course, available in the early 1990's when the Hubbard-model calculations were being developed. The main significance of the UM analysis is that it separates the dynamical spin-spin correlations ($\mathcal{K}_n$ from (2.5.17)–(2.5.18)) from the quasiparticle dynamical function (2.5.17) $\mathcal{X}_{at}(T) = \sum_q [\chi''(\boldsymbol{q}, \omega)/\omega]_{\omega\to 0}$. (This quantity is called $\tau_{\mathit{eff}}(T)$ in the UM analysis.) Given a theory for $\chi''(\boldsymbol{q}, \omega)$, then $\mathcal{K}_n$ and $\mathcal{X}_{at}(T)$ are straightforward quantities to evaluate. In the latter formulation, the ^{63}Cu and ^{17}O curves contain the same dynamical information, but very different correlation effects. The latter may be seen in Fig. 5.3a and b. The experimental dynamical function for YBCO7 is shown in Fig. 5.2b.
- Si et al. attributed the strong curvature of the ^{63}Cu $1/T_1$ curve to a *coherence temperature* [253]. This may be something of a misapprehension, since the curve is strongly affected by T-dependent spin-spin correlations. The true picture of dynamical behavior with temperature would be given by $\mathcal{X}_{at}(T)$ versus T. It would be most interesting to see whether this theory would provide an explanation for the flattening off of dynamical fluctuations at high temperatures, *since at the present time there is no theory for this available.* The large-U RPA model seems especially

suited to this, because the band-narrowing effect with hole dilution [254] is similar to what is observed in practice (see Table 4.2).

- We further suggest that the reason for the strong deviation from Korringa-like behavior for the ^{17}O in Fig. 5.15b is that the corresponding peak in Fig. 5.14 has much wider wings than the INS-measured peak in Fig. 4.21. This situation is a bit like the Lorentzian versus Gaussian models by MM [162], where the Gaussian model with steeper sides gives far less contribution to the $^{17}O(2,3)$ relaxation rate. The downward curvature of the plot of $1/T_1$ for the ^{17}O is probably mainly due to the dynamical correlation factor, leaving the impression that $\mathcal{X}_{at}(T)$ may be nearly linear in T.
- Nonetheless, an actual evaluation of $\mathcal{X}_{at}(T)$ for different carrier levels would answer a very interesting question about the large-U Hubbard model, namely whether it can account for the high-temperature transition in the dynamics seen in the experimental data, such as Figs. 5.2b and 5.4b.

5.6 Relaxation via Orbital Currents Proposed by Varma

Up to this point only spin-paramagnetic T_1 processes have been considered for the doped cuprate superconductors. However, in 1989 a somewhat radical idea was proposed for these systems, namely that for a certain class of charge excitations the polarizability has a term which obeys [159]

$$Im\tilde{P}(\boldsymbol{q},\omega) \sim \begin{cases} -N(0)(\omega/T)\ for\ \ |\ \omega\ | < T \\ -N(0)\ sgn(\omega)\ for\ \ |\ \omega\ | > T \end{cases}, \tag{5.6.1}$$

where N(0) is the one-particle density of states. Applying this marginal Fermi liquid (MFL) *ansatz* to magnetic processes at NMR frequencies $\omega_0 \ll k_B T/\hbar$, we have

$$\chi''_{mfl}(\boldsymbol{q},\omega) = \frac{\hbar^2 N(0)\omega_0}{k_B T}, \tag{5.6.2}$$

where we have added the factor $\hbar^2/k_B$ to make clear that this is dimensionally a correlation time.

To apply this *ansatz* to an HF perturbation process, we suppose that there is a quasiparticle interaction term $\gamma_n H_{mfl} \boldsymbol{I}\cdot\boldsymbol{f}(\boldsymbol{q})$ with a (dimensionless) $\boldsymbol{q}$-dependence factor $\boldsymbol{f}(\boldsymbol{q})$. Then, by the fluctuation-dissipation theorem ((3.5.8)–(3.5.9)) this generates a temperature-independent T_1 process

$$\frac{1}{T_1} = \frac{4\gamma_n^2 H_{mfl}^2 k_B T}{\hbar}\sum_{\boldsymbol{q}} f_\perp(\boldsymbol{q})^2 \frac{\chi''_{mfl}(\boldsymbol{q},\omega_0)}{\omega_0} = 4\hbar\gamma_n^2 H_{mfl}^2 N(0)\sum_{\boldsymbol{q}} f_\perp(\boldsymbol{q})^2, \tag{5.6.3}$$

where the second step used (5.6.2), and $\hbar N(0) \sim \hbar/E_F$ is a correlation time. Such processes in metals would be very strong. It has already been noted that such a dynamic would not apply to spin fluctuation processes in cuprates.

However, it was subsequently proposed by Varma [244] that the MFL scenario could be applied to magnetic fluctuations which result from circulating orbital currents, yielding a T_1 process for the Cu(2) sites. Here we review the arguments leading to an estimate given for ${}^{63}T_{1orb}$ [244]. The proposed T_1 mechanism would operate via the orbital HF interaction (see (2.2.1))

$$\mathcal{H}_{orb} = \boldsymbol{I} \cdot \boldsymbol{M}_{orb} = i\hbar \boldsymbol{I} \cdot \boldsymbol{L}/r^3 = i\hbar \boldsymbol{I} \cdot \boldsymbol{r} \times \nabla / r^3 , \tag{5.6.4}$$

where the local angular momentum operator $\boldsymbol{L}$ is associated with the Cu(2) site. $\mathcal{H}_{orb}$ could then act to scatter quasiparticles from $\boldsymbol{k} + \boldsymbol{q}/2$ to $\boldsymbol{k} - \boldsymbol{q}/2$, where the squared matrix element for the relaxation process would be summed over $\boldsymbol{k}$ and $\boldsymbol{q}$. However, since the occupied $3d$ wavefunction in the cuprates is thought to be purely $d_{x^2-y^2}$, the orbital HF matrix element

$$\langle \boldsymbol{k} + \boldsymbol{q}/2 \mid \boldsymbol{M}_{orb} \mid \boldsymbol{k} - \boldsymbol{q}/2 \rangle = \hbar \langle \mathrm{e}^{-i\boldsymbol{q}\cdot\boldsymbol{r}} \mathcal{U}^*_{\boldsymbol{k}+\boldsymbol{q}/2}(\boldsymbol{r}) \frac{\boldsymbol{L}}{r^3} \mathcal{U}_{\boldsymbol{k}-\boldsymbol{q}/2}(\boldsymbol{r}) \rangle , \tag{5.6.5}$$

($\exp[i(\boldsymbol{k} + \boldsymbol{q}/2) \cdot \boldsymbol{r}]\mathcal{U}_{\boldsymbol{k}+\boldsymbol{q}/2}(\boldsymbol{r})$ is a typical Bloch function) will have no intrasite contribution $\sim \hbar f_{orb} \gamma_n \mu_B \langle r^{-3} \rangle$, where f_{orb} is the orbital symmetry factor (see Fig. 2.9) and $\langle r^{-3} \rangle \sim 6$ a.u. for $3d$ electrons. Thus, for a $d_{x^2-y^2}$ ground state $f_{orb} = 0$. It is for this reason that the conventional orbital T_1 mechanism, which is often predominant in d-band metals, is absent in the cuprates.

The leading contribution to (5.6.5) will come from cross terms between the Cu(2) site and its oxygen neighbors. To show how this develops, the behavior of $\langle \boldsymbol{k} + \boldsymbol{q}/2 \mid \boldsymbol{M}_{orb} \mid \boldsymbol{k} - \boldsymbol{q}/2 \rangle$ was examined [244] for $\boldsymbol{k} \sim 0$ and small values of $\boldsymbol{q}$. In this limit, $\mathcal{U}_{\boldsymbol{k}+\boldsymbol{q}/2}(\boldsymbol{r}) \sim \mathcal{U}_0(\boldsymbol{r})$ can be written

$$\mathcal{U}_0(\boldsymbol{r}) = \frac{1}{\sqrt{N}} \sum_{i,\alpha} \mathcal{A}_{i,\alpha}(0)\, \phi(\boldsymbol{r} - \boldsymbol{R}_{i,\alpha}) , \tag{5.6.6}$$

where α is summed over the three sites in the CuO_2 unit cell, and i is summed over cells. The local wavefunctions are $\phi(\boldsymbol{r} - \boldsymbol{R}_{i,\alpha}) \sim d_{x^2-y^2}$ in a Cu(2) site and $\sim p_\sigma = p_{x,y}$ on the oxygen neighbors. The phase factors $\mathcal{A}_{i,\alpha}(0) = \pm 1$ must be adjusted appropriately (see [244], Fig. 1).

The principal HF fluctuations from orbital currents will be along the c axis, which we label z. Since $L_z \phi_d(\boldsymbol{r}) \sim d_{xy}$ for $\phi_{d_{x^2-y^2}}$ at the origin, one must find d_{xy}-like symmetry on the oxygen neighbors. For small $\boldsymbol{q}$ the relevant term (5.6.5) becomes

$$\langle \boldsymbol{k} + \boldsymbol{q}/2 \mid M_{orb} \mid \boldsymbol{k} - \boldsymbol{q}/2 \rangle \simeq i\hbar \sum_n \boldsymbol{q} \cdot \langle \boldsymbol{r} \phi_p(\boldsymbol{r} - \boldsymbol{R}_n) \frac{L_z}{r^3} \phi_d(\boldsymbol{r}) \rangle , \tag{5.6.7}$$

where n is summed over the neighbors. One can then choose the term from $\boldsymbol{r}\phi_p(\boldsymbol{r} - \boldsymbol{R}_n)$ which has xy symmetry to obtain a non-zero matrix element. There will evidently be a non-zero contribution from larger values of $\boldsymbol{q}$ as well. On the other hand it is argued that there is no corresponding non-zero Cu-O term if the origin is placed on the oxygen site. This is fairly obvious, because at $\boldsymbol{r} \sim \hat{i}x$, $L_z p_x \sim p_y$, and xp_y is orthogonal to $d_{x^2-y^2}$. So the orbital currents can produce a z-axis T_1 fluctuation, and possibly through the apical oxygens some smaller in-plane fluctuations at the Cu(2) site, but not at the O(2,3) sites.

To estimate the strength of the corresponding T_1 process, we can write the orbital dynamical structure factor, in analogy with (3.5.6)–(3.5.8), as

$$\frac{1}{T_{1orb}} = \frac{2k_BT}{\hbar\omega_0}\frac{\gamma_{63}^2\mu_B^2}{\hbar^2}\sum_{\boldsymbol{q}}\langle M_{orb}(\boldsymbol{q})M_{orb}(-\boldsymbol{q})\rangle_{\omega_0} = \frac{k_BT}{\hbar\omega_0}\,{}^{63}A_{orb}^2\sum_{\boldsymbol{q}}(qa)^2\chi''_{mfl}(\boldsymbol{q},\omega_0)\,, \tag{5.6.8}$$

where ${}^{63}A_{orb}$ is $\gamma_{63}\mu_B/\hbar$ times the intersite matrix element from the right-hand side of (5.6.7). To obtain the final expression for T_{1orb} the MFL *ansatz* correlation time (5.6.2) is modified by replacing the effective quasiparticle scattering rate $k_BT/\hbar$ in the denominator by $k_BT/\hbar + 1/\tau_0$, where $1/\tau_0$ is a disorder-generated scattering rate of the order of tens of degrees [244]. With this change, (5.6.8) then gives

$$\frac{1}{T_{1orb}} = \hbar\,{}^{63}A_{orb}^2N(0)\frac{k_BT}{k_BT + \hbar/\tau_0}\,. \tag{5.6.9}$$

In applying this to cuprate relaxation data, we note that for $T > 100\,\text{K}$, (5.6.9) assumes a nearly constant value $1/T_{1orb} \simeq \hbar\,{}^{63}A_{orb}N(0)$. The HF parameter ${}^{63}A_{orb}$ is very difficult to estimate from (5.6.7). The main contribution will come from the overlap region between the Cu(2) and O(2,3) sites, and is probably 1–2 orders of magnitude smaller than the $3d$ on-site orbital HF field (see Sect. 2.3.3 and Table 2.4) $H_{orb} = 2\mu_B\langle r^{-3}\rangle_{3d} \sim 125\,kOe/\mu_B$. From the UM analysis of Sect. 5.1 (Table 5.2) we estimate $\hbar N(0) \sim \hbar/k_BT_F \simeq 8\times 10^{-15}s$ for YBCO7. In this case, taking an optimal symmetry factor $f_{orb} = 0.6$, we find

$$\left.\frac{1}{T_{1orb}}\right]_{opt} = f_{orb}\gamma_{63}^2H_{orb}^2\hbar N(0) = 3.8\times 10^3\,s^{-1}. \tag{5.6.10}$$

This is $\sim$3 times the measured relaxation rate for YBCO7 at $T = 100\,\text{K}$. The rate in (5.6.9) is therefore estimated to be of the order of 10% of that at most.

Interestingly, the combination of (5.6.9) and spin-paramagnetic processes which vary as $1/T_1 \propto T$ have been shown to give a good account of ^{63}Cu(2) relaxation data for both YBCO7 and optimally-doped LSCO. These results are shown in Fig. 5.16, where they are compared with data from [143] and [239]. For the calculated curves in this figure $1/\tau_0$ is taken to be 25 K and 100 K for YBCO7 and LSCO, respectively. The quantity $\hbar\,{}^{63}A_{orb}^2N(0)$ is taken to be $\sim 3\times 10^3\,s^{-1}$, which is very near the estimated upper limit for an orbital contribution.

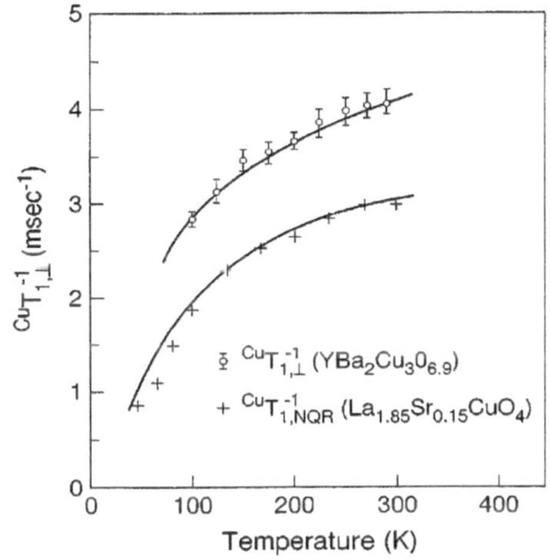

Fig. 5.16 Orbital relaxation rates estimated with (5.6.9) are compared with experimental data for YBCO7 [143] and optimally-doped LSCO [239] as shown. For the plotted curves $^{63}A^2_{orb}\,\hbar N(0)$ is taken to be $3200\,s^{-1}$ and $2800\,s^{-1}$, and $1/\tau_0$ is taken to be 25 K and 100 K for YBCO7 and LSCO, respectively. See text for discussion of these calculations

A final point has to do with how the orbital relaxation would fit into the scheme of T_1 analysis described in earlier subsections of this chapter. Since the UM analysis in Sect. 5.1, for example, is internally consistent and is consistent with the Mila-Rice, spin Hamiltonian parameters derived from shift measurements in Chaps. 3 and 4, a contribution from $1/T_{1orb}$ at the 10% level would pose no problem. To accommodate a larger contribution than that would pose something of a challenge to the internal consistency of our current thinking on T_1 in the cuprates.

The presence of orbital current fluctuations in cuprates has now been verified by neutron scattering studies [223], so there is no doubt about their existence. Further, they appear to be closely associated with the pseudogap effect. Whether these fluctuations are observable via NMR (other than the well-known NMR pseudogap effects reviewed in Chap. 4) remains to be established. An important theoretical step toward further elucidation would be a serious effort to evaluate the intersite orbital matrix element in (5.6.7).

5.7 Relaxation of Planar ^{63}Cu and ^{17}O at $T < T_c$

In Sect. 3.6 we saw how a straightforward adaptation of the RPA susceptibility model to the BCS formulation of nuclear relaxation by Monien and Pines [160] gave a good account of the NQR $T_1(T)$ for ^{63}Cu(2) at $T < T_c$ (YBCO7). Following that pioneering work there were important developments on the experimental side. As shown in Fig. 3.11b, there had been a serious discrepancy between high-field NMR T_1 data below T_c and that obtained with NQR. One suspected that, as was found for the V_3X compounds in Sect. 2.2.5 (see Fig. 2.13), relaxation by spin diffusion to the normal cores becomes appreciable in high fields. Thus, the precise scaling between

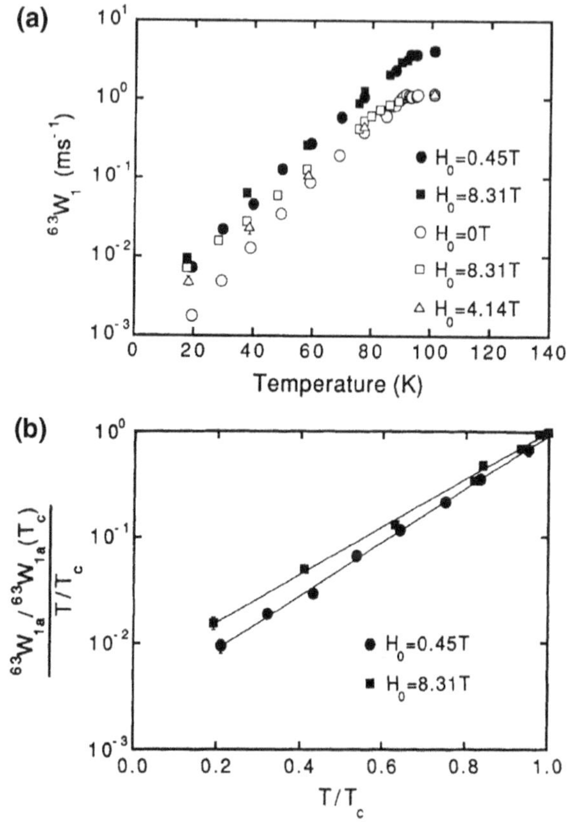

Fig. 5.17 ^{63}Cu(2) relaxation data from the work of Martindale et al. [258]. **a** Semilog plot of $1/T_1$ for ^{63}Cu(2) ($^{63}W_1$) versus T in the range $0 \leq T \leq 100$ K for YBCO7. Data are presented for different values of applied field, including NQR data for zero field. The open symbols show data with field (i.e., quantization axis) along the c axis, while filled symbols are for field $\perp$ the c axis. **b** Semilog plot of $1/T_{1ab}T$ for ^{63}Cu(2) ($^{63}W_{1a}/T$), normalized to its value at T_c, as a function of $T \leq T_c$. This plot illustrates the exponential behavior, as well as the weak field dependence of $1/T_{1ab}T$ below T_c

T_1 curves for ^{63}Cu(2) and ^{17}O(2,3) observed by Hammel et al. is strongly dependent on the applied field [156].

In an effort to determine the intrinsic relaxation times for ^{63}Cu(2), a careful study of the field dependence of T_1 below T_c in YBCO7 using oriented powder samples was conducted by Martindale et al. [258]. The results are as follows: For field (i.e. quantization axis) along the c axis, T_1 was measured in 8.31 T, 4.14 T, and 0 T (NQR). For field $\perp$ c axis, T_1 was measured in 8.31 and 0.45 T. Measurements were conducted at a series of temperatures $20\,\text{K} \leq T \leq 100$ K, with the results plotted semilogarithmically in Fig. 5.17a. For $H_0 \parallel c$ axis, the variation of T_1 with field is substantial, where at 20 K an 8 T field increases the relaxation rate by a factor ~5. The field dependence with the field in the ab plane is measurable, but much smaller.[4]

[4]The authors refer to the conclusion by Redfield and Genack [109] that relaxation by diffusion to the normal cores in the vortex state is not physically viable. Further analysis of this problem by Furman and Goren seems to have restored the validity of such a T_1 process [110]. Experimental evidence for it is, of course, very strong.

In the remainder of the discussion, then, NQR data or data taken in very low applied field will be preferred. In Fig. 5.17b $1/^{63}T_{1ab}T$, scaled to its value at T_c, is plotted semilogarithmically versus T. It is seen to behave very nearly as

$$\frac{1}{^{63}T_{1ab}T} \simeq \mathrm{e}^{-b(1-T/T_c)}, \qquad (5.7.1)$$

with $b \sim 6$. The c axis data exhibits similar behavior, except that it begins a bit more slowly and the coefficient b is slightly larger. Over a decline in relaxation rates of a factor ~300, the ratio $^{63}T_{1ab}^{-1}/^{63}T_{1c}^{-1}$ behaves as shown in Fig. 5.19a. To discuss the curves and data presented there, however, we must first review the calculations reported by Bulut and Scalapino [251, 252].

5.7.1 $\chi''(\mathbf{q}, \omega)$ *for the Small-U Hubbard Model at T < T$_c$*

The calculations by Bulut et al. [158] of T_1 behavior in the normal state based on

$$\chi(\boldsymbol{q}, \omega) = \frac{\chi_0(\boldsymbol{q}, \omega)}{[1 - U\chi_0(\boldsymbol{q}, \omega)]}, \qquad (5.7.2)$$

where U is the coulomb repulsion renormalized due to particle-particle correlations, is discussed at some length in Sect. 5.4. This theory was fitted to relaxation data by choosing $U = 2t$ and band occupation number $\langle n \rangle = 0.86$. Their treatment of the superconducting state with these same parameter values [251, 252] is discussed in this subsection.

To model the dynamic susceptibility for $T < T_c$, these authors used the RPA form (5.7.2) with the "irreducible BCS susceptibility" given by [252]

$$\begin{aligned}
\chi_0(\boldsymbol{q}, \omega) = \\
&\frac{1}{N}\sum_p \left[\frac{1}{2}\left[1 + \frac{\mathcal{E}_{p+q}\mathcal{E}_p + \Delta_{p+q}\Delta_p}{E_{p+q}E_p}\right] \frac{f(E_{p+q}) - f(E_p)}{\omega - (E_{p+q} - E_p) + i\Gamma}\right.\\
&+ \frac{1}{4}\left[1 - \frac{\mathcal{E}_{p+q}}{E_{p+q}} + \frac{\mathcal{E}_p}{E_p} - \frac{\mathcal{E}_{p+q}\mathcal{E}_p + \Delta_{p+q}\Delta_p}{E_{p+q}E_p}\right] \frac{1 - f(E_{p+q}) - f(E_p)}{\omega + (E_{p+q} + E_p) + i\Gamma}\\
&\left.- \frac{1}{4}\left[1 + \frac{\mathcal{E}_{p+q}}{E_{p+q}} - \frac{\mathcal{E}_p}{E_p} - \frac{\mathcal{E}_{p+q}\mathcal{E}_p + \Delta_{p+q}\Delta_p}{E_{p+q}E_p}\right] \frac{1 - f(E_{p+q}) - f(E_p)}{\omega - (E_{p+q} + E_p) + i\Gamma}\right] \qquad (5.7.3)
\end{aligned}$$

In (5.7.3) the dispersion relation is $E_p = (\mathcal{E}_p^2 + \Delta_p^2)^{1/2}$ and $\mathcal{E}_p = -2t\,(\cos p_x a + \cos p_y a) - \mu$, where μ is the chemical potential and Δ_p is the gap function. For Δ_p we will consider here the case of a d-wave form

$$\Delta_p(T) = [\Delta_0(T)/2](\cos p_x a - \cos p_y a), \qquad (5.7.4)$$

where a BCS temperature dependence is adopted for $\Delta_0(T)$. There is a finite broadening parameter $\Gamma(T)$ to be specified below. Finally, the superconducting transition temperature was taken to be $T_c = 0.1\,t$, where at $T = T_c$ the ^{63}Cu(2) relaxation rate has approximately the correct value and degree of AF enhancement to correspond to YBCO7 at 100 K.

In [252] this formalism was used to treat and discuss both s-wave and d-wave gap symmetries with different choices of parameter values. Since this work was completed several years before d-wave symmetry was confirmed for the cuprates through the observation of phase jumps at the boundaries of Josephson junctions [172, 173], at that time there were still many proponents of s-wave pairing. We shall not dwell on the s-wave case here, but will point out how convincingly the relaxation behavior argues in favor of d-wave pairing in this theory. The early work of Monien and Pines [160] discussed in Sect. 3.6 also weighed heavily on the side of d-wave.

We first look at the behavior of T_1 as treated with this formalism [252]. The usual expression (3.5.9)

$$\frac{1}{T_1} = \frac{k_B T}{4\hbar} \sum_{\boldsymbol{q}} \mid A(\boldsymbol{q}) \mid^2 \frac{\chi''(\boldsymbol{q}, \omega_0)}{\omega}\Bigg]_{\omega\to 0} \tag{5.7.5}$$

is employed with the Mila-Rice-Shastry HF tensors (3.3.11). To see how this model distinguishes s-wave and d-wave behavior, it is useful to consider how the quantity

$$\frac{\chi''(\boldsymbol{q}, \omega)}{\omega}\Bigg]_{\omega\to 0} = \frac{\chi_0''(\boldsymbol{q}, \omega)/\omega]_{\omega\to 0}}{[1 - U\chi_0(\boldsymbol{q})]^2} \tag{5.7.6}$$

behaves with temperature in different regions of $\boldsymbol{q}$-space. For the s-wave case, $\chi_0(\boldsymbol{q}, 0)$ declines for all $\boldsymbol{q}$ values below T_c, so that the enhancement factor $[1 - \chi(\boldsymbol{q}, 0)]^{-2}$ will decline for all HF tensors [252]. In contrast, for the d-wave case, $\chi_0(\boldsymbol{q}, 0)$ is only suppressed for $\mid \boldsymbol{q} \mid \leq 1/\xi_{sc}$, where ξ_{sc} is the superconducting correlation length. The region around $\boldsymbol{Q} = (\pi, \pi)$ is not suppressed because of the nodes in the gap. Thus, AFM fluctuations persist into the SC state for the d-wave case.

The other factor present in (5.7.6) is

$$D_0(\boldsymbol{q}) = \frac{\chi_0''(\boldsymbol{q}, \omega)}{\omega}\Bigg]_{\omega\to 0}. \tag{5.7.7}$$

Figure 5.18a is a plot of how $D_0(\boldsymbol{q})$ varies with temperature for a very small value of $\boldsymbol{q}$, and in part (b) it is shown how it varies for $\boldsymbol{q}$ near to $\boldsymbol{Q}$. In the s-wave case the two are very nearly the same, but for d-wave they are quite different. This sets up a diagnostic where if we look at the ratio of T_1 values for sites which depend very differently on $\boldsymbol{q}$, there will be a large expected difference between predicted behavior for s-wave and d-wave.

Comparison of RPA Calculations with T_1 Data

The authors have calculated and discussed T_1 behavior for a number of cases of parameter values with both s-wave and d-wave gap symmetry. The contrast between these two symmetries is best displayed by taking the ratio of T_1 values for cases which have different degrees of dependence on fluctuations near $\boldsymbol{Q}$, as shown in Fig. 5.18. As was clearly illustrated by Takigawa et al. [259], T_{1ab} for ^{63}Cu(2) is most strongly dependent on AFM fluctuations, with a lesser dependence for T_{1c}, and very nearly zero dependence for the oxygen ^{17}O(2,3) relaxation T_{1O} (for any axis).

To illustrate the foregoing effect, calculations and experimental data were presented for T_{1c}/T_{1ab} (^{63}Cu(2) in YBCO7) [251], which we reproduce in Fig. 5.19a. Over a range of ~3 orders of magnitude in relaxation rates, the variation of T_{1c}/T_{1ab} is relatively slight, but has a definite signature. The data shown are the low-field measurements of Martindale et al. [258] (squares) reviewed above, and similar low-field measurements of T_{1ab} combined with low-field NMR measurements (filled circles) and NQR (H = 0) measurements (open circles) of T_{1c} [259], which give equivalent results. The theory curves plotted for s-wave ($2\Delta_0/k_BT_c = 4$ (dotted line) and 8 (dashed line)) are seen to drop slightly, then remain flat, since there is no real distinction between the $D_0(\boldsymbol{q})$ curves (Fig. 5.18). For the d-wave case (solid curve), however, the drop and then rise reflect clearly the difference in behavior of the respective $D_0(\boldsymbol{q})$ curves in Fig. 5.18a and b. The observed ratio in Fig. 5.19a is a very clear sign in favor of d-wave gap behavior, which appeared several years before d-wave symmetry was widely accepted for the cuprates.

The ratio T_{1O} for ^{63}Cu(2) to T_{1c} in Fig. 5.19b is also clearly supportive of d-wave symmetry. The message here is somewhat muted, owing to the fact that no low-field

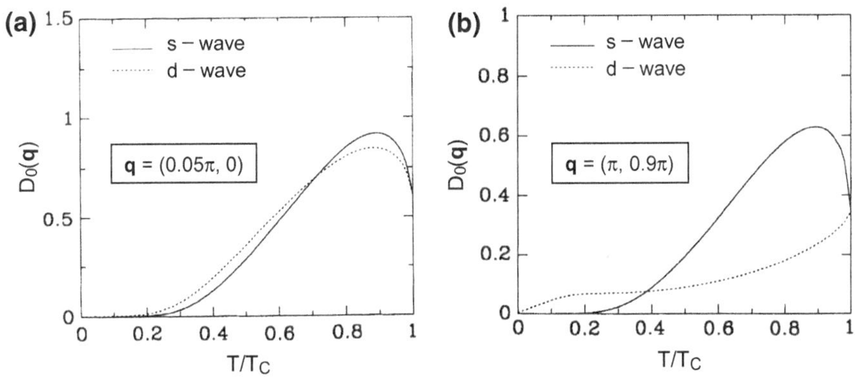

Fig. 5.18 **a** The quantity $D_0(\boldsymbol{q})$, defined in (5.7.7), characterizes the variation in intensity of low-energy fluctuations over the Brillouin zone. In part (**a**) this quantity is plotted versus T/T_c at $\boldsymbol{q} = (0.05\pi, 0)$ for s-wave and d-wave gaps as shown. Here and in part (**b**), $2\Delta_0 = 3.52k_BT_c$, and broadening $\Gamma = 0.3T_c$ has been used to control the coherence factor singularity. **b** The quantity $D_0(\boldsymbol{q})$ is plotted for the same parameters as in part (**a**), except that $\boldsymbol{q} = (\pi, 0.9\pi)$. A sharp contrast between s-wave and d-wave behavior is seen to prevail

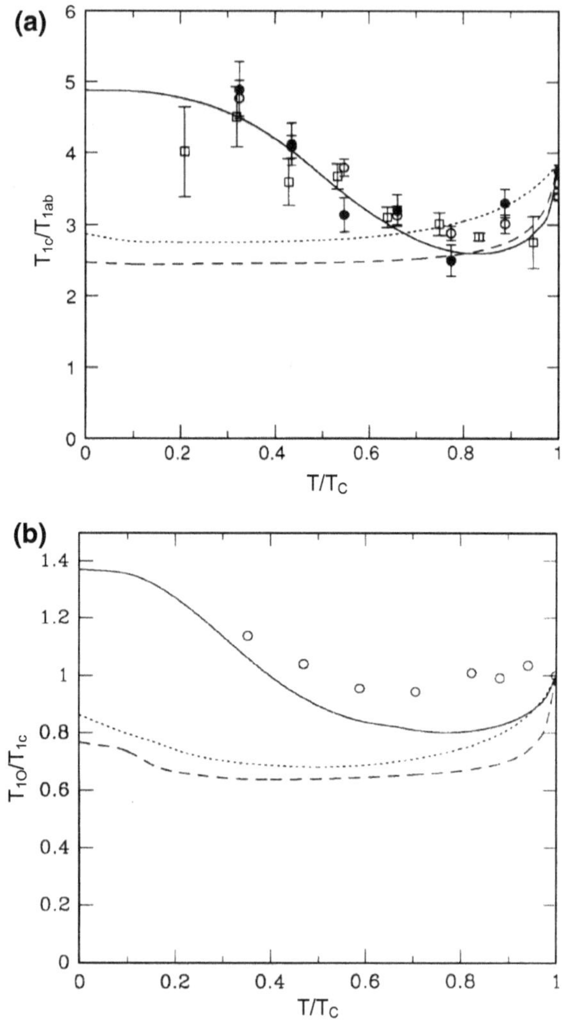

Fig. 5.19 **a** For ^{63}Cu(2) in YBCO7, a plot of the ratio T_{1c}/T_{1ab} versus T/T_c for temperatures in the superconducting state. The square symbols represent low-field data from Fig. 5.17 [258]. The circles are data from Takigawa et al. [259], where the open circles represent T_{1c} data taken using NQR (H = 0) and the closed circles represent data taken in a c-axis field of 0.44 T. The data for T_{1ab} were taken in a transverse field of 0.44 T. See the text for an explanatory note on these measurements. The solid line is the calculated curve for a d-wave gap as described in the text, using $2\Delta_0/k_BT_c = 8$. The other lines represent calculations with an s-wave gap, assuming $2\Delta_0/k_BT_c = 4$ (dotted) and 8 (dashed), respectively. **b** The ratio of the ^{17}O(2,3) relaxation time T_{1O} and T_{1c} (for ^{63}Cu(2)) is plotted versus T/T_c as in (**a**), but with plotted data scaled to their value at T_c. The data are from Hammel et al. [156] and were measured in fields of 7–8 T. The calculated curves correspond to the same set of cases as in (**a**)

^{17}O(2,3) T_1 data were available at the time, so the entire data base for the latter plot is 7–8 T NMR data [156]. The latter circumstance would only affect behavior at the low-temperature end, however, so the comparison with theory is for the most part justified.

It is interesting to note that a simple modification of the analysis would make the displayed effect in Fig. 5.19a somewhat more dramatic. This is suggested by the method of T_1 measurement employed by Takigawa et al. [259], in which they write

$$\frac{1}{T_{1ab}} = J_{ab} + J_c; \quad \frac{1}{T_{1c}} = 2\,J_{ab}\,, \tag{5.7.8}$$

where the J_α's are terms corresponding to HF fluctuations along the α axis.[5] With field applied $\perp$ to the NQR principal axis to measure T_{1ab}, a complex set of terms occurs in the relaxation process. Thus, $J_c(T)$ and $J_{ab}(T)$ are actually measured separately. Compared with T_{1c}/T_{1ab}, a plot of J_{ab}/J_c [259] would show a more striking contrast between s-wave and d-wave.

Finally, it is interesting to simply compare an RPA calculation of $T_{1c}(T)$ for $T < T_c$ [252] with NQR data attributed to Takigawa et al. [259]. These results are shown in Fig. 5.20a, where the experimental points are open circles and the calculation is shown as a solid line. The data points are seen to drop very steeply below T_c, then approach a T^3 behavior (dashed line), as expected for a d-wave gap. The RPA calculation gives a very good account of this behavior [252].

RPA Theory for the NMR Shift Below T_c

Bulut and Scalapino also calculated the NMR shift behavior with their model [251, 252], which is simply given by

$$K_s(T) \propto \chi(\boldsymbol{q} \rightarrow 0, 0) = \frac{\chi_0(\boldsymbol{q} \rightarrow 0, 0)}{1 - U\chi_0(\boldsymbol{q} \rightarrow 0, 0)}\,. \tag{5.7.9}$$

There is also a temperature-independent Van Vleck shift term, which renders the zero of $K_s(T)$ a bit uncertain. This has been subtracted off in the data plot of Fig. 5.20b. The data shown are for ^{63}Cu(2) by Barrett et al. [147] (crosses) and ^{17}O(2,3) by Takigawa et al. [178, 260] (circles). All data are scaled to their total estimated excursion between T_c and 0 K. Theoretical curves are shown for U = 2t and several values of the d-wave gap parameter $2\Delta_0/k_B T_c$. As the authors note [252], the lowest-temperature value is very important here. If the curves had been least-squares fitted to the two sets of data, it appears that the dashed curve ($2\Delta_0/k_B T_c = 6$) would have been an excellent fit.

The distinction between $K_s(T)$ for s-wave and that for d-wave is that for d-wave the curve approaches $T = 0$ with a finite slope, but with zero slope for s-wave [252]. The data are certainly consistent with a finite slope, although the scatter is large

[5] This is the same procedure as that used in the UM analysis of Sect. 5.1, where the U_α's defined there are the same as the present J_α's.

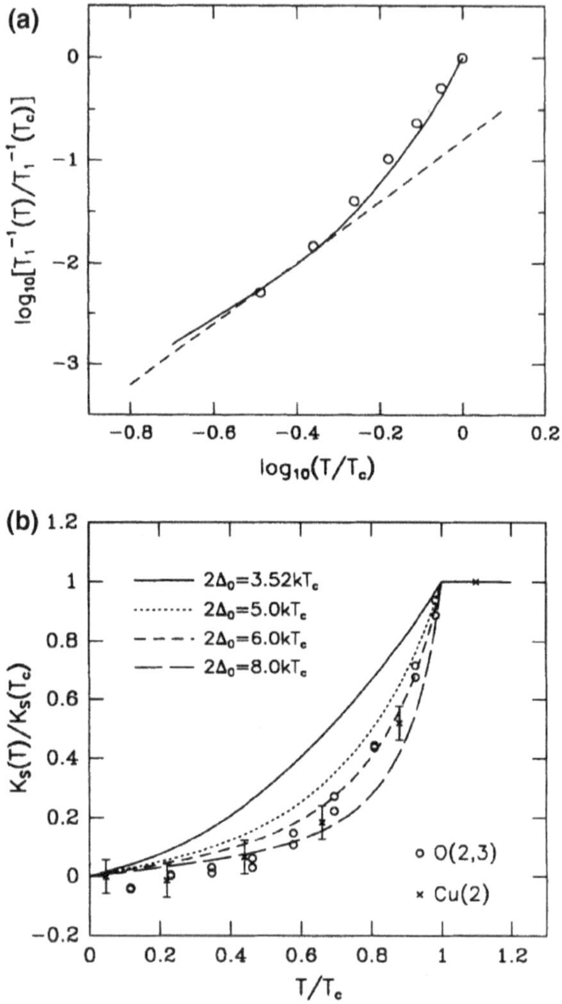

Fig. 5.20 **a** The c-axis relaxation rate $1/T_{1c}$ for ^{63}Cu(2), scaled to its value at T_c, is plotted logarithmically versus T/T_c. The data are attributed to Takigawa et al. [259]. The relaxation rate drops sharply below T_c, then falls very nearly as T^3 (dashed line). The solid curve shows the RPA theory calculation for the d-wave case with U = 2t and $2\Delta_0/k_BT_c = 8$. **b** The behavior of the NMR shift $K_s(T)$, scaled to its value at T_c, is plotted for $0 \leq T/T_c \leq 1$. The crosses with error bars are ^{63}Cu(2) data from Barrett et al. [147], and the open circles are ^{17}O(2,3) data from Takigawa et al. [178, 260]. The lines plotted are calculations for a d-wave gap with U = 2t and various values of Δ_0 as shown

enough so that s-wave is not ruled out. All in all, the shift results are a nice complement to the T_1 data. The two together make a very strong case for d-wave symmetry.

5.7.2 T_1 and Spin-Spin Correlations Below T_c: The UM Analysis

In this final subsection we consider the behavior of T_1 below T_c from another standpoint, one which does not employ any particular microscopic model of the spin dynamics, but relies only on the general formulation of the T_1 process involving

$\chi''(\boldsymbol{q}, \omega)$ and the Mila-Rice HF Hamiltonian (3.3.2) (see Sect. 3.5.3). This, of course, is the basic premise of the Uldry-Meier analysis of cuprate T_1 data presented in Sect. 5.1. These authors and co-workers have also employed this approach to discuss the behavior of spin-spin correlations above and below T_c and have come to two remarkable conclusions [261]. First, they find that in-plane spin-spin correlations vanish as $T \to 0$ in a fashion roughly symmetrical to how they vanish at high temperatures. Secondly, they give evidence that it is only the in-plane correlations which vanish, while the c-axis correlations remain basically unchanged below T_c.

To review the UM formulation, relaxation of the kth nuclear spin species quantized along the α axis is written (see Sect. 5.1)

$$\frac{1}{{}^kT_{1\alpha}(T)} = [{}^kV_\beta(T) + {}^kV_\gamma(T)]\,\tau_{eff}(T)\,, \tag{5.7.10}$$

where (α, β, γ) are a permutation of the axes (a, b, c) and ${}^kV_\alpha(T)$ is expressed in terms of the HF constants and the correlation functions $\mathcal{K}_n^\alpha(T)$. For example, for ^{17}O(2,3) one has

$${}^{17}V_\alpha(T) = \frac{1}{2\hbar^2} C_\alpha^2\,[1 + \mathcal{K}_1^\alpha(T)]\,, \tag{5.7.11}$$

where we emphasize that the $\mathcal{K}_n^\alpha$ were allowed to be anisotropic. For YBCO7 it was found, for example, that the $\mathcal{K}_n^c$ were ~20% larger than the $\mathcal{K}_n^{ab}$ (see Fig. 5.2a).

Behavior of ab Plane Correlations

In this same notation we also have (5.1.4)

$${}^{63}V_\alpha(T) = \frac{1}{4\hbar^2}[A_\alpha^2 + 4\,B^2 + 8\,A_\alpha B\mathcal{K}_1^\alpha(T) + 8\,B^2\mathcal{K}_2^\alpha(T) + 4\,B^2\mathcal{K}_3^\alpha(T)]\,. \tag{5.7.12}$$

Thus,

$$\frac{{}^{17}T_{1c}}{{}^{63}T_{1c}} = \frac{2\,{}^{63}V_{ab}(T)}{{}^{17}V_a(T) + {}^{17}V_b(T)} \underset{\mathcal{K}_n \to 0}{\Longrightarrow} \frac{A_{ab}^2 + 4\,B^2}{C_a^2 + C_b^2} \tag{5.7.13}$$

is purely a function of in-plane correlations, approaching the value on the right as they vanish.

Several sets of low-field data for ${}^{17}T_{1c}/{}^{63}T_{1c}$ are plotted in Fig. 5.21a as a function of T/T_c for $0 \le T/T_c \le 4.5$ for YBCO7 [206, 238, 263, 264] (see figure caption for details) and for Y248 [218]. Since the only temperature dependence for this ratio comes from the $\mathcal{K}_n(T)$'s, what the figure shows is that correlations are small at the highest temperature, grow as T is reduced until they reach a peak value $\mathcal{K}_1^{ab}(T_c) \sim -0.4$, and then decline again toward zero at low temperatures. The dashed line in this plot shows the value of ${}^{17}T_{1c}/{}^{63}T_{1c}$ for $\mathcal{K}_n^{ab} = 0$ (5.7.13), evaluated with numbers from the UM fit to YBCO7 data (Table 5.1). This line is seen to be in very good agreement with the high and low-temperature limiting values of the experimental T_1 ratio.

The shaded triangles in Fig. 5.21a show the same plot for Y248, taken from the results of Bankay et al. [218]. There we see that the correlation effects are more than

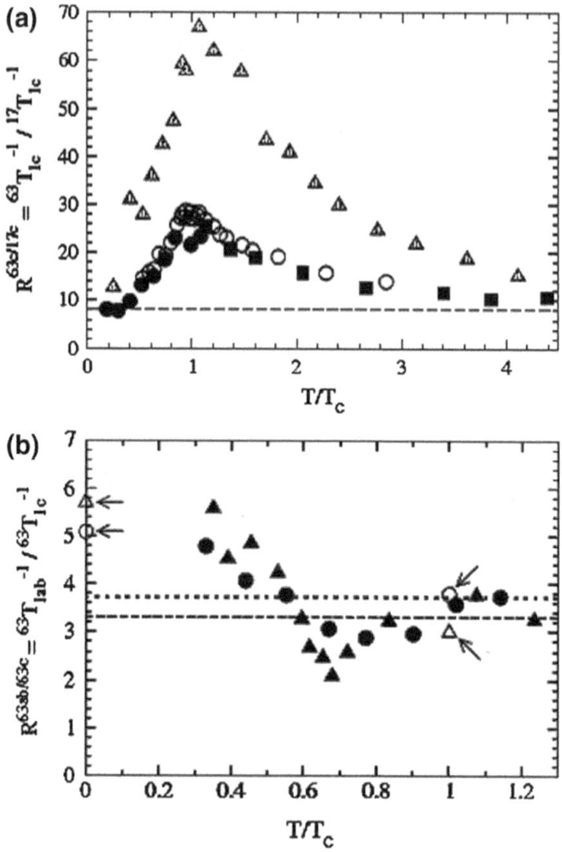

Fig. 5.21 **a** The ratio of $^{17}O(2,3)$ to $^{63}Cu(2)$ T_{1c} values is plotted versus T/T_c for YBCO (squares and circles) and for Y248 (shaded triangles). For YBCO7 the open circles are high-field data from Yoshinari et al. [238], and filled circles are low-field data from Martindale et al. [263]. The filled squares denote data points where $^{63}Cu(2)$ data from Barrett et al. [264] are combined with $^{17}O(2,3)$ data from Nandor et al. [206]. The Y248 data are high-field $^{17}O(2,3)$ data combined with NQR data for $^{63}Cu(2)$, both from Bankay et al. [265]. The dashed line is the model prediction for vanishing transverse AFM correlations ($\mathcal{K}_1^{ab} = 0$). **b** The ratio T_{1c}/T_{1ab} for $^{63}Cu(2)$ is plotted versus T/T_c for YBCO7 (filled circles) and for Y248 (filled triangles). The YBCO7 data are low-field data from Takigawa et al. [259] and the Y248 data are low-field data from Bankay et al. [265]. The dotted and dashed lines are normal state ratios from the same sources for YBCO7 and Y248, respectively. The open symbols are corresponding predictions from UM model calculations at T_c and at 0 K, which are obtained as described in the text

twice as great as in YBCO7 and do not seem to be very much affected by the spin pseudogap, which has quite a powerful effect on τ_{eff} for this compound. It follows that the $\mathcal{K}_n^{ab}(T)$'s are not much affected.

Behavior of c-axis Correlations

The foregoing results show that the ab-plane correlations $\mathcal{K}_n^{ab}(T)$ are strongly peaked at T_c in these systems and vanish at low temperatures $T \ll T_c$. However, if one looks at the ratio T_{1c}/T_{1ab} for ^{63}Cu(2), the value of T_{1ab} is strongly dominated by c-axis fluctuations, for which $^{63}V_\alpha(T)$ is markedly dependent on $\mathcal{K}_n^c(T)$. What they found is that $^{63}V_\alpha(T)$ behaves as though the $\mathcal{K}_n^c(T)$ do not change appreciably below T_c. Data for T_{1c}/T_{1ab} versus T/T_c are plotted in Fig. 5.21b, showing the variation of this ratio for both YBCO7 and Y248 [218]. Both sets of data were measured in low magnetic fields. In the normal state T_{1c}/T_{1ab} is fairly flat for both of these compounds, with a value $\sim$3.7 for YBCO7 and $\sim$3.3 for Y248. Below T_c the data decline slightly, then rise to new maximum values, Y248 slightly higher than YBCO7.

To discuss these data we use

$$\frac{T_{1c}}{T_{1ab}} = \frac{1}{2}\left[1 + \frac{^{63}V_c(T)}{^{63}V_{ab}(T)}\right] . \tag{5.7.14}$$

Data for T_{2G}, to be defined and discussed in Chap. 6, suggest that c-axis correlations do not change much below T_c [261]. If that were true, then the T_1 ratio in (5.7.14) will have the value $\frac{1}{2}[1 + {}^{63}V_c(T_c)/{}^{63}V_{ab}(T_c)]$ at T_c and the value $\frac{1}{2}[1 + {}^{63}V_c(0)/{}^{63}V_{ab}(T_c)]$ at $T = 0$. For YBCO7, these values are obtained from the fits and parameter plots from the original UM paper [195] (see Table 5.1 and Fig. 5.3). The resulting ratio values are plotted in Fig. 5.21b (open circles), where they are seen to be in good agreement with the YBCO7 data.

To evaluate T_{1c}/T_{1ab} for Y248 one needs a special procedure, because the magnitude of the correlations at T_c is not known from experimental data. The asymptotic, high-temperature value of $T_{1c}/T_{1ab} \simeq 3.3$ for ^{63}Cu(2) was obtained from the measurements of Bankay et al. [265]. However, it seems unlikely that this corresponds to $\mathcal{K}_n^\alpha \sim 0$. From fit results for YBCO7 [195], if $\mathcal{K}_n^\alpha = 0$, one has $T_{1c}/T_{1ab} \simeq 2.75$. For Y248, then, the values of $\mathcal{K}_1^\alpha$ were estimated from the fit values for YBCO6.63, arguing that these two compounds have similar carrier densities and pseudogaps. Taking $\mathcal{K}_1^{ab}(T_c) \simeq -0.53$ and $\mathcal{K}_1^c(T_c) \simeq -0.61$ from data fits for YBCO6.63 [195], they obtained

$$\left.\frac{T_{1c}}{T_{1ab}}\right]_{T=T_c} \simeq 3.00; \quad \left.\frac{T_{1c}}{T_{1ab}}\right]_{T=0} \simeq 5.78 . \tag{5.7.15}$$

These approximate values are plotted in Fig. 5.21b as open triangles, and are seen to correspond well with the data.

The point here is, of course, that these results give good evidence that the c-axis correlations do not decay at low temperatures. If the c-axis correlations decay to zero as $T \to 0$, then according to (5.7.14) $T_{1c}/T_{1ab} \to 2.75$ in Fig. 5.21b for both compounds. This value is below the dashed and dotted lines and is not in good accord with the data.

Comparing this discussion with the analysis given by Bulut and Scalapino in Sect. 5.7.1, their Hubbard model gives a functional form for $\chi''(\boldsymbol{q}, \omega)$ at low frequencies at $T < T_c$ which they have used *in effect* to evaluate the $\mathcal{K}_n^{\alpha}(T)$ (see (3.5.17) and (3.5.18)), and thus the behavior of T_{1c}/T_{1ab} versus T, etc. The major difference with the present analysis by Uldry et al. [261] is that $\chi''(\boldsymbol{q}, \omega)$ for the Hubbard model is *isotropic and would not predict the anisotropic correlation effects seen here*. Further, the Hubbard model with d-wave symmetry does not inhibit AFM correlations in the superconducting state. As we have seen from the analysis of this subsection, only the c-axis correlations survive below T_c, with correlations in the basal plane declining to zero as $T \rightarrow 0$. Thus, with the simple analytic method pioneered by Uldry and Meier, important effects have been uncovered which might never have seen the light of day through the usual process of theoretical modeling.

Chapter 6
Dynamic Susceptibility Studies via NMR for the Cuprates

In Chaps. 3 and 4 we saw how the dynamic susceptibility has come to play a pivotal role in interpreting cuprate spin-lattice relaxation data and rationalizing the sharp contrasts in T_1 behavior among sites in and near the CuO_2 planes. The principal driving force in all of the magnetic phenomenology of the cuprates is clearly the AFM fluctuation peak at or near $\boldsymbol{Q} = (\pi, \pi)$, which occurs in varying degrees of dominance depending in an inverse fashion on the degree of carrier doping. This scenario appears to be a unique property of doped Mott insulators.

A second consequence of strong AFM fluctuation effects is the indirect spin-spin couplings between neighboring nuclear spins which are thereby induced. This effect was recognized early on by Pennington et al. [139], shortly after which the Pennington-Slichter (PS) theory of indirect spin-spin coupling appeared, formulating this effect in terms of the static, $\boldsymbol{q}$-dependent susceptibility $\chi'(\boldsymbol{q}, 0)$. Equally important was the recognition that these couplings were responsible for a Gaussian spin-echo decay (T_{2g}) process which, with due care, could be measured experimentally [31]. In the first section of this chapter we review the PS formulation and survey a range of experimental cuprate results and interpretations.

The importance of T_{2g} for cuprate magnetism research was that now there were two methods of characterizing $\chi(\boldsymbol{q}, \omega)$, namely $T_1(T)$ and $T_{2g}(T)$. T_{2g} has been shown (and we document in Sect. 6.1) to be a fairly direct measure of the key parameter $\chi'(\boldsymbol{Q})$. It soon became *de rigueur* to measure the NMR shift, T_1 and T_{2g} in a cuprate NMR study. It will become clear in this chapter why this is the case.

The connection between $\chi''(\boldsymbol{q}, \omega)$ and $\chi'(\boldsymbol{q}, 0)$ is the rather tenuous one through the Kramers-Kronig (KK) relation. It is, of course, not tenuous in principle, but difficult to implement experimentally, because it requires the entire frequency spectrum of $\chi''(\boldsymbol{q}, \omega)$. In theoretical modeling, of course, this problem is handled through parameterization, and the KK relation becomes a constraint. For example, Bulut and Scalapino have extended their calculations of dynamical properties using a small-U

© Springer-Verlag GmbH Germany 2018
R.E. Walstedt, *The NMR Probe of High-T_c Materials and Correlated Electron Systems*, Springer Tracts in Modern Physics 276,
https://doi.org/10.1007/978-3-662-55582-8_6

Hubbard model to the case of indirect spin-spin coupling (T_{2g}) in the superconducting state. When compared with the data of Ohsugi et al. [279], the result was a dramatic confirmation of d-wave gap structure below T_c for Y248. We review these developments at the end of Sect. 6.1.

For a purely experimental implementation of the KK relation we turn to the case of optimally doped LSCO:x, x $\sim$ 0.15. For this system there has been an INS characterization of $\chi''(\boldsymbol{q}, \omega)$ for effectively all $\boldsymbol{q}$ values (in the first BZ) and for all frequencies for which there is a substantial contribution. Moreover, the latter characterization is in absolute units, leading to a direct comparison between INS and T_1 data with no adjustable parameters. The story of this first-ever quantitative juxtaposition of quantitative INS and NMR data is spelled out in detail in Chap. 4. A closely-related T_2 study for both ^{63,65}Cu and ^{17}O is included as well, where a successful implementation of the KK integral has been reported. The points where INS and NMR data are clearly incompatible is also an interesting part of the story.

An interesting interconnection between T_1, T_{2g}, and the dynamic susceptibility components has been created by the dynamic scaling theories of Pines and co-workers. This topic is addressed in Sect. 6.4, where in the limit of long AFM correlation lengths ξ the two parameters $\xi(T)$ and the spin fluctuation frequency $\omega_{sf}(T)$ appear to control the dynamical properties accessible to the NMR probe. In this context a high-temperature investigation of ^{63}Cu(2) static and dynamic NMR/NQR properties of $YBa_2Cu_4O_8$ (Y248) by Curro et al. [242] has given evidence for a sharply-defined crossover temperature ($\sim$500 K) for the dynamical exponent ($z = 1 \rightarrow 2$) in this system. These interesting developments are spelled out in detail, as well as suggestions for extension to other systems.

6.1 The Indirect Spin-Spin Coupling Theory of Pennington and Slichter

Indirect couplings between nuclear spins are very strong in the cuprates [139]. As was first demonstrated by Pennington and Slichter [31], these can be formulated in terms of the real part of the dynamic susceptibility $\chi'(\boldsymbol{q}, 0)$. $\chi'(\boldsymbol{q}, 0)$ can, in turn, be expressed in terms of $\chi''(\boldsymbol{q}, \omega)$, so the estimation of $\chi'(\boldsymbol{q}, \omega)$ ($\omega = 0$) through measurements of the spin-echo T_2 make the indirect couplings a powerful element in the NMR methodology for solids. In this section we give the derivation of the Pennington-Slichter result, which is straightforward, after which various applications of this result will be presented throughout the chapter.

There are indirect couplings between any pair of neighboring nuclei in (or near) the CuO_2 plane. We shall begin with the pertinent example of interactions among the ^{63,65}Cu nuclear spins. Although T_1 can be formulated in terms of $\chi''(\boldsymbol{q}, \omega)$ having the dimension of time (see discussion in Sect. 3.5), here we will use susceptibilities per mole, since $\chi'(\boldsymbol{q}, \omega)$ becomes a measured uniform, static susceptibility at $\omega = 0$. We calculate the indirect coupling of a single ^{63}Cu with surrounding ^{63,65}Cu spins, an effect which is transmitted via the bath of fluctuating electronic spins $\boldsymbol{S}_i$. It is driven by the HF coupling tensor

$$\mathcal{H}_{HF}(i) = \sum_{\alpha} \left[A_{\alpha} I_{i\alpha} S_{i\alpha} + B \sum_{j(nn)} I_{i\alpha} S_{j\alpha} \right], \tag{6.1.1}$$

where $\sum_{j(nn)}$ is taken over the four nearest-neighbor spins to the site i. We begin by using this HF Hamiltonian to calculate the spin polarization created in the local environment by a nuclear spin $\boldsymbol{I}_i$ located at $\boldsymbol{R}_i$. The effective field which this nucleus exerts on spin moments $-g_{\alpha}\mu_B S_{\alpha}(\boldsymbol{R})$ in the vicinity is

$$H_{i\alpha}(\boldsymbol{R}) = \frac{I_{i\alpha}}{g_{\alpha}\mu_B} F_{\alpha\alpha}(\boldsymbol{R}, \boldsymbol{R}_i), \text{ with } F_{\alpha\alpha}(\boldsymbol{R}, \boldsymbol{R}_i) = [A_{\alpha}\delta_{\boldsymbol{R},\boldsymbol{R}_i} + B \sum_{j(nn)} \delta_{\boldsymbol{R},\boldsymbol{R}_j}]. \tag{6.1.2}$$

The Fourier transform $H_{i\alpha}(\boldsymbol{q}) = \sum_k \exp(-i\boldsymbol{q}\cdot\boldsymbol{R}_k) H_{i\alpha}(\boldsymbol{R}_k)$ may be written

$$H_{i\alpha}(\boldsymbol{q}) = \frac{I_{i\alpha}}{g_{\alpha}\mu_B} \mathrm{e}^{-i\boldsymbol{q}\cdot\boldsymbol{R}_i} F_{\alpha\alpha}(\boldsymbol{q}), \text{ with } F_{\alpha\alpha}(\boldsymbol{q}) = [A_{\alpha} + 2B\,(cos\, q_x + cos\, q_y)]. \tag{6.1.3}$$

$H_{i\alpha}(\boldsymbol{q})$ then induces a $\boldsymbol{q}$-component of spin moment

$$\mathcal{S}_{\alpha}(\boldsymbol{q}) = -\frac{1}{N_A g_{\alpha}\mu_B} H_{i\alpha}(\boldsymbol{q}) \chi'_{\alpha}(\boldsymbol{q}), \tag{6.1.4}$$

where we divide $\chi'(\boldsymbol{q})$ by Avogadro's number for an atomic susceptibility. The inverse Fourier transform of $\mathcal{S}_{\alpha}(\boldsymbol{q})$ gives the resulting spin polarization in the vicinity of $\boldsymbol{R}_i$, i.e.

$$\mathcal{S}_{\alpha}(\boldsymbol{R}) = \sum_{\boldsymbol{q}}{}' \mathrm{e}^{i\boldsymbol{q}\cdot\boldsymbol{R}} \mathcal{S}_{\alpha}(\boldsymbol{q}) = -\frac{\hbar I_{i\alpha}}{N_A g_{\alpha}^2 \mu_B^2} \sum_{\boldsymbol{q}}{}' \mathrm{e}^{i\boldsymbol{q}\cdot(\boldsymbol{R}-\boldsymbol{R}_i)} \chi'_{\alpha}(\boldsymbol{q}) F_{\alpha\alpha}(\boldsymbol{q}), \tag{6.1.5}$$

where $\sum'_{\boldsymbol{q}}$ is a normalized sum over the Brillouin zone.

The interaction energy with a second nucleus $\boldsymbol{I}_j$ may now be obtained by inserting (6.1.5) into (6.1.1), where we use the notation of (6.1.2) to find

$$\sum_{\alpha} a_{\alpha}(i, j) I_{i\alpha} I_{j\alpha} = \sum_{\alpha} I_{j\alpha} \sum_{\boldsymbol{R}_k} \mathcal{S}_{\alpha}(\boldsymbol{R}_k) F_{\alpha\alpha}(\boldsymbol{R}_k, \boldsymbol{R}_j). \tag{6.1.6}$$

Substituting from (6.1.5) then gives the general result [241]

$$a_{\alpha}(i, j) = \frac{1}{N_A g_{\alpha}^2 \mu_B^2} \sum_{\boldsymbol{q}}{}' \mathrm{e}^{i\boldsymbol{q}\cdot(\boldsymbol{R}_j-\boldsymbol{R}_i)} \chi'_{\alpha}(\boldsymbol{q}) F_{\alpha\alpha}(\boldsymbol{q})^2. \tag{6.1.7}$$

A useful relation for calculating the second moment of the line as well as the Gaussian decay time T_{2g} of the spin echo may be derived with (6.1.7). This result, given independently by Takigawa [266] and by Thelen and Pines [267], is

$$\sum_j a_\alpha(i,j)^2 = \frac{1}{(N_A g_\alpha^2 \mu_B^2)^2}\left[\sum_{\boldsymbol{q}}{}'[\chi'_\alpha(\boldsymbol{q})F_{\alpha\alpha}(\boldsymbol{q})^2]^2 - \left[\sum_{\boldsymbol{q}}{}'\chi'_\alpha(\boldsymbol{q})F_{\alpha\alpha}(\boldsymbol{q})^2\right]^2\right]. \tag{6.1.8}$$

6.1.1 Some General Properties of the Indirect Spin-Spin Coupling

We adopt a notation where $\boldsymbol{q}$ is dimensionless and $\boldsymbol{R}_i = (n_{xi}, n_{yi})$ in units of a, where $n_{xi,yi}$ are integers. We shift the origin of the $\boldsymbol{q}$ sum to $\boldsymbol{Q}$, i.e. $\boldsymbol{q} = \boldsymbol{p} + (\pi, \pi)$. Then $\chi'_\alpha(\boldsymbol{p})$ is a symmetric function, where $\sum'_{\boldsymbol{p}} \rightarrow (1/2\pi)^2 \int_\pi^\pi \mathrm{d}p_x \int_\pi^\pi \mathrm{d}p_y$, and $F_{\alpha\alpha}(\boldsymbol{p}) = [A_\alpha - 2B(cos\ p_x + cos\ p_y)]$. The exponential factor in (6.1.7) reduces to $cos\ (n_x p_x)\ cos\ (n_y p_y)$, since the sine terms cancel by symmetry, where $n_x = n_{xj} - n_{xi}$, etc. Thus, (6.1.7) becomes

$$a_\alpha(i,j) = \frac{(-1)^{(n_x+n_y)}}{N_A g_\alpha^2 \mu_B^2} \sum_{n'_x=n_x-2}^{n_x+2}\ \sum_{n'_y=n_y-2}^{n_y+2} C_\alpha(n'_x, n'_y) G_\alpha(n'_x, n'_y)\,, \tag{6.1.9}$$

where

$$G_\alpha(n'_x, n'_y) = \frac{1}{\pi^2}\int_0^\pi \mathrm{d}p_x \int_0^\pi \mathrm{d}p_y\, \chi'_\alpha(\boldsymbol{p})\, cos(n'_x p_x)\, cos(n'_y p_y), \tag{6.1.10}$$

and where the $C_\alpha(n'_x, n'_y)$ are functions of A_α and B.

Some general properties of $a_\alpha(i,j)$ are as follows:

- Terms in $\chi'(\boldsymbol{q}, 0)$ which are independent of $\boldsymbol{q}$ give no contribution to $a_\alpha(i,j)$.
- The essential range function of a(i, j) is the Fourier transform of the peak function $\chi'_\alpha(\boldsymbol{q})$.
- A sign reversal on moving between *nn* sites results from the AFM nature of the fluctuations.
- The range function depends only on the magnitudes of the $n_{x,y}$ and is invariant on exchange of n_x and n_y.
- Therefore $a_\alpha(i,j)$ is symmetric about the x and y axes, and also about the $x = \pm y$ axes.

6.1.2 The Gaussian Model of Indirect Spin-Spin Coupling in YBCO

In the MM follow-up to MMP [162], a Gaussian model has been suggested for $\chi''(\boldsymbol{q}, \omega)$. Citing this example, Imai et al. [268] made use of this model in their interpretation of data for T_1 and T_{2g} (see below for definition of T_{2g}). More recently, Auler et al. [245] employed this model for similar purposes, citing INS studies which showed a Gaussian-like shape with a width essentially independent of temperature. Their interpretation of T_1 data on this basis was presented and discussed in Sect. 5.3.

INS data provide an important characterization of $\chi''(\boldsymbol{q}, \omega)$, which is then used to interpret NMR data. An important question regarding INS data is to what extent they are dependent on the frequency scale. Does a peak in $\chi''(\boldsymbol{q}, \omega)$ at or near $\boldsymbol{Q}$ with a certain measured width at $\omega = 10\,\text{meV}$ have the same shape and width at an NMR frequency four orders of magnitude lower? With the limitations of current spectroscopic techniques one cannot answer this question experimentally. Many discussions of NMR T_1 interpretation tacitly assume that the INS results will still be valid at NMR frequencies. We address this question in various contexts throughout this chapter.

For the purposes of the present discussion we adopt the Gaussian shape and width for the AFM peak given by [245] (ξ = correlation length in units of a):

$$\chi'_\alpha(\boldsymbol{q}) = \frac{g_\alpha^2}{4}\chi'_{AF}\mathrm{e}^{-(q_x^2+q_y^2)\xi^2/4)}\,, \tag{6.1.11}$$

with $\xi \simeq 1.3$, where we place the origin at $\boldsymbol{Q} = (\pi, \pi)$ and we denote $\chi'(\boldsymbol{Q}) = \chi'_{AF}$. Inserting this into (6.1.10), a simplification occurs, where we can write $G_\alpha(n'_x, n'_y) = (g_\alpha^2/4)\chi'_{AF}g(n_x)g(n_y)$, with

$$g(n) = \frac{1}{\pi}\int_0^\pi \mathrm{d}p\mathrm{e}^{-p^2\xi^2/4}cos(np) \approx \frac{1}{\sqrt{\pi}\,\xi}\mathrm{e}^{-n^2/\xi^2}\,. \tag{6.1.12}$$

In (6.1.12) the result on the far right is obtained by extending the upper integration limit to ∞. Since $\exp(-\pi^2\xi^2/4) \simeq 0.015$, we suggest that this is a reasonable approximation. Inserting this expression for $G_\alpha(n'_x, n'_y)$ into (6.1.9), we find, with $n_x = n_{xj} - n_{xi}$, etc.,

$$\begin{aligned} a_\alpha(i,j) = (-1)^{(n_x+n_y)}&\frac{\chi'_{AF}}{\pi N_A\xi^2 g_\alpha^2\mu_B^2}\mathrm{e}^{-(n_x^2+n_y^2)/\xi^2}\{A_\alpha^2 + 4B^2 \\ &- 4A_\alpha B\,\mathrm{e}^{-1/\xi^2}[cosh(2n_x/\xi^2) + cosh(2n_y/\xi^2)] \\ &+ 8B^2\mathrm{e}^{-2/\xi^2}cosh(2n_x/\xi^2)\,cosh(2n_y/\xi^2) \\ &+ 2B^2\mathrm{e}^{-4/\xi^2}[cosh(4n_x/\xi^2) + cosh(4n_y/\xi^2)]\}\,. \end{aligned} \tag{6.1.13}$$

Equation (6.1.13) is used below to calculate indirect coupling energies for YBCO6.9 and to discuss results which have appeared in the literature. Before that, however, we make a brief digression to introduce spin echo decay measurements.

6.1.3 Measuring Indirect Spin-Spin Coupling via Spin-Echo Decay

Under suitably restrictive assumptions, it is possible to get good estimates of the indirect spin-spin coupling (6.1.7) by measuring spin-echo decay times. We give here a brief review of some of the essential points of spin-echo decay, followed by a discussion of results from the literature in the next subsection.

We consider spin-echo decay for the specific problem of ^{63,65}Cu(2) NMR in the CuO_2 planes of a cuprate superconductor. If the ^{63,65}Cu isotopes are labelled A and B, where the A spins are under observation, then the nuclear spin-spin coupling has the general form

$$\mathcal{H}_{II} = \mathcal{H}_{II}^{AA} + \mathcal{H}_{II}^{AB} + \mathcal{H}_{II}^{BB} . \tag{6.1.14}$$

For cases of orthorhombic symmetry or higher, the latter terms can be written

$$\mathcal{H}_{II}^{AA} = \sum_{i>j} \left[\alpha_{ij} I_{zi}^A I_{zj}^A + \beta_{ij} I_{xi}^A I_{xj}^A + \gamma_{ij} I_{yi}^A I_{yj}^A\right], \text{ and } \mathcal{H}_{II}^{AB} = \sum_{i>j} \alpha_{ij} I_{zi}^A I_{zj}^A , \tag{6.1.15}$$

where we have taken z to be the quantization axis. For $\mathcal{H}_{II}^{AB}$ the x and y terms have been omitted, because they are non-secular and have no measurable effects. Whether or not the x and y terms for $\mathcal{H}_{II}^{AA}$ are secular, i.e. physically viable, is a matter of some discussion.

In cases where there is no detuning of neighboring A spins, i.e., where locally, the Zeeman splitting $\gamma_A(1+K)H$ as well as quadrupolar splittings are homogeneous, then the x and y terms are technically viable. However, if the zz coefficient α_{ij} is much larger than β_{ij} and γ_{ij} (a realistic case as we shall see below), then "flip-flop" processes such as $| m_A, m_B\rangle \rightarrow | m_A + 1, m_B - 1\rangle$ may become severely inhibited and the transverse terms become de facto non-secular. Below, a simple test is suggested to determine the effectiveness of the dynamical terms.

For the moment we shall assume the flip-flop terms to be inoperative. Approximating the decay function with a Gaussian $\exp(-(2\tau)^2/2T_{2g}^2)$, then the echo decay time constant is found to be [26]

$$\frac{1}{T_{2g}} = sin(\theta_2/2)\left[\frac{c_{63}}{8}\sum_j \alpha_{ij}^2\right]^{1/2} , \tag{6.1.16}$$

where $c_{63} \simeq 0.69$, and the sum is over all neighbors. The dependence on θ_2 suggests an interesting test for the spin dynamics of the sample. If the spins are completely static on the time scale of the echo sequence, then T_{2g} should become correspondingly long when θ_2 is small. If T_{2g} does not change a great deal over a range of values for θ_2, it probably means that there is an appreciable degree of flip-flop motion taking place among the spins.

In applying (6.1.16) to actual measurements of $\sum_j \alpha_{ij}^2$, it is clearly necessary to be able to flip over nearly all of the spins with the refocusing pulse. This is an important constraint on experimental technique. For cuprates this means that one will probably have to employ only the $(-1/2 \leftrightarrow 1/2)$ transition, and possibly work at low fields in addition in order to narrow the line.

6.1.4 Behavior and Interpretation of Indirect Couplings in YBCO and Y248

There have been a number of measurements of T_{2g} for both YBCO and Y248 in the literature, with much discussion of its behavior as compared with T_1. Since they are both driven, at least in part, by AFM fluctuations, these relaxation times are reflections of the same underlying quasiparticle physics. Thus, the advent of T_2g as an additional characterization of cuprate hole dynamics was a most welcome development.

Results Given by Pennington and Slichter

In their original paper formulating the indirect couplings, expounded in the previous subsection, PS [31] presented calculations of coupling coefficients $a_\alpha(i, j)$ as well as of T_{2g} for fully oxygenated YBCO7. They calculated their range function using the MMP Lorentzian form

$$\chi'(\boldsymbol{q}) = \chi_0 \left[1 + \frac{(\xi/\xi_0)^2)}{1 + (\boldsymbol{q} - \boldsymbol{Q})^2 \xi^2} \right] \tag{6.1.17}$$

centered on $\boldsymbol{Q}$. The ξ^2 factor gives only a rough normalization for $\chi'(\boldsymbol{q})$, since $\sum_q \chi'(\boldsymbol{q})$ depends on $ln(\xi)$. The resulting relaxation time is found to vary $1/T_{2g} \propto \xi$ [31]. The parameters and assumptions for various calculations of T_{2g} discussed here are shown in Table 6.1. Adopting the MMP estimate [161] $\xi = 2.5\,a$, the half-width of the AFM peak in the PS calculations with (6.1.17) is 0.4. Also from (6.1.17), $\chi'(\boldsymbol{Q}) = 1.66 \times 10^{-3}$ emu/mol Cu(2). With values of A_c and B as shown in Table 5.1, PS find a value of $1/T_{2g}$ which is just slightly smaller than their experimental number, which clearly validates their formulation of the indirect spin-spin coupling.

Effects of Transverse Coupling and of Dipolar Fields

Before proceeding with discussion of experimental data, we note in passing that transverse and dipolar coupling contributions to $\mathcal{H}_{zz}$ are much smaller and can be ignored for most purposes. The dipolar coupling coefficient for quantization axis

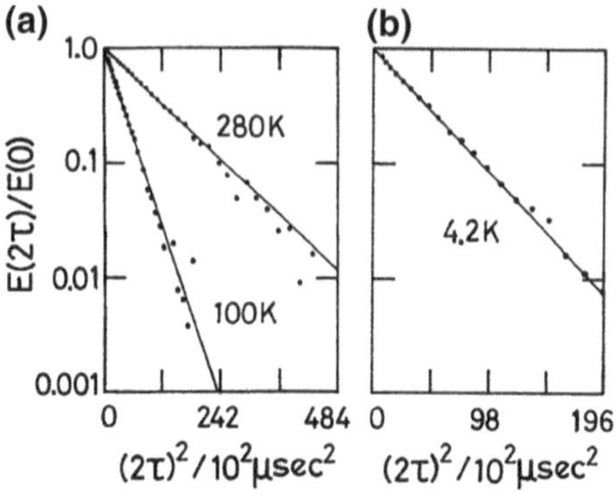

Fig. 6.1 **a** and **b** Spin echo decay curves, corrected for T_1 effects, are plotted semilogarithmically versus $(2\tau)^2$, showing nearly perfect Gaussian behavior over two orders of magnitude of decay at several temperatures. These data for $^{63}Cu(2)$ in $YBa_2Cu_40_8$ are from Itoh et al. [272]

$\perp$ CuO_2 plane is simply $\alpha_c^{dip}(i, j) = \hbar\gamma^2/R_{ij}^3$, which has the value 908 s^{-1} for *nn* $^{63}Cu(2)$. The corresponding second moment value is less than 6% of the estimated indirect coupling value calculated above. Moreover, it has a constant positive sign for all in-plane neighbors, so that its effects tend to cancel when combined with the staggered indirect coupling. We shall ignore the dipolar terms from this point forward. The transverse indirect coupling terms are also much smaller than the c-axis contribution, giving a second moment only 15% of the latter effect using $A_{ab} \sim 0.31$ in the units of Table 6.1, with B having the same value as for the c-axis terms. As far as we are aware, no effort has been made to measure the transverse terms, and we shall not discuss them further.

Comparison of the PS Results with Other Models and Data

Throughout the derivation of indirect spin-spin coupling given above [31], it is assumed that the resulting spin-echo decay process will be Gaussian in form or nearly so. Itoh et al. have tested this hypothesis for the case of NQR studies of $^{63}Cu(2)$ in Y248 [272]. Their results are plotted in Fig. 6.1a, b for several temperatures. After corrections for T_1 effects (see Appendix A.1), it is seen that these decay curves exhibit nearly perfect Gaussian character over two orders of magnitude. Some discussion of the origin of such Gaussian behavior, which is very widespread in T_{2g} studies, is presented in Sect. 6.3.

In the second line of Table 6.1 we list the calculated result of Bulut and Scalapino (BS) based on their small-U Hubbard model calculations [271] along with experimental numbers from Itoh, Yasuoka et al. [272]. BS calculated the indirect coupling using the same model and parameters they had applied to the T_1 problem [251, 252], finding good agreement with experimental data with an AFM peak width ~25% greater than that used by PS. Itoh et al. also tested the proportionality of $^{63}Cu(2)$ second moment to the concentration of ^{63}Cu by measuring a sample which contained 100% 63Cu. The result, shown on the third line of Table 6.1, is within experimental error of the expected increase $1/T_{2g} \propto (c_{63})^{1/2}$. The BS calculated result scaled in this fashion is shown for comparison. Itoh et al. [272] also measured the behavior of T_{2g} in the superconducting state of Y248. The latter results and the BS model calculations for this case are discussed in the next subsection.

Table 6.1 Models, data, and results for c-axis indirect coupling values of $1/T_{2g}$ are presented from various sources for different charge states of YBCO. The first line gives the original results of Pennington and Slichter with a Lorentzian model. HF parameters vary only slightly between calculated results given. Experimental data in the final column corresponds generally to 100 K. Under "Source", Imai and Auler are first authors, as are Bulut and Itoh. The Takigawa reference is [266]. For Imai et al. and Auler et al., the estimated values of $1/T_{2g}$ have been made using (6.1.13) and direct summation to give $\sum_j a_c(i, j)^2$. For the Takigawa estimate, which was made with a Lorentzian model, the PS result is simply scaled to the larger value of $\chi'(\boldsymbol{Q})$ shown. The units of parameters given are as follows: A_c and B ($10^{-6} eV$); $\chi'(\boldsymbol{Q})$ ($10^{-3} emu/mol$); $1/T_{2g}$ (ms^{-1}). These results are discussed extensively in the text

System	Source	$\chi'(\boldsymbol{q})$ form	H.W.(a^{-1})	A_c	B	$\chi'(\boldsymbol{Q})$	$(1/T_{2g})_{est}$	$(1/T_{2g})_{exp}$
YBCO7	PS [31]	Lorentzian	0.40	−1.53	0.38	1.66	5.26	7.7
YBCO7	Bulut/Itoh	Hubbard RPA	0.5	−1.53	0.38	1.23	9.0	11.1
YBCO7	Itoh($c_{63} = 1$)	Hubbard RPA	0.5	−1.53	0.38	1.23	10.8	12.9
YBCO6.9	Imai [268]	Gaussian	0.37	−1.77	0.45	3.28	9.2	9.6
YBCO6.9	Auler [245]	Gaussian	1.28	−1.56	0.39	1.41	7.5	11.5
YBCO6.63	Takigawa	Lorentzian	0.40	−1.53	0.38	6.14	19.5	18.4

We make further comparisons with the results in Table 6.1 with calculations using (6.1.13) based on the parameters provided by Auler et al. [245]. Results obtained with a Gaussian peak for $\chi'(\boldsymbol{q})$ and other parameters from Auler et al. [245] are listed in line five of the table. The INS-derived peak width is more than three times that used by PS. The value of $(1/T_{2g})_{est}$ is substantially larger than the PS value with a slightly smaller value for $\chi'(\boldsymbol{Q})$. Although this is a satisfactory result, it is puzzling that it does not give the rather larger relaxation rate which they quote, and from which the value given for $\chi'(\boldsymbol{Q})$ was supposedly derived [245]. The origin of this discrepancy is not understood.

In line four of Table 6.1 are listed the results of a similar calculation based on (6.1.13) as well as the HF coupling, $\chi'(\boldsymbol{Q})$, and peak width parameters given by Imai et al. [268], on which the calculation is based. The estimated value of $1/T_{2g}$ is in good agreement with the experimental value given. A result reported by Takigawa [266] is given in the bottom line of the table. Parameter values were stated in [266] for the Lorentzian model with a range of peak widths. Choosing the peak width used by PS for the same model with the same HF parameter values, we simply scale the PS result to the larger value of $\chi'(\boldsymbol{Q})$ given in [266] to yield the value of $(1/T_{2g})_{est}$ stated in the table, which is in good agreement with the experimental value given for YBCO6.63.

Before continuing to discuss the significance of these results, we address the question of how the results for $1/T_{2g}$ scale with the correlation length.

Scaling of $\mathbf{T_{2g}}$ *with* $\chi'(q)$ *Parameters*

There appears to be an interesting difference between Lorentzian and Gaussian models of $\chi'(\boldsymbol{q})$ which has to do with their dependence on the correlation length parameter ξ. For the Gaussian model as prescribed by Auler et al. [245], we find that the second moment $\propto \sum_j \alpha(i,j)^2$ varies roughly as ξ^{-1}. From (6.1.16) it follows that $T_{2g} \sim \xi^{1/2}$. In contrast, PS note that $T_{2g} \propto \xi^{-1}$ for their Lorentzian model. However, the latter model (see (6.1.17)) included a prefactor $\propto \xi^2$ multiplying a peak function of unit amplitude, whereas the Gaussian has none. *The* ξ^2 *prefactor by itself accounts for the stated variation of* T_{2g} *with* xi. Without the prefactor effect, then, the PS result is that T_{2g} is essentially independent of ξ.

We next consider whether it is possible to account for the foregoing behaviors using the expression (6.1.8) for the second moment. While under some circumstances it may be possible to deduce how $\sum_j \alpha_{ij}^2$ scales with ξ using (6.1.8), there is no simple answer to this question in general for two reasons. First, if $F_{\alpha\alpha}(\boldsymbol{q})$ contains transferred HF couplings, the cosine factors do not scale in any simple way. Second, in the case of a Lorentzian form for $\chi'(\boldsymbol{q})$, the sum $\sum_{\boldsymbol{q}}\chi'(\boldsymbol{q})$ does not converge at the boundaries, so that the integration limits become ξ–dependent. This problem is particularly severe for small values of ξ.

Nonetheless we shall consider what sort of results (6.1.8) gives for $\boldsymbol{q}$-independent HF couplings. Such a calculation can be carried out straightforwardly, with the result

$$\sum_j \alpha_{ij}^2 = \frac{1}{2\pi\xi^2}\left[1 - \frac{2}{\pi\xi^2}\right]. \quad (6.1.18)$$

Thus, as ξ declines, the leading term gives the $T_{2g} \sim \xi$. However, at small values of ξ the correction term in the brackets becomes increasingly important. For $\xi \sim 1.3$, the correction term increases T_{2g} by ~25%. We suggest that the $\boldsymbol{q}$-dependent HF couplings also contribute to the behavior mentioned for YBCO, where $T_{2g} \propto \xi^{1/2}$.

For the Lorentzian case, with $\chi'(\boldsymbol{q}) = \chi'(\boldsymbol{Q})/(1 + q^2\xi^2)$, the leading term in (6.1.8) gives $T_{2g} \propto \xi$ as with the Gaussian case. However, the second term gives a somewhat larger correction than in (6.1.18), so that $\sum_j \alpha_{ij}^2$ levels off and may even diminish at low values of ξ. It is plausible that over a modest range of ξ values $\sum_j \alpha_{ij}^2$ may be rather flat, leading to the ξ-independent behavior noted, in effect, by PS. Therefore, it appears that the contrast between Gaussian and Lorentzian susceptibilities may be accounted for in this simple fashion.

A second and equally important conclusion which we draw from this discussion is that the results for T_{2g} do not depend strongly on ξ or on the choice of Gaussian or Lorentzian for the shape of the AFM peak in $\chi'(\boldsymbol{q})$. For this reason it will also be difficult to extract information from T_{2g} data on either of these points. At the same time, the results in Table 6.1 show that T_{2g} can be analyzed to yield semi-quantitative results for $\chi'(\boldsymbol{Q})$, even if the width and detailed character of the associated AFM peak are not well established.

Comparing the Temperature Variation of T_{2g} with that of T_1

The relatively weak dependence of T_{2g} on the width of the AFM peak in $\chi'(\boldsymbol{q})$ leads to an interesting contrast between echo decay times and T_1 processes. While $1/T_1T$ is given by sums of $\chi''(\boldsymbol{q}, \omega)$ over all of $\boldsymbol{q}$-space, $1/T_{2g}$ is very nearly proportional to $\chi'(\boldsymbol{Q})$, i.e. to the staggered susceptibility. T_1 is, of course, also sensitive to the staggered susceptibility, since it measures the area under the AFM peak in $\chi''(\boldsymbol{q}, \omega)$ at $\boldsymbol{Q}$. But T_1 for ^{63}Cu(2) has major contributions from the rest of the Brillouin zone as well, much of which collapses in the presence of a pseudogap. We see the result of this in Fig. 6.2 from Takigawa [266]. The curve for $1/T_{2g}$ indicates that the staggered susceptibility is rising fairly rapidly in a Curie-Weiss-like fashion. T_1 for the ^{63}Cu(2) rises below 300 K at a similar rate. When the spin pseudogap sets in around 200 K, however, $1/T_1T$ goes over a maximum and then begins a steep descent which indicates collapsing behavior for $\chi''(\boldsymbol{q}, \omega)$ over much of the BZ. The fact that $1/T_{2g}$ is immune to the pseudogap decline suggests that $\chi'(\boldsymbol{Q})$ is made up from parts of the Fermi surface which are not gapped. A related result is presented in the next subsection, where $1/T_2$ for Y248 does not decay in the superconducting state, presumably because contributions to $\chi'(\boldsymbol{Q})$ are not affected by the superconducting gap either.

Curie-Weiss behavior for $1/T_{2g}$ is also found in YBCO7 [245, 268], though with rather smaller increases between 300 K and T_c. It would be, of course, most enlightening to be able to compare such results with INS data on the same systems, including temperature dependences. While the INS characterizations are not easy to perform, systems such a YBCO provide an ideal test bed for comparison of results

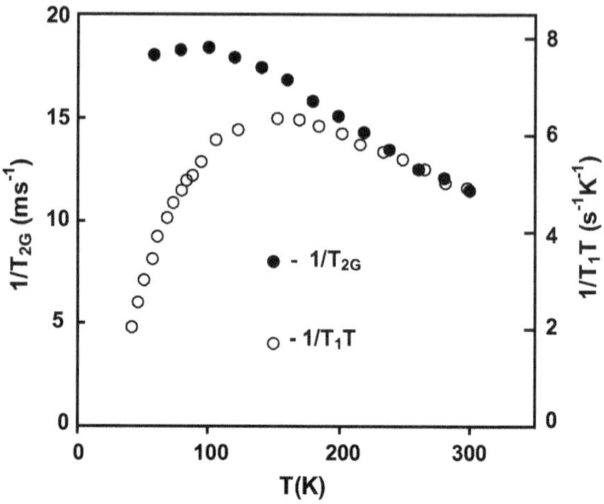

Fig. 6.2 Measured values of $1/T_{2g}$ and $1/T_1T$ for ^{63}Cu(2) in YBCO6.63 are plotted versus temperature. The data have been scaled to coincide near room temperature, where they show similar behavior. This plot is from Takigawa. [266]

for the dynamic susceptibilities. We shall see this in more detail in Sect. 6.3 with the system LSCO.

Takigawa also made an effort to apply the scaling theory of Sokol and Pines [273] to the susceptibility behavior of YBCO6.63 [266]. The scaling effect may be seen in Fig. 6.2, where just at the upper end of the temperature range the ratio T_1T/T_{2g} approaches a constant value. Effects such as this have become a major branch of the study of cuprate magnetism and will be reviewed at some length in Sect. 6.4.

6.2 The Bulut-Scalapino Calculation of $1/T_{2g}$ for T < T_c

Bulut and Scalapino (BS) used their small-U Hubbard model of spin dynamics to treat both T_1 (Sect. 5.7) and T_{2g} in the superconducting state [271]. In the latter case they found a major distinction between the cases of s-wave and d-wave pairing. We first describe their calculation, and then review the experimental T_{2g} data of Itoh, Yasuoka et al. [272] on $YBa_2Cu_4O_8$ (Y248) which has been compared with the BS theory.

Model Calculations of Bulut and Scalapino: T_{2g} in the Normal State

The BS calculations of T_{2g} used much the same parameter values as employed in related work on T_1 (see Sect. 5.4) [251, 252], where they used the RPA formula

$$\chi'(\boldsymbol{q}) = \frac{\chi_0(\boldsymbol{q})}{1 - U\chi_0(\boldsymbol{q})} \tag{6.2.1}$$

for the real part of the dynamic susceptibility $\chi'(\boldsymbol{q})$ at $\omega = 0$. They give for the noninteracting susceptibility $\chi_0(\boldsymbol{q} \to 0) = 0.214\,t^{-1}$, where t is the hopping integral giving band energies $\mathcal{E}_q = -2t[cos\, q_x + cos\, q_y] - \mu$. The chemical potential was set so that the band occupation $\langle n \rangle = 0.86$, very near the critical value for instability (see [252] and Sect. 6.4).

The enhanced curve of $\chi'(\boldsymbol{q})$ from the BS model is shown in Fig. 6.3a for the path in $\boldsymbol{q}$-space shown in the inset. We determine the molar value of $\chi'(\boldsymbol{Q})$ as follows. From discussion of NMR shift calculations in [252] the $\boldsymbol{q} = 0$ Stoner enhancement is given as 1.75. Equating their c-axis susceptibility result to $\chi_c = 1.18 \times 10^{-4}$ emu/mol Cu(2) from Table 2.4, we find $\chi_0(0) = 6.74 \times 10^{-5}$ emu/mol Cu(2). Relative to that value, the peak value in Fig. 6.3a is enhanced by a factor $\sim(3.8/0.214)$, giving the estimate $\chi'(\boldsymbol{Q}) \simeq 1.23 \times 10^{-3}$ emu/mol Cu(2). The latter value is within $\sim$15% of the result obtained by taking $t = 0.1$ eV [252]. The AFM peak half-width of $\sim$0.5 (Table 6.1) is also taken from Fig. 6.3. The latter results apply in the vicinity of 100 K.

BS also calculated the temperature variation of $1/T_{2g}$, both in the normal state up to 300 K and below T_c down to essentially zero temperature. Their results as given are shown as the dashed line in Fig. 6.3b, where they are seen to fall just below the YBCO7 data from Itoh et al. [272] (filled circles). The open circles in the same figure are a plot of measured values of $1/T_{2g}$ for ^{63}Cu(2) in Y248 also taken from [272]. The latter data have been compared with the BS calculations (solid lines) from [271], which have been scaled to coincide with the experimental data in the region around 100 K. Since the Y248 data were obtained using NQR, the original (dotted) curve was first rescaled by $\sqrt{2}$ to reflect the enhanced NQR relaxation rate. Following that, rescaling with the Stoner factor has been done using (6.2.1) as follows. It is presumed that the temperature dependence of $\chi'(\boldsymbol{Q})$ is driven by that of $\chi_0(\boldsymbol{Q})$. From [271] it was determined that the enhancement at the peak of the curve in Fig. 6.3a is $\simeq$5.4 at T = 100 K. Using that fact and (6.2.1) to recover $\chi_0(\boldsymbol{Q})$ as a function of T, the dotted curve with an adjusted value of U was plotted as the solid curve in Fig. 6.3b. For the region below T_c, the original curves from [271] were simply rescaled to meet the normal-state curve at T = 82 K.

The Superconducting State: A Dramatic Success for D-Wave

The normal-state calculation gives a reasonable account of the Y248 data for $1/T_{2g}$. In the superconducting state, these data, combined with the BS calculations, give a very dramatic confirmation of d-wave pairing symmetry. The effect of superconducting pairing on the susceptibility curve is shown in Fig. 6.3a where calculated susceptibility curves at $T = 0.8T_c$ are shown for both s-wave (dashed line) and d-wave (dotted line). The contrast between these two is attributed by BS to nodes in the d-wave energy gap which simply allow contributions to $\chi'(\boldsymbol{Q})$ to persist into the superconducting state. This is very likely related to the continued c-axis spin correlations below T_c which were demonstrated by Uldry et al. [261], as was discussed in Sect. 5.7. It may also be related to the fact that $\chi'(\boldsymbol{Q})$ is immune to the pseudogap in the normal state, exhibiting Curie-Weiss behavior down to T_c. Thus, the pseudogap and the superconducting gap have similar character in this respect.

The present results along with the dramatic T_1 effects discussed in the latter Section consitituted a very strong case for d-wave pairing several years before it was widely accepted by the high-T_c community. To summarize, we have seen in this section

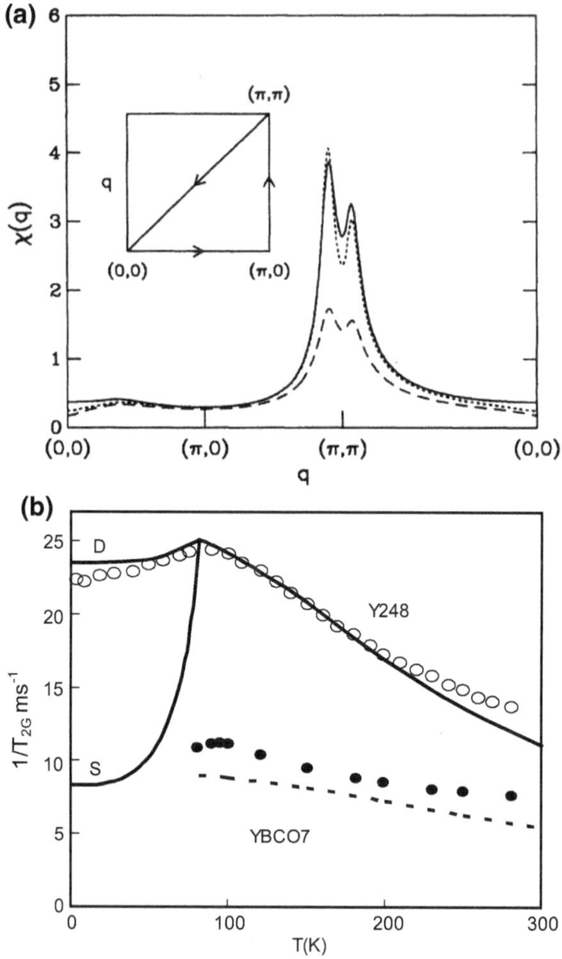

Fig. 6.3 **a** The enhanced susceptibility $\chi'(\boldsymbol{q})$ obtained with (6.2.1) [271] is plotted for the path in $\boldsymbol{q}$-space shown in the inset. The solid curve shows behavior at $T \sim 100\,\mathrm{K}$, while the dashed and dotted curves show behavior for s-wave and d-wave pairing, respectively, at $T = 0.8\,T_c$. See text and part (b) of this figure for discussion of this effect. (**b**) NMR measurements of $1/T_{2g}$ for YBCO7 (dots) and NQR measurements for Y248 (open circles) by Itoh et al. [272] are plotted versus temperature. The dotted line shows the normal-state calculations of Bulut and Scalapino [271] for ^{63}Cu(2) NMR in YBCO7. The solid curves are determined as follows. For the normal state the dotted line is scaled by $\sqrt{2}$ for the NQR case, then the enhancement is re-calculated using (6.2.1) to match the experimental curve near 100 K. For $T < T_c$ the curves from [271] are simply scaled to meet the normal-state curve at $T_c = 82\,\mathrm{K}$. See text for additional discussion

and the last that the formulation of indirect spin-spin coupling in terms of $\chi'(\boldsymbol{q})$ by Pennington and Slichter opened up a very important new avenue of approach for the characterization of the AFM enhancement peaks as well as of $\chi'(\boldsymbol{Q})$.

6.3 Coordinated Interpretation of NMR and INS Data for LSCO

In Sect. 4.1 we discussed data for the NMR shift, uniform susceptibility, and spin-lattice relaxation time T_1 for ^{63}Cu in LSCO:x, for x ranging over the superconducting region, $0.06 \leq x \leq 0.25$. This system was seen to exhibit a peculiar sort of pseudogap effect, which strongly affects the susceptibility even at the peak of the T_c curve, but allows the ^{63}Cu T_1 curves to follow a Curie-Weiss-like behavior down to temperatures very close to T_c. Here, we look at another aspect of the LSCO system which came about, because this system emerged as being unique among the cuprates in having received a very thorough characterization of $\chi''(\boldsymbol{q}, \omega)$ by means of INS. The detailed geometry of the four mildly incommensurate peaks near (π, π) has been scanned in great detail over a wide range of frequencies. As a consequence, LSCO became a classic "test bed" for the theoretical relationship between NMR and $\chi''(\boldsymbol{q}, \omega)$. We first review the theoretical basis for such an investigation.

The fluctuation-dissipation theorem relation between spin-fluctuation-generated T_1 processes in a metallic environment and the dissipative term $\chi''(\boldsymbol{q}, \omega)$ of the dynamical susceptibility, as recounted in Sect. 3.5, had been known for many decades [58] before it was tested using experimental INS data for $\chi''(\boldsymbol{q}, \omega)$. The latter test was conducted via T_1 measurements for ^{63}Cu and ^{17}O (both planar and apical) in LSCO:0.15 [239] and was based on the INS data of Cheong et al. [274] and Mason et al. [275]. Such an enterprise—to correlate two pieces of experimental data using an "exact" theoretical relationship[1]—may not seem well motivated at the outset. However, this relationship had not been tested experimentally before. Furthermore, it led to the uncovering of some interesting limitations, among which is the clash between INS and NMR data for the case of planar ^{17}O [239], which we review below.

In this section we review the circumstances of the latter test and suggest a very simple resolution, namely that the INS result may not apply at NMR frequencies, which are lower by a factor $\sim 10^4$. Additional INS data which have appeared since that time may throw some new light on the situation, as well. We shall also review indirect spin-spin coupling measurements for LSCO:0.15, interpreted with values of $\chi'(\boldsymbol{q})$ which were calculated from the measurements of $\chi''(\boldsymbol{q}, \omega)$ using the Kramers-Kronig relation [5]

$$\chi'(\boldsymbol{q}) = \frac{2}{\pi} \int_0^\infty d\omega \chi''(\boldsymbol{q}, \omega)/\omega \, . \tag{6.3.1}$$

[1]This relationship is very nearly exact under the assumption that spin hyperfine processes are predominant. However, it has not yet been determined how important the orbital relaxation rates are for the cuprates (see Sect. 5.6).

From (6.3.1) one sees that such a procedure requires data for $\chi''(\boldsymbol{q}, \omega)$ over a wide range of frequencies. Since that time, such high-frequency data have actually appeared in the literature [277]. The latter data will be taken into account in reviewing this calculation. For the T_2 processes we come to a similar conclusion as with T_1, namely a successful account of the ^{63}Cu data with (or without) discommensurations, but not for the planar ^{17}O (see Sect. 6.3.2).

6.3.1 Review of INS Data for Optimally-Doped LSCO

We first set the stage with a brief review of relevant INS data for LSCO:x, with x ~ 0.15. It seems that of all the cuprate compounds, LSCO alone was available in large, uniform crystals with homogeneous oxygen content. Conveniently enough, $\chi''(\boldsymbol{q}, \omega)$ is a quantity which can be measured directly by means of INS. The principal difficulty is that INS measurements can only be made at frequencies several orders of magnitude higher than those at which NMR data are recorded. We examine the available INS data with that problem in mind.

Some relatively low-frequency INS data for $\chi''(\boldsymbol{q}, \omega)$ are shown in Fig. 6.4. In part (a) we see data for scans through two of the four incommensurate peaks located at $(\pi, \pi) \pm \delta(0, \pi)$ and $(\pi, \pi) \pm \delta(\pi, 0)$, where $\delta = 0.245$ for $La_{1.86}Sr_{0.14}CuO_4$ [275]. The peaks are seen to be well resolved, with relatively little difference between behavior near $T_c = 35$ K and that at 4.6 K. There is, however, an increase in intensity at the point between the peaks at higher energy. That trend is also visible in the lower-energy scans from Thurston et al. [276] in part (b) of Fig. 6.4, where at $\omega = 1.5$ meV the peaks are completely resolved, while at 4.0 meV there is already some filling in of the space between peaks. The incommensurability varies with composition, $\delta \simeq 2x$ [274], but it shows no measurable variation with energy at the lowest energies scanned [276].

From available INS results for LSCO, then, it does not appear that discommensurations will vanish at lower (i.e. at NMR) frequencies, although the point remains at issue. At higher frequencies, however, early work showed the discommensurations fading into a single broad peak by $\omega = 15$ meV [274]. That this trend would continue to higher energies was verified by studies to above 200 meV carried out at ISIS [277]. In that study, results for LSCO:x, x = 0.14, were compared with the undoped antiferromagnet La_2CuO_4, both $\boldsymbol{q}$-scans and energy spectra. For LSCO:0.14, discommensurations are not visible in the lowest energy scans (25–50 meV) reported [277], nor any above that. The resolution in $\boldsymbol{q}$-space may be somewhat marginal for this purpose, although the AFM (Bragg?) peak at 295 K for La_2CuO_4 showed a width $\Delta \boldsymbol{q} \sim 0.25$ r.l.u.

We display three composite data plots from Hayden et al. [277], namely the equal-time spin correlation function $S(\boldsymbol{q}) = \int_{-\infty}^{\infty} d\omega\, \chi''(\boldsymbol{q}, \omega)$, the zero-frequency susceptibility $\chi'(\boldsymbol{q}) = (2/\pi) \int_0^{\infty} d\omega \chi''(\boldsymbol{q}, \omega)/\omega$, and the local susceptibility $\chi''(\omega) = \sum_{\boldsymbol{q}}' \chi''(\boldsymbol{q}, \omega)$. These quantities have been computed from the results of the spallation–

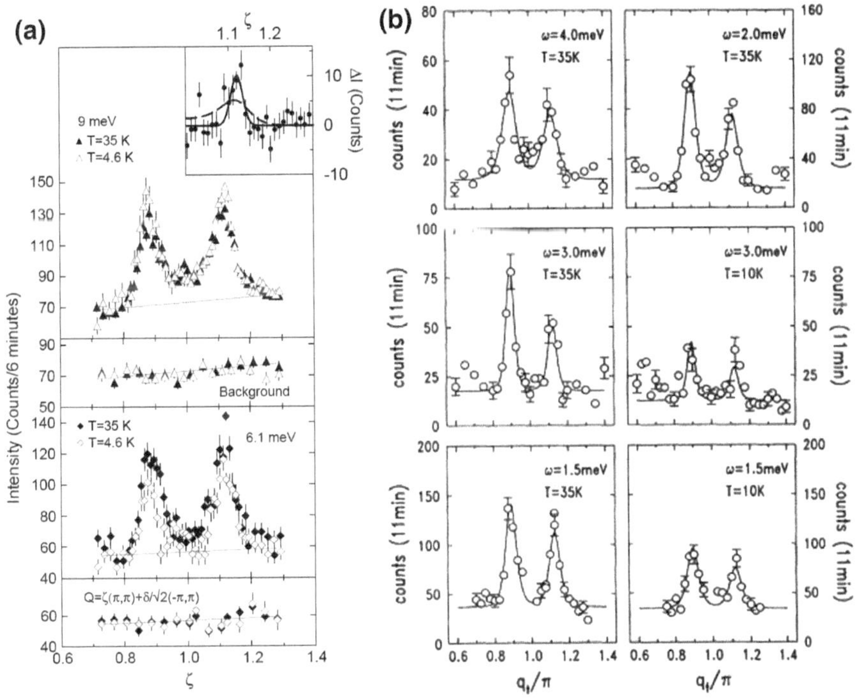

Fig. 6.4 **a** INS $\boldsymbol{q}$-scans along the path shown at the bottom of the figure are plotted for $T \sim T_c$ and for $T \ll T_c$ for the two scattering energies shown [275]. These data show the well-resolved incommensurate peaks and a relatively small difference in scattering intensity for the two temperatures. The inset shows the difference in intensity measured for the two temperatures at 9 meV. Note the small, but definite increase in intensity between the peaks between 6 and 9 meV. **b** INS data at energies $\omega \leq 4.0$ meV, showing very well-resolved incommensurate peaks with no measureable variation of the discommensuration δ with ω. These authors find $\delta = 0.22$ for $La_{1.85}Sr_{0.15}CuO_4$. They also find relatively small variation of intensity with temperature below T_c

source study [277]. The local susceptibilities are shown in Fig. 6.5a, b, where it is noteworthy that the results are plotted in absolute units, and where in (b), reactor-based results are included below ~25 meV. Below we shall compare these data with the absolute susceptibility results which were used to calculate T_1 in [239] (see Sect. 4.1). The results in Fig. 6.5a, b show that the effect of doping is to concentrate $\chi''(\boldsymbol{q}, \omega)$ at energies below 50 meV (as compared with La_2CuO_4), and that the peak seen at 22 meV is a "new energy scale" for this superconductor [277]. This energy is also a rough upper limit for the appearance of discommensurations. The solid line plotted in panel (b) is

$$\chi'' = \frac{\Gamma\omega}{(\Gamma^2 + \omega^2)}, \qquad (6.3.2)$$

with $\Gamma = 22 \pm 5$ meV.

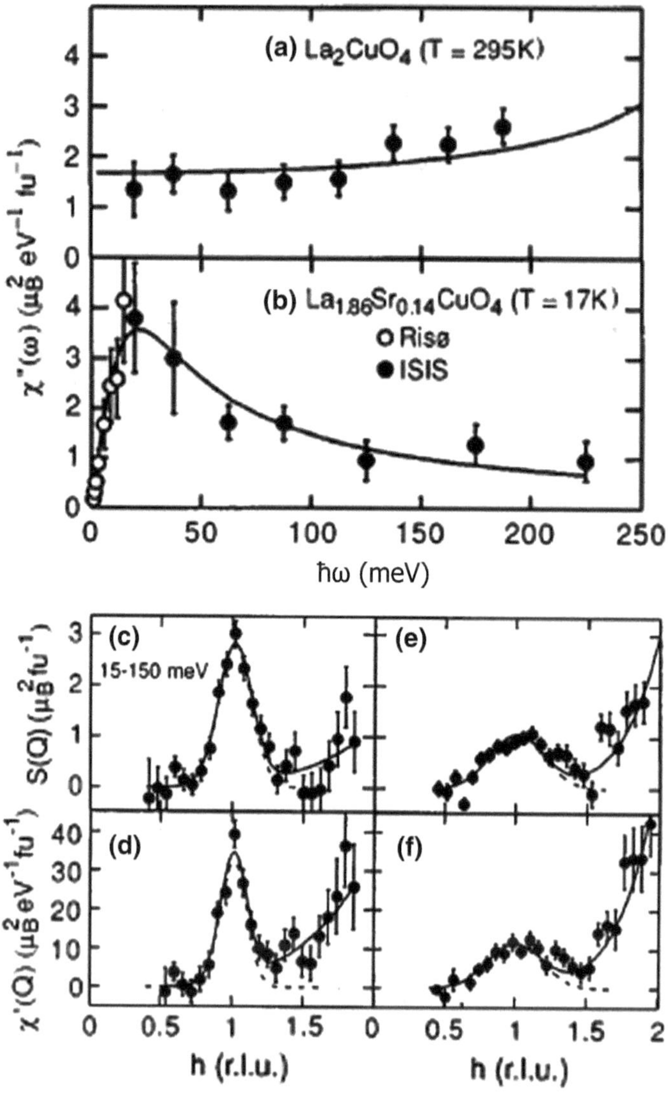

Fig. 6.5 **a** and **b** The local susceptibilities $\chi''(\omega)$ for La_2CuO_4 (295 K) and LSCO:0.14 (17 K), derived from the data presented in [277], are plotted as a function of energy $\hbar\omega$. As compared with La_2CuO_4, the effect of doping is seen to concentrate the amplitude at low energies, giving a peak at ~22 meV. Below that point are plotted data from earlier reactor-based studies [275], which serve to determine the initial slope used for T_1 calculations. **c** and **e** are the equal-time spin correlation functions $S(\boldsymbol{q})$ defined in the text, for La_2CuO_4 and LSCO:0.14, respectively. **d** and **f** are zero-frequency susceptibilities $\chi'(\boldsymbol{q})$ obtained with (6.3.1) in a similar fashion. For LSCO:0.14 ((**e**) and (**f**)) these quantities are nearly indistinguishable, since the width of $\chi''(\boldsymbol{q},\omega)$ shows very little dependence on energy [277]. In (**c**)–(**f**) the rise in intensity to the right is caused by phonons, where the suggested behavior of the quasiparticle properties is indicated by dashed lines. Although the spallation source resolution width is not as good as in reactor-based studies, it is clearly well below the peak widths found for LSCO:0.14 in (**e**) and (**f**)

The equal-time correlation function $S(\boldsymbol{q})$ and the zero-frequency susceptibility $\chi'(\boldsymbol{q})$ obtained with (6.3.1) in [277] are shown in Fig. 6.5 for the undoped ((**c**) and (**d**)) and the doped ((**e**) and (**f**)) cases, respectively. Owing to the near-constant width of $\chi''(\boldsymbol{q}, \omega)$ in $\boldsymbol{q}$-space for LSCO:0.14 [277], there is very little difference between $S(\boldsymbol{q})$ and $\chi'(\boldsymbol{q})$. The associated energy integrals have been carried out for ω ranging from 15 to 150 meV. The central peak in Fig. 6.5e, f has an HWHM of $\kappa = 0.26(5)\,\text{Å}^{-1}$, which is just slightly larger than the incommensurate splitting. Since the Kramers-Kronig integral favors low energies, one wonders whether there would be visible incommensurate peaks in $\chi'(\boldsymbol{q})$ if the reactor-based intensity (e.g. that which gives the open circles in Fig. 6.5b) were included in the calculation of $\chi'(\boldsymbol{q})$. As noted below, such a contribution might be appreciable for indirect spin-spin coupling with the planar ^{17}O.

6.3.2 *Interpretation of ^{63}Cu and ^{17}O T_2 Data with INS Data for $\chi'(\mathbf{q})$*

In Sect. 6.1 the interpretation of spin-echo decay (T_{2g}) measurements with the purely static indirect coupling term $\mathcal{H}_{zz}$ (see also Appendix A.1.4) was discussed in some detail. This technique has been applied widely to cuprate NMR/NQR measurements. Here we consider the case of LSCO:0.15, for which we shall use the INS-derived calibration of $\chi''(\boldsymbol{q}, \omega)$, along with the Kramers-Kronig relation (6.3.1), to obtain quantitative estimates of $\chi'(\boldsymbol{q})$. The latter data, combined with HF constants from Table 4.1, are used for the indirect coupling calculation.

Spin echo decay in LSCO:0.15 presents a number of complications which did not occur in the "static" cases discussed in Sect. 6.1. The effects of T_1 fluctuations, both direct and indirect (i.e., of neighbors), are quite important. Moreover, there is strong evidence that like-spin ^{63,65}Cu neighbors in the $m = \pm\frac{1}{2}$ states are fluctuating among themselves on the time scale of T_2, changing the nature of the problem in a fundamental way.

The planar ^{17}O also present a different and unique situation, in that T_2 is dominated by the $^{63,65}Cu$-^{17}O indirect spin-spin coupling term, which is time-modulated by the ^{63,65}Cu T_1 processes. The ^{17}O-^{17}O coupling term is reckoned to be negligibly small. The somewhat different methodology needed to treat these cases was reported in [26]. The techniques used there will be reviewed briefly here in the course of discussing the T_2 data for LSCO:0.15 and its interpretation.

^{63,65}*Cu T_2 Data and Simulations of Nuclear Spin Fluctuations*

Spin-echo decay measurements for ^{63,65}Cu were carried out on a partially oriented powder sample of LSCO:0.15 containing ~30% ^{17}O [26] at $T \simeq 100$ K. The measurements were made on the $(-\frac{1}{2} \leftrightarrow \frac{1}{2})$ transition in a field of 7 T. Decay curves were measured for ^{63}Cu with several levels of refocussing pulse as well as for ^{65}Cu with $\theta_2 \simeq \pi$. The echo decay data obtained are shown in Fig. 6.6a [26].

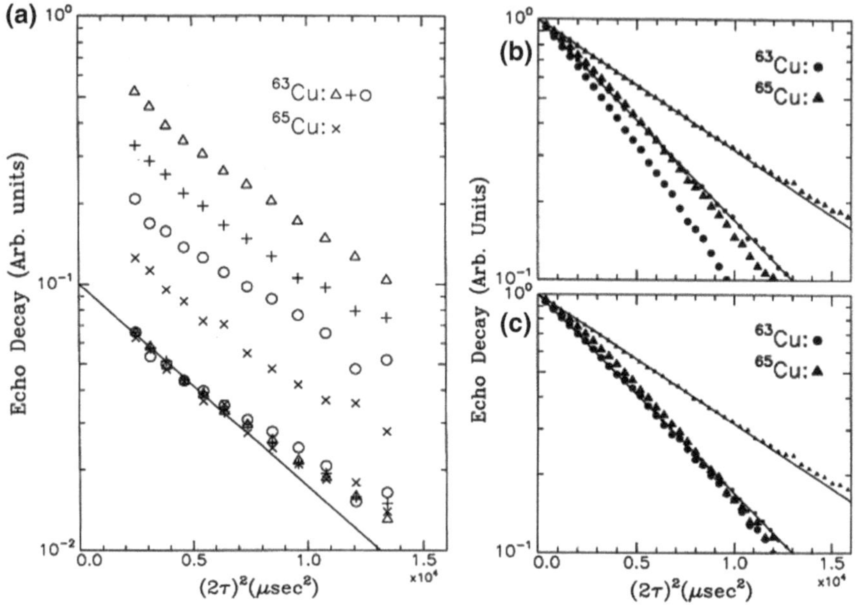

Fig. 6.6 **a** Spin echo decay curves for ^{63,65}Cu taken on the $(-\frac{1}{2} \leftrightarrow \frac{1}{2})$ transition in a field of 7 T at a temperature of 100 K. Data for different refocussing pulses have been displaced vertically for clarity. Conditions are as follows: Full power, $\theta_2 \simeq \pi$ (^{63}Cu–Δ). Power down 10 db, $\theta_2 \simeq 3\pi/4$ (^{63}Cu–+). Power down 10 db, $\theta_2 \simeq \pi/4$ (^{63}Cu–$\circ$). Full power, $\theta_2 \simeq \pi$ (^{65}Cu–$\times$). The curvature is thought to be caused by a distribution of indirect coupling strengths. The bottom curve is a composite of all the data. The solid line is the estimated Gaussian decay with $T_{2g} = 75.6\,\mu$s. These data have all been corrected for direct T_1 effects as described in the text. **b** and **c** Simulated decay curves obtained with random-number selection techniques as described in [26]. Small dots and triangles show static decay curves for ^{63}Cu and ^{65}Cu, respectively, where the decay is caused by a single π pulse. The solid lines show pure Gaussian decay. For (**b**) the large triangles and dots show how the decay curves are modified if random transitions simulating T_1 relaxation effects are included. For (**c**) the large triangles and dots show results for when both T_1 fluctuations and flip-flop transitions between the $\pm\frac{1}{2}$ levels for ^{63,65}Cu are included as described in the text

The decay data shown have been corrected for direct T_1 relaxation effects using results from Appendix A.2. The essence of the procedure is as follows. Direct T_1 relaxation effects result in a factor $\exp(-2\tau/T_{1E})$ multiplying the spin echo decay function. The value of T_{1E} depends on the T_1 process of the nuclei observed, i.e. on its anistropy and on the transition being excited. Defining W_α to be the contribution to $1/T_1$ from fluctuations along the α axis, then the relaxation rate with field along the z axis becomes $1/T_{1z} = W_x + W_y$. For arbitrary half-integral spin the $(-\frac{1}{2} \leftrightarrow \frac{1}{2})$ transition echo decay rate for isotropic spin-lattice relaxation is $1/T_{1E} = (I + \frac{1}{2})^2/T_1$ [283]. For the anisotropic case this becomes

$$\frac{1}{T_{1Ez}} = \frac{[I(I+1) - 3/4]}{T_{1z}} + \frac{1}{2T_{1x}} + \frac{1}{2T_{1y}}, \tag{6.3.3}$$

in terms of T_1 values measured along the three axes. For axial symmetry with $I = \frac{3}{2}$ this becomes $1/T_{1Ez} = 3/T_{1z} + 1/T_{1xy}$ in an obvious notation. With known values of $T_{1\alpha}$ for LSCO:0.15, this formula gives $T_{1Ec} = 90\,\mu$s for ^{63}Cu and 80 μs for ^{65}Cu. The corresponding correction factors have been applied to the echo decay data in Fig. 6.6a. The decay curves in Fig. 6.6a have been shifted vertically for clarity. At the bottom of the figure is a composite plot of all four curves, which shows that they are very nearly indistinguishable. This is a remarkable result in the sense that the decay is independent of θ_2 and is also the same for both isotopes, whose γ values differ by ~7%. The first effect shows that there is considerable amount of "flip-flop" motion taking place, and the second can be rationalized by means of the latter effect combined with the effect of T_1 fluctuations [26]. We give here a brief account of these mechanisms, which were presented in detail in [26].

In order to take account of the foregoing effects in a calculation of spin echo decay, the decay process has been simulated with a model where the spin in question is coupled to neighbor spins with a set of randomly chosen coefficients. The coefficients are scaled to give a second moment adjusted at first to produce the observed value of $1/T_{2g}$ (line drawn in Fig. 6.6a) in a static decay process for ^{63}Cu. In a statistical simulation of such a static echo decay this yields the small dots in Fig. 6.6b for ^{63}Cu and small triangles for ^{65}Cu. Straight lines representing precise Gaussian decay functions are drawn for comparison. Note that the ^{65}Cu spin echo decays much more slowly than the ^{63}Cu. This is because the 31% abundant ^{65}Cu has only half as many "like-spin" neighbors with which to relax the echo [26].

The next step is to introduce the "indirect T_1 effects" by allowing the neighbor spins to execute random transitions between m levels at the correct rate dictated by magnetic dipole coupling T_1 processes. The result of this is shown as large dots (^{63}Cu) and large triangles (^{65}Cu) in Fig. 6.6b. The decay constant $1/T_{2g}$ for ^{63}Cu is increased by ~15% by the T_1 fluctuations, but for ^{65}Cu the increase is nearly three times as much. This is because the number of neighbor spins relaxing the ^{65}Cu is increased by about a factor of six, whereas the increase for ^{63}Cu is much smaller. Now the ^{63}Cu are relaxing ~10% faster than the ^{65}Cu.

The final step is to initiate flip-flop transitions between the $\pm\frac{1}{2}$ levels of both isotopes. That these are actually taking place is evidenced by the absence of change in the decay curve with θ_2 in Fig. 6.6a. Since the $(\pm\frac{1}{2} \leftrightarrow \pm\frac{3}{2})$ transitions undergo severe first-order quadrupolar broadening, flip-flop transitions are only allowed for the $\pm\frac{1}{2}$ levels. Estimates from the indirect interaction calculations suggest that flip-flops are present on a time scale of about 50 μs. Introducing these into the simulation then gives the decay curves shown in Fig. 6.6c (large dots and triangles). The ^{65}Cu decay time increases slightly, but the ^{63}Cu increases by more than 10%, so that the two now very nearly coincide. Evidently, the flip-flops give a slightly greater "narrowing" effect for like spins than for unlike spins. In any case, now the decay times are nearly the same for both isotopes, as was observed. Quite coincidentally, they are both now equal to the initially estimated static decay time for ^{63}Cu within just a small error. We do not believe that there is any simple way to obtain this result except by simulation [26].

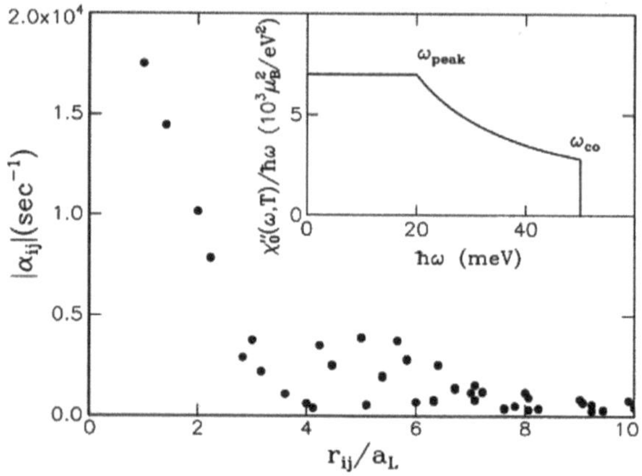

Fig. 6.7 The calculated range function α_{ij} for the "zz" indirect coupling terms among the ^{63,65}Cu is shown as filled circles versus r_{ij}. Only coefficients greater than 2% of the *nn* value are included in the plot. The inset shows the coefficient from (6.3.4) which occurs in the KK integral, adjusted to yield the measured value of T_2

Indirect Coupling Calculations for* ^{63}*Cu

Having determined a reasonable experimental estimate of $1/T_{2g}$ for the ^{63}Cu in LSCO:0.15, we could now compare that number with calculations based on an evaluation of $a_\alpha(i, j)$ with (6.1.7), the HF parameters from Table 6.1, and $\chi'(\boldsymbol{q})$ calculated with INS data for $\chi''(\boldsymbol{q}, \omega)$ using the Kramers-Kronig relation (6.3.1). This kind of calculation was reported in [26], using the form (4.1.3) for $\chi''(\boldsymbol{q}, \omega)$ at all energies. Below, we display the latter results and then make a critical commentary on how this evaluation should be modified in the light of more recent INS data, particularly the high-energy spallation source results [277].

In [26] the same basic expression (6.1.9) was used for the coefficients, where the Kramers-Kronig integral of the expression

$$\chi''(\boldsymbol{q}, \omega) = \frac{\chi_0''(\omega, T)\kappa^4(\omega, T)}{[\kappa^2(\omega, T) + R(\boldsymbol{q})]^2}, \tag{6.3.4}$$

in the same notation as (4.1.3), is used for $\chi'_\alpha(\boldsymbol{q})$. In this expression, one has $\kappa(\omega, T) = \kappa_0^2 + a_L^2[(k_B T/E_T)^2 + (\hbar\omega/E_\omega)^2]$, with $\kappa_0 = 0.034\,\text{Å}^{-1}$ and $E_T = E_\omega = 47$ meV [278]. For the low-frequency region ($\omega < 20$ meV), the function $\chi''(\omega, T)$ can be represented as linear in ω and declining with temperature as $1/T^2$ [278]. It was found that the KK integral up to 20 meV was insufficient to account for $1/T_{2g}$, so the integral was extended, with $\chi''(\omega, T)$ = constant, up to a cutoff for an evaluation at T = 100 K. Adjusting the cutoff to obtain the experimental value of $1/T_{2g}$, it was found to be $\omega_{co} \simeq 50$ meV.

The range function α_{ij} for the ^{63}Cu coupling ($\mathcal{H}_{zz}$) calculated as described is displayed in Fig. 6.7a, where the KK integrand $\chi''(\omega, T)/\omega$ is shown in the inset. Since there are usually four neighbors at each value of r_{ij}, this figure shows that the indirect coupling second moment is dominated by 15-20 neighbor nuclei.

It is interesting to consider how this analysis might be changed in the light of high-frequency $\chi''(\boldsymbol{q}, \omega)$ data [277] (Fig. 6.5), which were not available at the time [26] was submitted. There are two main differences if one envisions using the data of Fig. 6.5. First, to the extent that discommensurations survived in $\chi'(\boldsymbol{q})$, they would no doubt be greatly diminished in importance with data from Fig. 6.5, which would be represented by a single peak with amplitude varying as $\Gamma\omega/(\Gamma^2 + \omega^2)$ with $\Gamma = 22{\pm}5$ meV. This is probably of little significance for the ^{63,65}Cu, but it will figure in our discussion of ^{17}O below. Second, the Lorentzian gives a broader "spectrum" of contributions, extending out to 150 meV and beyond. Interestingly, if one performs the KK integral for these two cases, the results agree within a few percent. Thus, the new data [277] corroborate the earlier result for ^{63,65}Cu [26].

T_2 Decay of the Planar ^{17}O at 100 K

The planar ^{17}O spin-echo decay data at 100 K are displayed in Fig. 6.8. This decay curve cannot be characterized in a simple way, a circumstance resulting, we believe, from the nature of the sample. Whereas only a modest percentage of field-oriented sample material contributed to the Cu NMR signals, the ^{17}O second-order quadrupolar shifts are negligibly small. As a result, essentially all of the poorly-aligned sample material contributes to the planar ^{17}O line. We therefore treat the system as a random powder. Furthermore, quite unexpectedly, the dipolar spin-spin coupling alone accounts for the decay curve in Fig. 6.8. Its peculiar character, with a long tail of low amplitude, is caused by the "magic angle" region of phase space where $3\,cos^2\theta - 1 \sim 0$. The corresponding subset of nuclei relax very slowly.

There is, however, more to the story than just those simple points. On evaluating the indirect coupling terms for the planar ^{17}O, it was found that the ^{17}O–^{17}O terms were small, but that there was a sizeable ^{63,65}Cu –^{17}O term, which gives an unlike spin indirect second moment $\langle\Delta\omega^2\rangle^I_{AB} = 1.5 \times 10^7\,s^{-2}$ [26], larger than the corresponding dipolar term $\langle\Delta\omega^2\rangle^{dip}_{AB} = 6.25 \times 10^6\,s^{-2}$ (98% of which is contributed by the two *nn* ^{63,65}Cu spins). This unlike-spin second moment is expected to dominate the ^{17}O spin-echo decay, because of its time-modulation by the ^{63,65}Cu T_1 processes. Thus, for a crystal of unique orientation one has $1/T_{2O} = \langle\Delta\omega^2\rangle_{AB} T_{1Cu}$.

The calculation of the powder-average decay curve, including indirect couplings evaluated as for the ^{63,65}Cu case above, leads to the decay curve plotted with "×" symbols in Fig. 6.7b. This is widely at variance with the experimental data. The difficulty, we surmise, is again related to the assumed discommensurations at NMR frequencies. If $\delta \to 0$ in $R(\boldsymbol{q})$ (4.1.4), the indirect coupling drops by a large factor and its contribution to $\langle\Delta\omega^2\rangle_{AB}$ becomes negligibly small. *This inconsistency with discommensurations is on the same scale as the estimate of* T_1 *in Fig.* 4.9b. It again suggests that the discommensurations are not "seen" at NMR frequencies.

There is yet another small surprise in the interpretation of Fig. 6.8. If one takes the dipolar second moment in the foregoing formula, i.e. $1/T_{2O} = \langle\Delta\omega^2\rangle^{dip}_{AB} T_{1Cu}$, averaged over all orientations, one finds the curve plotted with small triangles in Fig. 6.7b, not quite in quantitative agreeement with the data. The final curve, plotted with filled squares, is only obtained when one includes the 50 μs–rate flip-flops

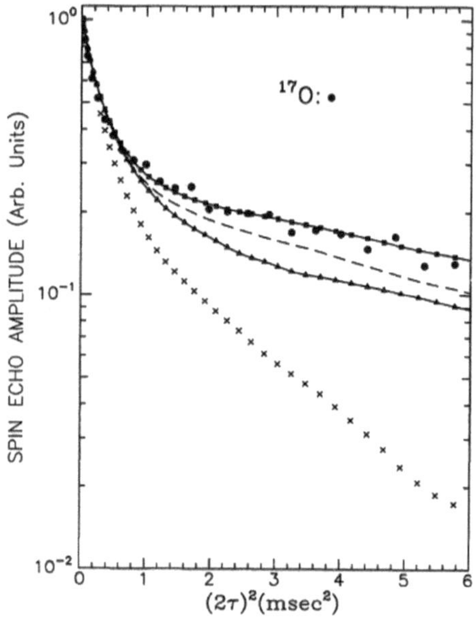

Fig. 6.8 Experimental data for the ^{17}O spin echo decay at 100 K (filled circles) are plotted versus $(2\tau)^2$. The field is oriented along the c axis. A simulated decay curve with the combined effects of dipolar and indirect coupling calculated as for the ^{63,65}Cu is shown as "×" symbols. A simulated decay curve with purely dipolar couplings modulated by T_1 processes alone is plotted with filled triangles connected by a solid line. The same calculation with both T_1 modulation and dynamic flip-flop transitions between the $\pm\frac{1}{2}$ levels of the ^{63,65}Cu neighbors, shown as filled squares connected by a solid line, is seen to be in good accord with the data [26]

between the $\pm\frac{1}{2}$ levels. These, we recall, were important in interpreting the ^{63,65}Cu T_2 decay curves as well.

In the end, however, the most astonishing result is the total abnegation of indirect spin-spin coupling for the planar oxygen nuclei.

6.4 The Sokol-Barzykin-Pines Scaling Theory of $\chi(\mathbf{q}, \omega)$

A large body of work has grown up around the discussion of scaling theories in the normal state of YBCO, Y248, and LSCO by Pines and co-workers [240, 273] and Moriya et al. [163]. We offer here a brief summary of NMR/NQR results and interpretation in terms of the theories mentioned. Although a great deal has been done, there seem to be important questions which remain unresolved as well. The details of these theoretical models go beyond the scope of this book, so we shall focus here on predicted behavior and experimental tests thereof. In the interest of presenting a compact account of this subject, many interesting experimental details have been omitted. The interested reader is referred to the literature cited for a more detailed account of these matters.

6.4.1 *Phase Diagram and Basic Relationships for* T_1, T_{2g}

To facilitate discussion we display the basic phase diagram presented by Barzykin and Pines (BP) [240] in Fig. 6.9, which refers specifically to YBCO:y (see also the paper by Sokol and Pines (SP) [273]). For the superconducting phases ($6.45 \le y \le 7.0$) the important regions are labelled "quantum critical" (QC), "quantum disordered" (QD), and "no scaling", which was formerly termed the "overdamped" (OD) region [273]. These terms refer, of course, to the behavior of the AFM fluctuations in the system, which are generally assumed to control the static and dynamic magnetic properties. One of the key parameters is a temperature-varying correlation length ξ, which is dimensionless in units of the lattice constant a. In the QC region, $\xi \gg 1$ but grows shorter with increasing temperature, until at $\xi \sim 2$ there is a crossover labelled T_{cr} to the OD region, in which ξ is of the order of 1 or less. At higher temperatures and doping levels the AFM fluctuations cease to be a controlling factor and the system reverts to a normal Fermi liquid. The temperature T_* which marks the boundary between QC and QD will be defined below.

It is useful to note that the phase boundary in Fig. 6.9 is similar to the phase boundary between regions I and II of the phase diagram proposed by Varma (see Fig. 4.34). The implied physical significance of these phase boundaries may also be quite different in detail [222].

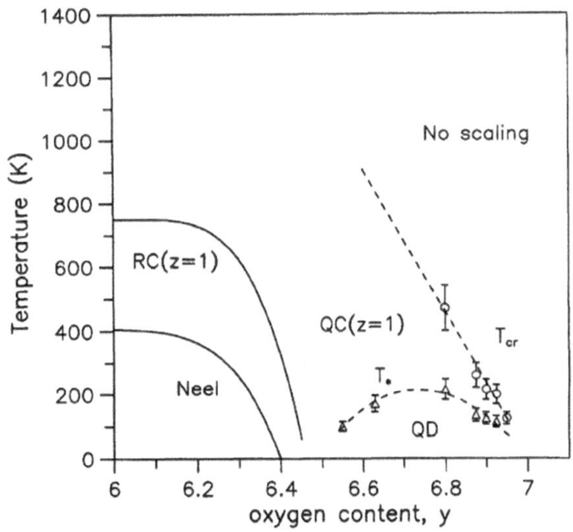

Fig. 6.9 Phase diagram from Barzykin and Pines [240] for magnetic scaling behavior in the YBCO family of cuprates. In the superconducting phases there are three regions identified, the quantum critical (QC: z = 1), the quantum disordered (QD: z = 1), and the "no scaling" (or overdamped OD: z = 2). Definition and measurement of the phase boundaries is discussed in the text

In Fig. 6.9, T_{cr} has also been identified with a maximum in the uniform susceptibility by BP, and data points shown in Fig. 6.9 for T_{cr} have derived from NMR shift data. Large error bars reflect the flatness of such maxima. Below, we review an effort to locate T_{cr} more precisely with NQR studies of T_1 and T_2 of ^{63}Cu(2) in Y248. From the behavior of ^{89}Y NMR shift data, Han et al. [292] have determined that the equivalent oxygen level of Y248 in the YBCO:y family is at YBCO6.8. On that basis, BP suggest that T_{cr} for Y248 should occur at $\sim$ 500 K.

Although drawn for YBCO, the phase diagram structure shown in Fig. 6.9 is presumed to apply to all cuprates. The superconducting region lies within the QD region. A system doped to the right of the T_{cr} line has no pseudogap and lies in the OD region at all temperatures. If there are AFM fluctuations in the OD region, they are assumed to obey mean-field behavior [163] as opposed to QC behavior. We now define the behavior of $\chi(\boldsymbol{q}, \omega)$ according to the BP model [240, 273].

To consider the SP-BP model, AFM correlations are characterized by the correlation length ξ and the low-frequency fluctuation energy scale ω_{sf}, both of which may vary with temperature. These energy and length scales are related by the *dynamical exponent* z, thus $\omega_{sf} \sim \xi^z$. In this connection it is also useful to employ the MMP model susceptibility [161, 242, 273]

$$\chi(\boldsymbol{q}, \omega)_{MMP} = \frac{\xi^2 \chi_0}{1 + q^2\xi^2 - i\omega/\omega_{sf}} \tag{6.4.1}$$

as was employed by PS in their indirect coupling calculation [31]. Then, taking the simplified forms

$$\frac{1}{T_1 T} \sim \int d^2q \left.\frac{\chi''(\boldsymbol{q}, \omega)}{\omega}\right]_{\omega \to 0}, \quad \frac{1}{T_{2g}} \sim \int d^2q\, \chi'(\boldsymbol{q}, 0)^2, \tag{6.4.2}$$

one finds the relations

$$\frac{1}{T_1} \sim \frac{T}{\omega_{sf}}, \quad \frac{1}{T_{2g}} \sim \xi. \tag{6.4.3}$$

Without paying much attention to the approximations involved, we apply (6.4.3) to different parts of the phase diagram. In the overdamped, z = 2 region, we have $\omega_{sf}^{-1} \sim \xi^2$, leading to $T_1 T / T_{2g}^2 \sim$ constant, which was the scaling prediction of Moriya et al. [287]. On this basis it was suggested that $T_1 T / T_{2g}^2$ is approximately constant for YBCO7 [266, 273] (however, see further discussion of this point below). At high temperatures BP suggest $\omega_{sf} \sim T$, so that $T_1 \sim$ constant and $T_{2g} \sim T^{1/2}$, which is at least a semiquantitative description of the data of Imai et al. [268].

The point here is that across the T_{cr} line in Fig. 6.9 the behavior is expected to be qualitatively different. Moreover, the approximations for (6.4.2) and (6.4.3) are better in that $\xi \gg 1$. Thus, with z = 1, $\omega_{sf}^{-1} \sim \xi$ and one expects that $T_{2g}/T_1 T \sim$ constant. Again, at high T, $\omega_{sf} \sim T$, so with (6.4.3) on again expects $T_1 \sim$ constant [240]. An interesting example of the latter behavior is seen in the LSCO data of Ohsugi et al. [279] shown in Fig. 6.10. The behavior of $T_1 T$ versus T is such that the

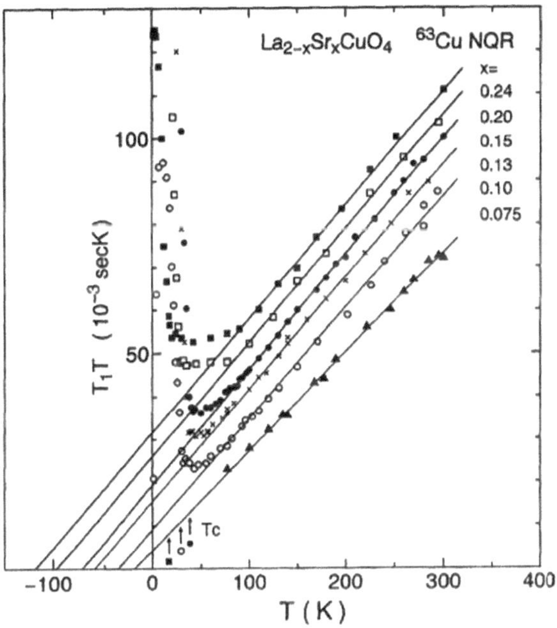

Fig. 6.10 Plot of T_1T data versus T for compositions of LSCO:x, with x ranging over the supeconducting phase (measurements of Ohsugi et al. [279]). These data exhibit the high-temperature limiting behavior of the SP model [273], where T_1T goes asymptotically as $\omega_{sf} \sim T$

curve is linear at high temperatures for all compositions of LSCO:x up to $x = 0.24$. Lest one suppose that $x = 0.24$ is "overdoped" (or OD), however, it is noteworthy that at low temperatures $1/T_1T$ for LSCO:0.24 is more than twice as great as for YBCO7 [279]. What is quite remarkable, however, is that for $x = 0.075$ the linear behavior of T_1T in Fig. 6.10 extrapolates nearly to the origin; T_1 is almost perfectly constant from ~75 K up to 300 K.

The latter figure also illustrates the QC-QD boundary T_*, which is the point at which the data for T_1T break away from the linear high-T behavior. T_* is seen to range from ~50 K to ~100 K over the range of compositions tested.

We note in passing that the behavior of T_1T in Fig. 6.10 suggests that the ^{63,65}Cu relaxation process over this range of temperature and concentrations is dominated by a single mechanism, namely the contribution of the AFM peak(s) in the vicinity of (π, π). That viewpoint, however, is not quite consistent with the behavior seen in Fig. 4.9a, where it appears from the INS data employed there that by room temperature the relaxation will be dominated by a "background" term not resolvable with INS. This discrepancy remains to be analyzed and understood.

As for the behavior of T_{2g} in LSCO, the variation of T_1T/T_{2g}^n, $n = 1$ or 2, was examined in some detail in a study where the indirect T_1 corrections to spin echo decay discussed in Sect. 6.3 were carefully evaluated [204]. It was found for LSCO:0.15 that T_1T/T_{2g} is constant well within the errors and T_1T/T_{2g}^2 was not, clearly placing this system in the QC category. Interestingly, a similar evaluation for YBCO7 found that QC behavior for this system was favored as well [204], contrary to the conclusion mentioned earlier [273]. In much of the literature, the indirect T_1 corrections to T_{2g} have not been taken account of. Note, however, the work of Keren et al. [289].

6.4.2 *The Dynamical Crossover Study by Curro et al: Y248*

A very interesting piece of experimental work regarding the $z = 1$ to $z = 2$ crossover is that of Curro et al. [242], who performed a temperature scan of the dynamics on the system Y248. To our knowledge, no other such test of the T_{cr} boundary has been reported. Y248 is an excellent system for such a test, because it is stable (i.e. does not lose oxygen) for temperatures well beyond 700 K. This study reported T_1 and T_{2g} data up to the latter temperature. In order to perform such a study it was necessary to use a random powder sample, since no known epoxy resin would be stable at such high temperatures. Thus, T_1 and T_{2g} were measured using NQR and the NMR shift was measured using the powder-pattern NMR spectrum. Stray magnetic fields had to be shielded out of the sample space to a fraction of a Gauss, because they disturb the NQR spin echo decay function. Finally, the authors examined and corrected for the indirect T_1 modifications of T_{2g} described in Sect. 6.3 for LSCO [26].

One stimulus for the Y248 study discussed here was the conjecture by Barzykin and Pines [240] that the Y248 system would exhibit a crossover transition at $T \sim$ 500 K. The ^{63}Cu(2) T_1 data from Y248 exhibited a curve of T_1T versus T very similar to one of the lightly-doped compounds in Fig. 6.10 with $T_* \sim$200 K, but showed no feature at 500 K [242]. The T_{2g} data, on the other hand, showed a definite kink near 500 K, as may be seen in Fig. 6.11. Also shown in that figure are data (open circles) from a previous study by Corey et al. [290] which are in excellent accord, but extend only to ~450 K.

In Fig. 6.12 T_1 and T_{2g} data for ^{63}Cu(2) are combined to make plots of both T_1T/T_{2g} (upper panel) and T_1T/T_{2g}^2 (lower panel), using the data described. The upper panel data show that once past $T_* \sim 200$ K, T_1T/T_{2g} remains very nearly constant up to the crossover $T_{cr} \simeq 500$ K, as expected for QC behavior. Above that point it veers upward very slightly. In the lower panel, the ratio T_1T/T_{2g}^2 is plotted with the same data. Up to 500 K the latter ratio is seen to execute a downward path, levelling off at 500 K just where the putative entry into the OD phase should occur. This seems about as clear-cut a crossover as one could expect. Moreover, these data show that the transition from QC to OD behavior is relatively sharp.

The foregoing identification of the crossover was also corroborated by NMR shift data, which indicate (Fig. 6.13) a flat maximum in the uniform susceptibility in the vicinity of 500 K. These data do not locate the crossover as precisely as the dynamical plots, but they are consistent with the results in Fig. 6.12.

The authors have also developed a procedure to observe the crossover directly in terms of $\xi(T)$. Using, first, (6.4.3) with the coefficient adjusted to give the BP predicted value 2.0Å at the crossover, they find the bottom curve in Fig. 6.14 (triangles). Secondly, they have adopted the exact solution for the susceptibility of a Heisenberg AFM given by Sokol, Singh, and Elstner (SSE) [291] to give a second estimate of the

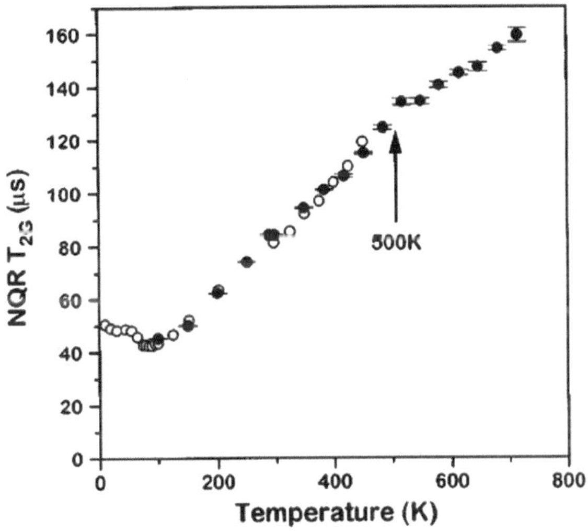

Fig. 6.11 Plot of measured T_{2g} for ^{63}Cu(2), taken on the NQR line of Y248. Data from the study of the dynamical crossover in Y248 by Curro et al. [242] are shown as solid circles. The open circles are from a previous study by Corey et al. [290]. Note the abrupt change of slope at 500 K

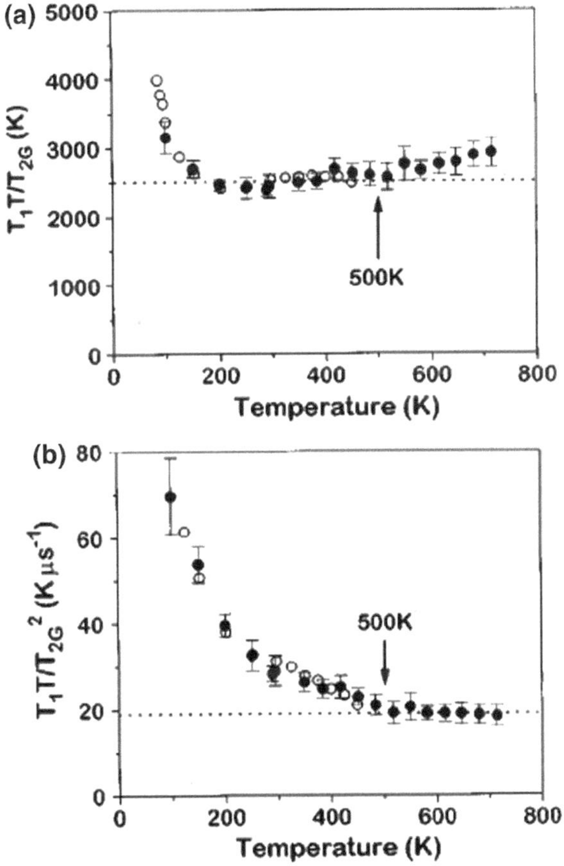

Fig. 6.12 Data for T_1 and T_{2g} for ^{63}Cu(2) in Y248 (Curro et al. [242]) plotted as T_1T/T_{2g} versus T (**a**) and as T_1T/T_{2g}^2 versus T (**b**) (closed circles). The (**a**) panel shows the scaling (QC: z = 1) regime extending from $T_* \sim$200 K to $T_{cr} \simeq 500\,K$. The (**b**) panel shows the onset of the non-scaling (OD: z = 2) regime at $\simeq$ 500 K. The earlier data of Corey et al. [290] are shown as open circles

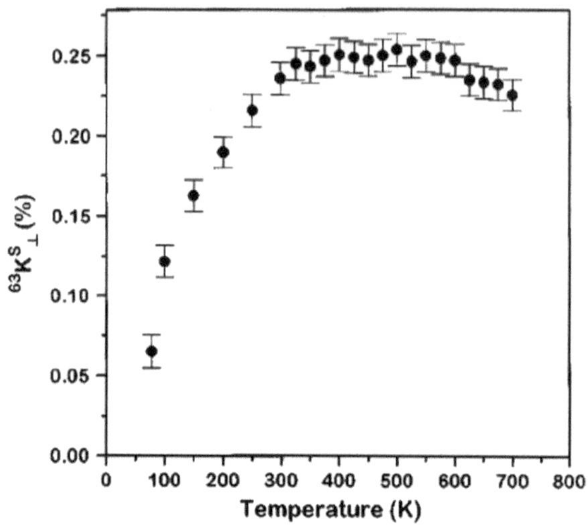

Fig. 6.13 The ^{63}Cu(2) NMR shift is plotted versus temperature, showing the maximum of the uniform susceptibility at $T_{cr} \sim 500$ K, in agreement with the plot of T_{2g} (Fig. 6.11. From Curro et al. [242]

$\xi(T)$ curve, where the latter workers also suggested that this would be a reasonable approximation to the behavior of lightly-doped cuprates at high temperatures [242]. The latter susceptibility is given by

$$\chi(\boldsymbol{q}, \omega, T)_{SSE} = \frac{\chi_0(T)[1+\theta(T)]}{1+\theta(T)\gamma(\boldsymbol{q})+i\Gamma_q\omega}, \tag{6.4.4}$$

where $\gamma(\boldsymbol{q}) = \frac{1}{2}[cos\, q_x + cos\, q_y]$, $\theta(T) = 4\xi^2(T)/[1+4\xi^2(T)]$, and Γ_q is a fluctuation rate which is not required for $\chi'(\boldsymbol{q}, 0)$ (and thus T_{2g}). In using (6.4.4) to interpret the T_{2g} data, the value of $\chi_0(T)$ was taken from the NMR shift measurements, the known HF constants were used, and the second moment was calculated using the full expression (5.1.8). The only unknown parameter $\xi(T)$ is then extracted from the T_{2g} data in Fig. 6.11. These values are plotted in Fig. 6.14 as open circles. According to this determination, ξ = 1.4 at the crossover temperature, somewhat less than the BP estimate. If the determination using (6.4.3) is adjusted to give $\xi = 1.4$ at 500 K, then the curve with filled squares results, which is remarkably similar to the exact solution by SSE.

In all three of the plotted curves for $\xi(T)$ versus T there is a definite point very near to 500 K which one would naturally associate with a transition from one type of behavior to another, even if there had been no theoretical prediction of such a point. With the SSE susceptibility (6.4.4), however, there is a remarkably large change in the slope of the curve at T_{cr}, giving very dramatic evidence of the dynamical crossover.

In conclusion, it seems that the QC, z = 1 behavior for AFM fluctuations is clearly established for Y248 as well as for LSCO [204]. For Y248, the present work finds a clear-cut crossover transition at $T \simeq 500$ K. In the case of LSCO there is evidence for a crossover curve from susceptibility [189, 232] as well as Hall effect data [240, 293]. It appears quite possible that a crossover curve for LSCO could also be established

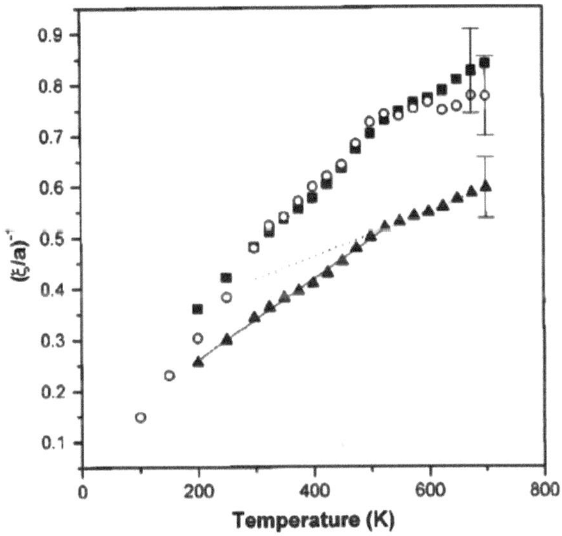

Fig. 6.14 Values of correlation length $\xi(T)$ derived from the data for T_{2g} in Fig. 6.11 as follows: Triangles are from (6.4.3) with coefficient adjusted to give $\xi = 2.0$ at $T = 500$ K. Open circles are from the SSE exact solution [291], where $\chi_0(T)$ is derived from the shift data in Fig. 6.13. Closed squares are from (6.4.3) with the coefficient adjusted to give $\xi = 1.4$ at $T = 500$ K. See text for interpretation of this result

using T_1T/T_{2g}^n plots as was done with Y248. Measurements would probably have to be done with single-crystal samples and conducted in an oxygen atmosphere in order to maintain sample integrity at elevated temperatures (e.g. see [119]). It would be interesting and important to investigate the *sharpness of the crossover transition* in this fashion, since the present work has shown that this could be a very interesting issue [242].

Chapter 7
NMR on Actinide Compounds

7.1 Introduction

There are relatively few laboratories with the specialized facilities required to conduct NMR research on compounds with uranium or with the transuranium elements Np, Pu and Am, which we refer to collectively as the actinides. In the last two decades, however, there has been a substantial increase in the volume of research in general and in NMR in particular on actinide compounds. Beginning with the first observation ever of the NMR (i.e., the AFNMR) of ^{235}U in the AFM state of UO_2 in 1998 [296], there followed sustained activity that has continued up to the present time. It must be noted that because the abundant isotopes of uranium are only very weakly radioactive, research on uranium compounds is far more widely pursued than on all other actinides combined. Since ^{235}U, with a half-life of only 71 million years, is required for NMR studies, shielding and handling facilties are a much greater problem than with the abundant isotope ^{238}U, the half–life of which is nearly two orders of magnitude longer. The natural abundance of $^{235}U \sim 0.72\%$. In this chapter we will discuss ^{235}U and ^{17}O NMR studies in UO_2, while more exotic heavy–fermion compounds such as URu_2Si_2 are deferred to future chapters. In addition to the original papers referred to here, there was also more recently a review of actinide oxide NMR studies [298].

Next, we consider ^{239}Np in NpO_2, a compound with a somewhat dramatic history. In the end it has been identified as having an octupolar ground state, the first actinide identified to have a multipolar ground state. It was also found to exhibit a nuclear cross–relaxation effect from ^{237}Np to ^{17}O, leading to identification of a new indirect coupling mechanism between nuclei. A similar cross–relaxation effect was later identified in another actinide compound, $NpPd_5Al_2$ [314]. Results for these effects are unique among NMR studies and are presented in some detail.

The NMR of the ^{239}Pu nucleus mysteriously eluded discovery for many years, but was finally reported for the compound PuO_2 in 2012 [299]. This is only the second actinide NMR to be observed. Whereas for most actinides the HF coupling

© Springer-Verlag GmbH Germany 2018

R.E. Walstedt, *The NMR Probe of High-T_c Materials and Correlated Electron Systems*, Springer Tracts in Modern Physics 276,
https://doi.org/10.1007/978-3-662-55582-8_7

is too powerful and T_2 is unmeasurably short, in the case of PuO_2 the compound is magnetically inert and there is no obvious T_1 mechanism. This makes the NMR signal very hard to detect. The ^{239}Pu NMR results are discussed briefly.

Finally, we touch upon the difficult case of ^{17}O in AmO_2. With each step across the actinide series the half-life decreases by about two orders of magnitude. For AmO_2, just beyond Pu, it is so short that it is no longer possible to grow a crystal and verify its properties before recoil damage from α–particle decay renders it as disordered as a glass. The NMR results will be seen to reflect such a circumstance.

7.2 The Study of ^{235}U and ^{17}O in UO_2

There are many interesting NMR–related features to the UO_2 compound. Here we present a selection of highlights and refer the interested reader to [297] for the full details. The gyromagnetic ratio $\gamma_{235}/2\pi \simeq 0.78$ MHz/T [296] of ^{235}U is one of the smallest of all detectable nuclei. At the same time, this circumstance can be said to make it possible to observe the AFNMR of ^{235}U in the AFM ordered state. The point here is that f–electron orbital HF interactions are so much larger than any encountered in spd–band solids that one requires a small value of γ as well as rather low absolute temperatures for T_1 and T_2 to be long enough for experimental resolution with pulsed NMR techniques. Thus, ^{235}U and ^{239}Pu are the only actinide nuclei are for which there is any prospect of observing their NMR directly, and that only because ^{239}Pu is available in the non–magnetic host compund PuO_2 [299] (see below).

The actinide oxides AnO_2 all crystallize in the cubic CaF_2 structure. In such a crystal the An^{4+} ions form a *fcc* lattice, while the O^{2-} ions form a *sc* lattice with half the lattice constant of the *fcc* host, with sites located at the centers of An^{4+} tetrahedra. In addition to the 'smallness' of γ_{235}, there are two other circumstances that render the AFNMR of ^{235}U in UO_2 observable, namely the possibility of isotopic enrichment to as much as 90% or more and the occurrence in AFM UO_2 of a HF field at the ^{235}U nucleus of $\sim$252 T, yielding for the $\pm 1/2$ transition an AFNMR frequency of 198 MHz. Along with AFM magnetic ordering, it has been shown with resonant X-ray scattering that there is also quadrupolar ordering among the 5f-electrons [301], generating an electric field gradient at the ^{235}U with its principal axis parallel to the magnetic moment. The result is an AFNMR spectrum with a first-order nuclear quadrupole splitting of about 14 MHz as is shown in Fig. 7.1. With I $= 7/2$, there are seven roughly equally spaced transitions in the AFNMR spectrum.[1]

In addition to scanning the AFNMR spectrum of ^{235}U it is possible to measure $^{235}T_1$ as a function of temperature, as shown in Fig. 7.2. We note in passing that, since $I_{235} = 7/2$, the magnetization recovery curve can be written

[1]In [298] the value of I for ^{235}U is given at first as 7/2, then later stated to be 5/2. The correct value is 7/2.

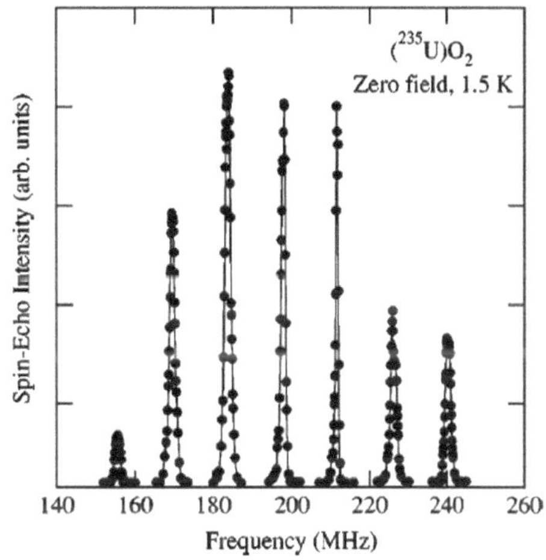

Fig. 7.1 ^{235}U spin echo spectrum in the AFM state of UO_2 at 1.5 K

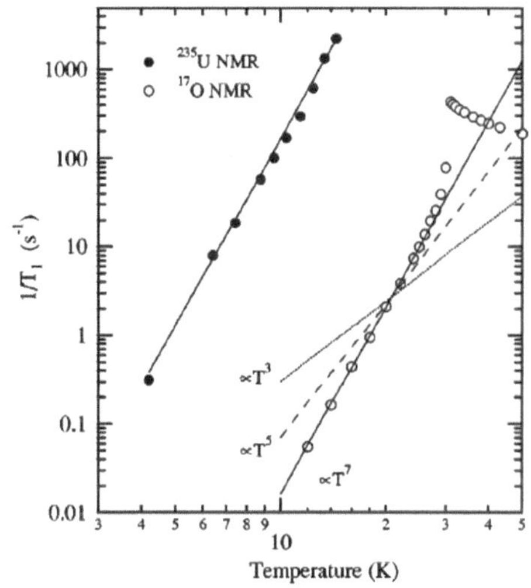

Fig. 7.2 Spin–lattice relaxation rates $1/T_1$ for ^{235}U (filled circles) and ^{17}O (open circles), measured in the ordered state of UO_2, are plotted versus T in a log–log plot. For ^{17}O the data extend beyond $T_N = 30.5$ K up to 50 K showing the paramagnetic peak of critical fluctuations just above the first–order transition. The solid lines through the data plots are fits to a T^7 law. See text for discussion

$$M_0 - M(t) = M_0[Aexp(-t/T_1) + Bexp(-6t/T_1) + Cexp(-15t/T_1) + Dexp(-28t/T_1)], \quad (7.2.1)$$

where M_0 is the equilibrium magnetization. If the central transition ($m = 1/2 \leftrightarrow -1/2$) is completely saturated at t = 0, then it can be shown that (A, B, C, D) = (1/84, 3/44, 75/364, and 1225/1716). See Sect. 2.1.5 for further discussion of these modes.

The data in Fig. 7.2 illustrate why it was only possible to observed the AFNMR of ^{235}U up to $T = 14$ K or so, because above that point $^{235}T_1$ becomes just a few 100 μs, i.e. no longer than the duration of the two-pulse spin echo sequence. This hastens the loss of available signal strength with increasing T. Also displayed in Fig. 7.2 is a plot of $^{17}T_1$ data for ^{17}O in the sample, which has been isotopically enriched for this purpose. The T–dependence of $^{17}T_1^{-1}$ is seen to follow the same T^7 law as for the

^{235}U, except that it is four orders of magnitude smaller. Near T_N it is slightly enhanced by critical fluctuations on both sides of what is actually a first–order phase transition. We finally note that the T^7 variation for both the ^{235}U and ^{17}O nuclear relaxation rates in Fig. 7.2 has been rationalized as a phonon–like excitation with magnon–like couplings to the nuclear moments [297]. At low temperatures the phonon–Raman scattering process that is believed to be at work here varies with temperature as T^7 [302].

^{17}O NMR in the Paramagnetic State

First, we look at the NMR shift and relaxation of the ^{17}O in the paramagnetic state, as depicted in Fig. 7.3. The small and very flat NMR shift is suggestive of a magnetic source that is isotropic and little affected by short–range AFM order such as an orbital susceptibility. Indeed, the dipolar HF effects that are quite powerful at the O^{2-} sites in the ordered state have far less effect in the paramagnetic state because of the high symmetry of the lattice. The very–nearly symmetric line profile shown in the inset to part (a) of the figure shows the absence of any anisotropic shift that would lead to anisotropic broadening. Meanwhile, the FWHM of about 30 G (<20 kHz) is clearly beyond any nuclear dipolar broadening, and could very well represent stray shift

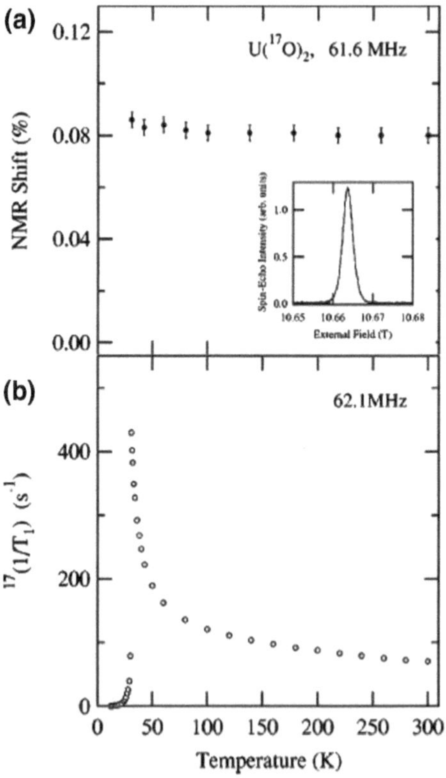

Fig. 7.3 **a** Temperature dependence of the NMR shift at the ^{17}O sites at 61.6 MHz in the paramagnetic state, which indicates a rather small transferred HF coupling constant. The inset shows the ^{17}O spin–echo spectrum in the paramagnetic state. Data were taken with pulse conditions $4.2\,\mu s - \tau - 8.4\,\mu s$. **b** Temperature dependence of $^{17}(1/T_1)$ at 62.1 MHz, which stands in sharp contrast with the constancy of the ^{17}O NMR shift in the paramagnetic state

effects from the relatively huge dipolar HF fields that would cancel under conditions of perfect cubic symmetry. The bulk of such broadening must be static in order to permit the observation of spin echoes. Thus, T_1 broadening would be less than 1 kHz at any temperature.

The T_1 process is, however, another matter entirely. As short–range AFM order develops on approaching T_N, the cancellation of dipolar HF fields that is so effective with uniform polarization begins to break down, and $1/T_1$ exhibits a spectacular critical fluctuation peak. Interestingly, that is eventually cut short and the system executes a first–order jump into the AFM ordered state. The relaxation rate suffers an order–of–magnitude collapse at T_N. At $T = T_N$ the contribution of quadrupolar fluctuations to the T_1 process is unlikely to be significant, as the authors point out [297].

AFM Ordering in the fcc UO_2 Lattice

To understand the static and dynamic AFNMR effects in UO_2, we need to examine the basic features of the ordered AFM state.

It is easy to divide the simple cubic and body–centered cubic lattices into a two–sublattice AFM arrangement of static spin orientations. Not so for the fcc lattice, in which each site has twelve nearest neighbors. Thus, it was only after the AFNMR studies of the fcc UO_2 that the so-called '3k' ordering was confirmed. The term '3k' refers to the fact that there are components of ordered magnetism along all three spatial dimensions. A planar rendition of both *2k* and *3k* ordering is shown in Fig. 7.4, where *3k* order appears to comprise great complexity. It is, however, straightforward to describe verbally: The fcc lattice is composed of four interpenetrating sc lattices.

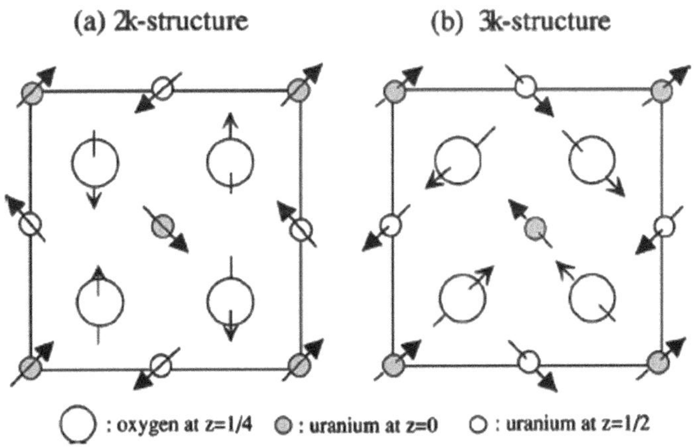

Fig. 7.4 The (001) projections of the fluorite structure for the *2k* and *3k* ordering schemes. The dipolar fields at the O^{2-} sites point along the (101) directions for the *2k* ordering and along the (111) directions for the *3k* ordering. The Jahn–Teller distortions are discussed at length in [297], but are not shown in this figure. These distortions do not significantly modify the dipolar field at the ^{17}O sites, but are entirely responsible for the observed EFG splittings. See text for details

Assign to each sc sublattice one of the four [111] axes of the crystal and you are very near to the '3k' prescription. The one complication is that there are many ways to assign the 'sense' of the magnetic vectors, so that needs to be specified in detail. In Fig. 7.4 we are looking down along the z axis. Assigning in–plane x (horizontal) and y (vertical) axes in the usual way, we see that there are two sublattices in each [100] and [010] plane. The simplest way to specify the entire ordering scheme is to focus on the U^{4+} moments at the eight vertices of the fcc unit cell, four of which are visible in Fig. 7.4. The typical U^{4+} moment in the lower left corner of the figure is seen to be oriented along the [111] axis that intersects the neighboring O^{2-} site, whose local field (see below) is seen to be oriented along the same [111] axis. The four nn U^{4+} moments to that lower–left O^{2-} site are seen to be pointed along all four possible [111] orientations in the crystal. *That pattern repeats for every fcc vertex site in the lattice, and it specifies the entire 3k magnetic structure.* It is also seen that for each O^{2-} site, one of the four nn moments is oriented along a [111] axis that passes through the O^{2-} neighbor. The local O^{2-} field ($\mathbf{H}_{int}$) is always oriented along that line. These two statements specify all of the details of the *3k* ordering.

As noted by the authors, there is a substantial dipolar magnetic field at the oxygen sites. This field always points along one of the [111] axes, as noted above. As a result of the small Jahn–Teller motion of the O^{2-} sites, there is also a small EFG and quadrupolar splitting of the ^{17}O NMR line. The local field and EFG both have a considerable effect on the ^{17}O NMR spectrum in the powder sample. At an NMR frequency of ν_0, the resonance condition can be met at field values over the range $\nu_0/\gamma - H_{int} \leq H_0 \leq \nu_0/\gamma + H_{int}$. Here H_{int} is the internal field at the ^{17}O site, which $\simeq 0.74\,\mathrm{T}$ at low temperatures. At any point on this scale the applied field will point at some angle θ relative to the internal field, so that the resonant field $\simeq H_0 \pm H_{int}\, cos\theta$. At such a point the first–order quadrupole splitting will be $\nu_Q(3\,cos^2\theta - 1)$. Such a splitting will only shift the NMR line by a few kHz, but will affect the excitation conditions for the spin echo. Moreover, the spin echoes will oscillate according to these quadrupole splittings in their decay processes, as was shown in an early classic paper on the subject [303].

To illustrate these points we present in Fig. 7.5 an ^{17}O NMR spectrum taken at a temperature where $H_{int} \sim 0.72\,\mathrm{T}$, and at a large enough value of 2τ so that echo oscillations have dissipated. Also shown is a plot of the quadrupole splitting derived from analysis of spin echo oscillations at points across the spectrum. The NMR spectrum is the somewhat noisy scan extending from 10.1 to 11.45 T. Quadrupole frequencies are shown as solid dots, and the solid line is a fitted curve of $|3\,cos^2(\theta) - 1|$ scaled in height and width to fit the data. Two critical points are, of course, the "magic angles", where $cos(\theta) = \pm 1/\sqrt{3}$. Coincidence between the principal axis of the electrical field gradient tensor and field data verifies that these share the same [111] axis in the lattice.

Lastly, we address the discrepancy between the measured value (0.74 T) and the direct dipolar estimate (0.42 T) of H_{int}. We attribute the difference to a small hybridization transfer of ordered moment from the four nn U^{4+} into p–states in the oxygen. Ordinarily the s–like hybridization might be dominant, but here this contribution is zero, because the vector–sum moment of the nn sites equals zero.

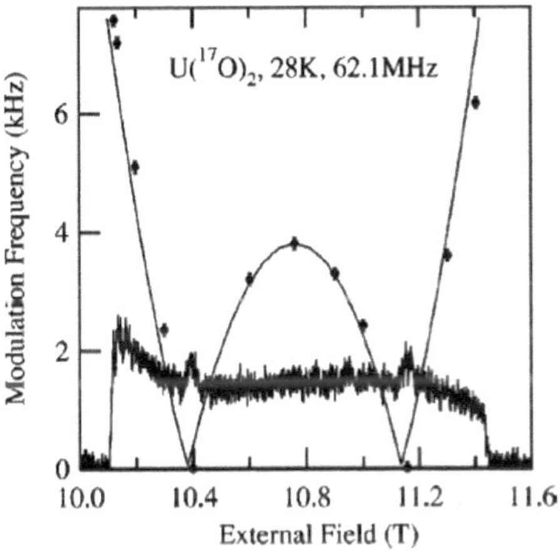

Fig. 7.5 The ^{17}O NMR spectrum (noisy trace) in UO_2 taken below T_N in a scan from 10 T to above 11 T. The echo amplitude was recorded at a value of pulse separation τ large enough so that the first–order quadrupolar oscillations have died out. Black dots represent values of the quadrupolar oscillation frequency recorded using the echo decay oscillations. The solid line fit to the latter data was determined by least-squares fit to the form $|3cos^2(\theta) - 1|$

The ^{17}O $1/T_1$ curve below T_N (see Fig. 7.2) mainly follows the same T^7 behavior as the ^{235}U in the ordered state, executing a noticeable upward deviation owing to critical fluctuations as T_N is approached from below. However, the inhomogeneous broadening, shift, and oscillatory spin echo decay present a challenging array of complex phenomena to be analyzed for the ^{17}O in the ordered state. First, as noted above, owing to the Jahn–Teller motion of the O^{2-} ions in the ordered state of UO_2 there is a small EFG at the ^{17}O site. Careful studies conducted by the authors demonstrate that both $\mathbf{H}_{int}$ and the EFG principal axis lie along the [111] axis that passes through every O^{2-} site. Moreover, another simplifying feature is that the EFG appears to be axial in nature. Under these conditions it has been shown [303] that if the first–order quadrupole splitting is α, then the spin echo decay will have the form[2]

$$E(2\tau) = c_0 + c_1 f_1(2\tau)cos[2\alpha\tau + \delta_1] + c_2 f_2(2\tau)cos[2\alpha\tau + \delta_2] + c_3 f_3(2\tau)\times \\ cos[2\alpha\tau + \delta_3] + c_4 f_4(2\tau)cos[2\alpha\tau + \delta_4], \quad (7.2.2)$$

where the factors $f_n(2\tau)$ represent the approximately exponential decay of the oscillations over several 100 μs.

Next, we note that the EFG principal axis coincides with $\mathbf{H}_{int}$ throughout the crystal, but the orientation of $\mathbf{H}_{int}$ varies randomly throughout the powder sample

[2]This is similar to Eq. (8) from [297], except that decay factors $f_n(2\tau)$ have been inserted to represent the roughly exponential decay of the spin echo oscillations.

used. The ^{17}O resonance condition selects only a particular orientation of the $\mathbf{H}_{int}$ axis over the sample relative to the applied field $\mathbf{H}_0$. We denote that angle θ, so that the resonance condition may be specified as

$$\nu_0 = (1+K)\gamma_{17}[H_0 + H_{int}cos\theta], \tag{7.2.3}$$

where the perpendicular component of $\mathbf{H}_{int}$ is neglected, because $H_0 \gg H_{int}$. As the field H_0 is swept upward, the resonance condition in (7.2.3) is met over a range of θ values from 0 to π, over which H_0 varies from $\nu_0/((1+K)\gamma_{17}) - H_{int}$ up to $\nu_0/((1+K)\gamma_{17}) + H_{int}$. This results in the NMR spectrum shown across the bottom of Fig. 7.5, extending at $T = 28$ K from 10.1 to 11.45 T, so that $H_{int} \sim 0.67$ T. Since the first–order quadrupole splitting is only a few kHz, it does not play a significant role in the resonance condition in (7.2.3). Nonetheless, the orientation of H_{int} can be monitored over this field scan using the spin echo decay oscillations given by (7.2.2) to monitor the first–order quadrupole frequency $|\Delta\nu_Q| = |\nu_Q(3cos^2\theta - 1)/2|$. Measurements of $|\Delta\nu_Q|$ across the NMR spectrum are plotted as dots in Fig. 7.5. A fitted curve of $|\Delta\nu_Q|$ versus θ is drawn as a solid line, using θ values derived from (7.2.3), where the end points of the spectrum determine the locus of $\theta = 0$ and $\theta = \pi$ on the field scale. The correspondence is excellent. The data in Fig. 7.5 can also be used to determine accurate values of the ordered-state ^{17}O NMR parameters as follows. The solid line functional fit to the quadrupole frequencies measured across the broad spectrum is seen to intersect the horizontal axis at two points, known as the "magic angles", where $cos(\theta) = \pm 1/\sqrt{3}$. Thus, $|\Delta\nu_Q|$ given in the previous paragraph vanishes at these points, which are labeled $H^{\pm}_{ma}$. These two points are determined quite accurately by the fit, and the resonant field that is halfway between them determines the shift at that temperature. Thus, we have

$$\begin{aligned} \nu_0 &= \gamma(1+K)[H^{+}_{ma} + H^{-}_{ma}]/2 \\ H_{int} &= (\sqrt{3}/2)(1+K)[H^{+}_{ma} - H^{-}_{ma}], \end{aligned} \tag{7.2.4}$$

where ν_0 is the NMR frequency. The first of these may be solved for K, the second gives H_{int}, and ν_Q is extracted from the scale of the solid line in Fig. 7.5. Data for H_{int} and ν_Q are plotted in Fig. 7.6, while K versus T data are shown in the inset.

The data obtained for the NMR properties of UO_2 offer a substantial characterization of this interesting compound. See [297] for more details and for comments on the f-electron hyperfine mechanism, which goes beyond the scope of our discussion here. We finally note that the T^7 variation for both the ^{235}U and ^{17}O nuclear relaxation rates in Fig. 7.2 has been rationalized as a phonon–like excitation with magnon–like couplings to the nuclear moments [297]. At low temperatures the phonon–Raman scattering process that is believed to be at work here varies with temperature as T^7 [302].

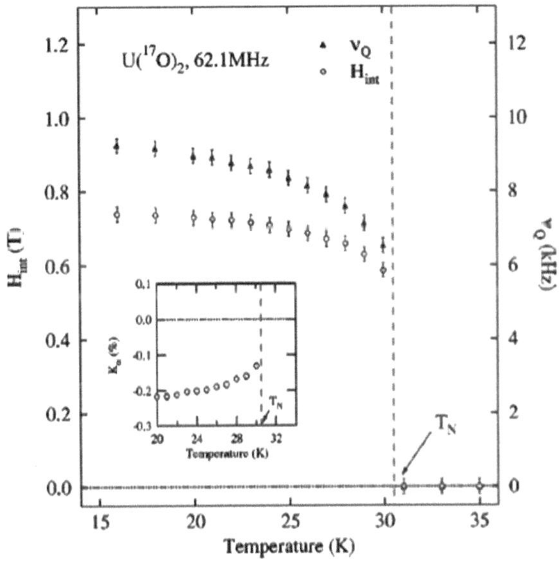

Fig. 7.6 Temperature dependence of the internal field H_{int} and the nuclear quadrupole frequency ν_Q for ^{17}O spins in the AFM state of UO_2. Over a similar temperature range, the shift deduced as described in the text is plotted in the inset

7.3 Octupolar Ordering in NpO_2: ^{17}O and ^{237}Np NMR Parameters

Early studies of specific heat and magnetic properties of NpO_2 and UO_2 showed were antiferromagnets [304, 305]. NpO_2 showed strong Curie–Weiss behavior with a fluctuating moment greater than $3\mu_B$. It was rather surprising, then, when somewhat later both Mössbauer [306] and neutron scattering [307] studies found no evidence for any nonzero magnetic moments in the "ordered state" of this compound. Eventually, nearly fifty years after the initial studies of this system, Santini and Amoretti summarized current experimental evidence and proposed a Γ_2 ground state that carries an octupolar moment [308].

The latter proposal was followed by a seminal resonant X-ray scattering paper and analysis, which showed that [308] was correct in principle, but incorrect in detail [310]. The X-ray results showed that the correct ground state was actually an ordering of octupoles of Γ_5 symmetry, leaving a ground state of zero dipolar moment. The interesting point about this result is that it was not motivated directly by resonant X-ray scattering data, which showed strong quadrupole scattering, but essentially zero octupolar scattering. The only evidence for a ground state that breaks time–reversal symmetry at the time was μSR data [309] that showed interstitial magnetic fields below $T_0 \simeq 26$ K of more than 700 G. Such fields could presumably only arise from a "magnetic" ground state, i.e. from octupolar (or higher multipolar) order. In retrospect, it seemed clear that octupolar X-ray scattering was not observed only because the relevant matrix elements were too weak [310].

It then remained for other probes to verify the octupolar nature of the NpO_2 ground state. NMR was an obvious candidate, and we review here results reported for both NMR shift measurements on ^{17}O that served to verify both the $3k$ quadrupolar ordering and the octupolar terms in the shift data. The observed quadrupolar order [310] was suggested to set up as a $3k$ charge ordering scheme quite similar to the magnetic $3k$ order found for the AFM state of UO_2. Again, each of the four *sc* lattices that make up the *fcc* lattice of the Np^{4+} ions is assigned one of the four [111] axes as symmetry axes for quadrupolar ordering. That is the essential result in a nutshell.

7.3.1 Analysis of ^{17}O NMR Shift Data in the AFO/AFQ Ordered State of NpO_2

In the paramagnetic state the crystal has the cubic space group Fm$\bar{3}$m, while in the triple-q ordered state the symmetry is lowered to Pn$\bar{3}$m. In the latter space group two of the eight O^{2-} sites in the unit cell retain cubic symmetry in their NMR shift tensors, while the other six have shift tensors with axial symmetry, two each for the x, y and z axes. None of the ions in the crystal is displaced from cubic symmetry in the (quadrupolar/octupolar) ground state, so that no EFG's or local magnetic fields occur at the O^{2-} sites. Confirmation of the local symmetries of the various ^{17}O sites has been obtained from a study of the corresponding NMR shifts that occur in the ordered state. Shift results at a series of temperatures are shown in Fig. 7.7. Frequency scans are shown for a series of temperatures beginning above $T_0 = 26\,\mathrm{K}$ at the top, progressing down to 14 K at the bottom, for each of two field orientations on a single crystal sample of NpO_2. For $\mathbf{H}\|\langle 111\rangle$ (part (a)) the six axial site axes were all at the same angle relative to $\mathbf{H}$ and were thus combined into a single NMR line with three times the intensity of the two cubic sites, which have an isotropic shift. Then when the field is aligned with one of the cubic axes (i.e., $\langle 100\rangle$), four of the axial sites (i.e., with y and z symmetry axes) will combine into a single line while those with x symmetry axes will form another line with a strongly positive NMR shift (Fig. 7.7). Meanwhile, the cubic sites have the same (negative) shift as in part (a), since their shift is isotropic. Thus, the ^{17}O NMR spectra show the expected shift behavior corresponding to the $3k$ configuration of quadupolar ordering axes.

There are other ^{17}O NMR effects in the ordered state of NpO_2 that are quite surprising. One is the behavior of the corresponding linewidth as a function of applied field. This is shown in Fig. 7.8. There we see a striking contrast in the behavior of linewidth versus field H at T = 17 K. For the cubic symmetry line the linewidth extrapolates quite precisely to zero as $H \rightarrow 0$. For the composite axial symmetry line, however, the linewidth extrapolates to a very substantial non–zero value. The cause of this effect is clearly some form of vestigial magnetic ordering in the crystal that can produce local magnetic fields when there is no applied field present. Interestingly, this effect was anticipated by Santini and Amoretti in their pioneering PRL on octupolar order in this system [308]. These authors suggested that

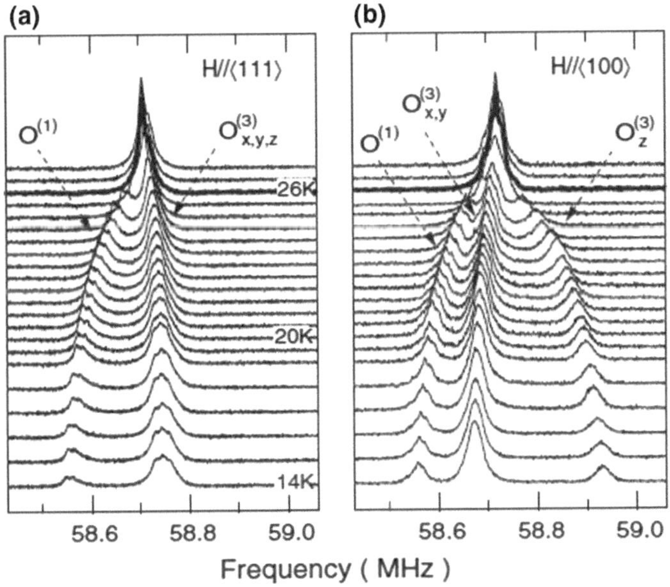

Fig. 7.7 Frequency–scanned NMR spectra of ^{17}O in a single crystal of NpO_2 for two orientations of the applied magnetic field as shown. Each data set consists of a series of scans at temperatures beginning just above the ordering temperature $T_0 \simeq 26$ K and progressing at intervals of first 0.5 K, then 1.0 K, ending at 14 K. As described in the text, the spectral features behave as expected for a *3k* AFO/AFQ ordered state

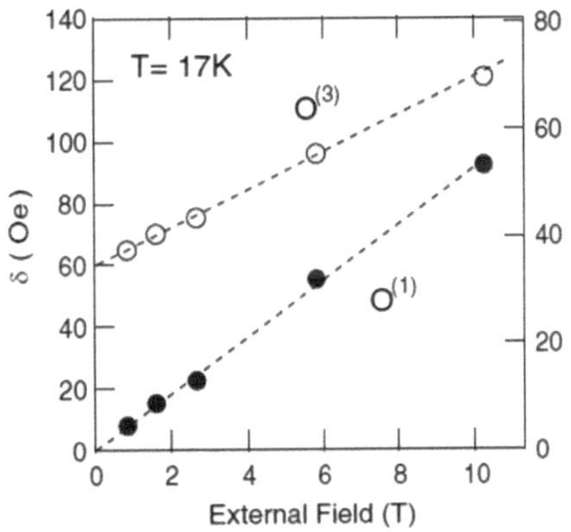

Fig. 7.8 ^{17}O line broadening parameters δ are plotted versus applied field in Tesla for the cubic sites (O^1) and for the composite line of axial sites with $\mathbf{H} \parallel [111]$. The axial site extrapolated linewidth extrapolates to a value of $\simeq 60$ G, confirming the presence of "tiny moment AFM ordering" in their vicinity [308]

minuscule distortions of the crystal, particularly along the [111] axis, would result in "tiny moment antiferromagnetism". Thus, the occurrence of trigonal strains in the crystal were suggested to account for zero–field Mössbauer line broadening in crystals of NpO_2. These stray AFM ordering phenomena also clearly account for zero-field ^{17}O NMR line broadening effects, which, interestingly, only occur for sites with axial symmetry in the octupolar/quadrupolar state.

Unfortunately, NMR is unable to probe octupolar moments in the ground state of systems like NpO_2 and $Ce_{1-x}La_xB_6$. This is because octupolar order produces no magnetic fields at lattice points of high symmetry [308]. While μSR sensed a substantial field in NpO_2 [309], it is not clear that this is a suitable probe of octupolar order. That is because the muon occupies an interstitial position in the lattice, disrupting the local crystalline symmetry. One cannot be entirely sure that locally a dipolar moment is not formed as a result of such a perturbation. Thus, there may not be a suitable probe of octupolar order available for this system.

On the other hand there are field–induced (FI) octupolar effects in NpO_2. Sakai, Shiina and Shiba have contributed a symmetry–based analysis of FI multipole effects in this system [312]. A summary of these results is shown in Table 7.1, which we now describe in detail.

The top line of Table 7.1 gives the EFG axis, then FI terms for different multipoles. The second line gives the coefficients (C.C.) that are defined in the source paper. Lines 3–6 give the form of expected results for the four oxygen sites. In particular, the far–right column gives the form for two types of octupolar shift term allowed by symmetry. The field–induced EFG terms described in the fourth column have been too weak to be identified in the NMR spectra.

We now describe a fitting of the FI–AFM and FI–AFO terms to ^{17}O NMR shift data from an angular scan of field from the [001] axis to the [110] axis in the [110] plane. The data are shown in Fig. 7.9. The scan is described by the polar angle of the field, which begins at value $\theta = 0$ with the field along [001] and extends down to $\theta = \pi/2$ with the field aligned along the [110] axis. The scan was carried out at $T = 17$ K, with field components $(H_x, H_y, H_z) = \mathrm{H}(sin\theta/\sqrt{2}, sin\theta/\sqrt{2}, cos\theta)$, where H = 10.17 T. The NMR shift data were fitted to the terms under FI–AFM and the two kinds of terms under FI–AFO in Table 7.1. The expressions fitted to the three curves in Fig. 7.9 are

Table 7.1 HF interactions with ^{17}O nuclei in the longitidal, triple-q AFO/AFQ ordered state from [312]. The definitions of coupling constants (C.C.) are given in [312]. q_z is the principal axis of the EFG tensor

Item $\mathbf{H} = (H_x, H_y, H_z)$				
Multipole	AFQ	FI – AFM	FI-AFQ	FI–AFO (T^β, T_{xyz})
C.C.	$C_{2,2}$	$2C_{1,2}$	$C_{2,5}$	$-2C_{1,3}/C_{1,4}$
O^1	0	(H_x, H_y, H_z)	0	(0,0,0)/(H_x, H_y, H_z)
$O_x^{(3)}$	$q_z \| [001]$	$(0, 0, -H_z)$	$[O_{yz}H_x + O_{zx}H_y]$	$(H_x, H_y, 0)/(-H_x, -H_y, H_z)$
$O_y^{(3)}$	$q_z \| [010]$	$(0, -H_y, 0)$	$[O_{xy}H_z + O_{yz}H_x]$	$(H_x, 0, H_z)/(-H_x, H_y, -H_z)$
$O_z^{(3)}$	$q_z \| [100]$	$(-H_x, 0, 0)$	$[O_{zx}H_y + O_{xy}H_z]$	$(0, H_y, H_z)/(H_x, -H_y, -H_z)$

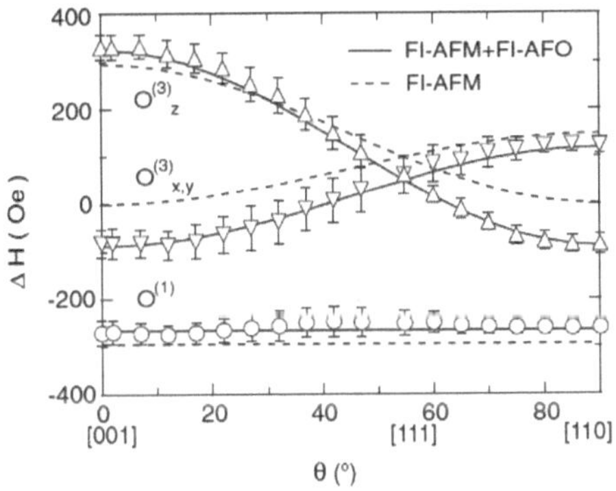

Fig. 7.9 A constant–temperature, constant–field plot of ^{17}O NMR shifts (in field units) for a rotational scan of the applied field from alignment along the c axis (polar angle $\theta = 0$) to alignment along the [110] axis in the basal plane. Field alignment in the [110] plane is maintained throughout. The spectrum consists of three distinct lines, similar to the spectrum in Fig. 7.7. The solid lines drawn are a fit to the data using possible HF interactions at ^{17}O nuclei as listed in Table 7.1 as based on a AFO–AFQ ground state from [312]. The latter analysis includes field–induced octupolar shift effects that make an important contribution, as indicated by the noticeably poorer fit that results when they are omitted (dashed lines)

$$\begin{aligned} \Delta H_1 &= [2C_{1,2} + C_{1,4}]H \\ \Delta H_{3xy} &= [C_{1,2}sin^2\theta - C_{1,3}(1 + cos^2\theta) - C_{1,4}cos^2\theta]H \\ \Delta H_{3z} &= [-2C_{1,2}cos^2\theta - 2C_{1,3}sin^2\theta + C_{1,4}(2cos^2\theta - 1)]H. \end{aligned} \quad (7.3.1)$$

A really excellent fit to the three shift curves is provided by these expressions with only three adjustable parameters. Their fitted values are $C_{1,2}H = -143.7$ Oe, $C_{1,3}$ $H = C_{1,4}H = 29.5$ Oe. Moreover, the (induced) octupolar terms make an important contribution to these fits. Thus, the quality of the fits is grossly degraded if the latter terms are omitted (dashed lines, Fig. 7.9). The measured octupolar terms are good evidence that the ordered state of NpO_2 is at least very much like the *3k* AFO/AFQ state upon which the calculations in [312] are based. It is interesting to note that according to recently published calculations the rank 5 term is the dominant multipole in NpO_2 [313].

7.4 Cross Relaxation in NpO_2

Cross relaxation (CR) is a relatively rare phenomenon in NMR, that is, a spin–lattice relaxation process that is driven by coupling between the nuclear spins under study with a set of unlike nuclear spins in the same compound. And thus, it may not always be instantly recognizable. We show in Fig. 7.10 a comprehensive set of curves showing $1/T_{1O}$ versus T for the ^{17}O nuclear spins in NpO_2. Data are plotted

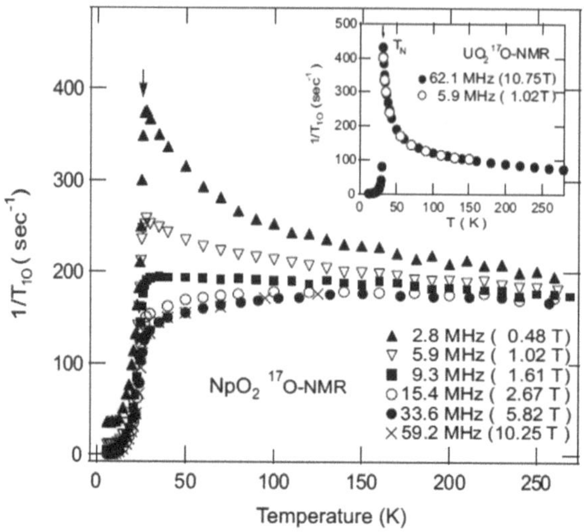

Fig. 7.10 The nuclear relaxation rate $1/T_{1O}$, measured at a series of applied field values given in the figure, is plotted as a function of temperature from He temperatures up to 260 K or so. T_{1O} becomes independent of field at frequencies above $\simeq$ 30 MHz or so, revealing a background relaxation rate from exchange–driven dipolar fluctuations. We argue that the increase of relaxation rate observed at frequencies of 15 MHz and below is caused by cross relaxation from the fluctuating ^{237}Np, which are 100% abundant. See text for discussion and detailed modeling of the CR process. The inset shows a similar plot for ^{17}O in UO_2, showing the complete absence of any CR effect for that system

from He temperatures up to nearly room temperature for a series of applied fields ranging from 0.48 T up to 10.25 T. There is a marked—and slightly peculiar—field dependence to the relaxation rate. For comparison, data for UO_2 at two extreme field values (1.02 and 10.25 T) are shown in the inset, and are seen to coincide very precisely, where the data for NpO_2 differ widely. In this section we argue and build a case for a CR process in which the ^{17}O are being relaxed through contact with the ^{237}Np nuclear spins, which are 100% abundant in this sample. What is peculiar is that $1/T_{1O}$ increases *as the field is reduced*, where the CR effect is visible up to NMR frequencies of 15 MHz. One only rarely encounters nuclear T_1 processes fast enough to have such a broad spectrum. Note, however, that the fluctuation density increases markedly at lower fields (i.e., lower ^{17}O NMR frequencies), as well as *at lower temperatures*, which we make sense of below with a simple model.

What is also evident is that there is another T_1 process for the ^{17}O, toward which the measured values in Fig. 7.10 asymptotically approach at high fields and frequencies. This is the process driven by exchange fluctuations of the Np^{4+} $5f$ moments, to which the ^{17}O are coupled by a transferred HF process. This latter T_1 process is unaffected by applied field or NMR frequency and is found in many systems containing exchange–coupled localized moments [58]. We do not analyze it further, but subtract the 10.25 T curve in Fig. 7.10 from each of the others in order to isolate the CR process more quantitatively. The results for CR thus quantified are shown in Fig. 7.11. With these

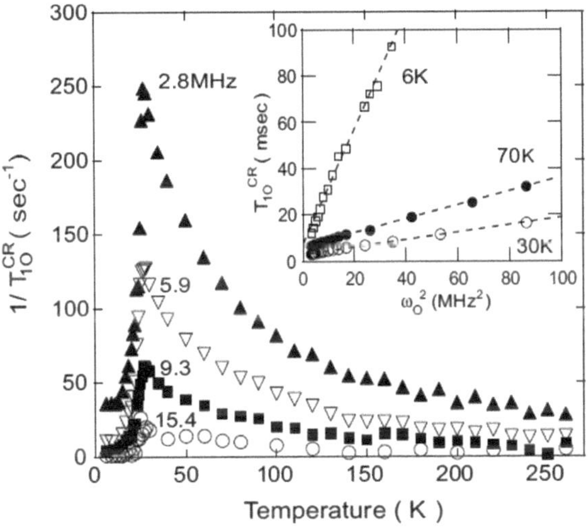

Fig. 7.11 The data from Fig. 7.10 for NMR frequency 15.4 MHz and below are replotted with the data taken in a 10.4T applied field subtracted off as a field–independent background. What remains is presumably 100% cross–relaxation in origin. Note that $1/T_{1CR}$ drops nearly to zero at T_0, where the dipolar magnetism collapses. At 2.8 MHZ the relaxation rate drops to a finite value. This effect and its field dependence are discussed in detail in the text. In the inset, T_{1CR} is plotted versus ω_{17}^2 (see text for details), illustrating the Lorentzian form of the high-temperature T_{1CR} process

data we can study the variation of T_{1CR} with frequency $\omega_O = \gamma_{17}H$. Abragam gives some examples of CR and states the expected variation with field and with T_{1Np} ([5], Chap. VIII) as

$$\frac{1}{T_{1CR}} = \frac{\langle \Delta^2\omega\rangle_\alpha T_{1Np}}{[1+(\omega_O-\omega_{Np})^2T_{1Np}^2}] + \frac{\langle \Delta^2\omega\rangle_\beta T_{1Np}}{[1+\omega_O^2T_{1Np}^2}] \tag{7.4.1}$$

where the two terms represent processes driven by squared matrix elements of $I_{\pm Np}I_{\mp O}$ and $I_{zNp}I_{\mp O}$, respectively. Both of these terms behave in a very similar fashion with frequency, i.e. with H. It is argued that because of the very large shift for ^{237}Np in NpO_2, the α term is bound to be less important. Thus, we complete the analysis using just the β term. With that term alone, it is clear that a plot of T_{1CR} versus H^2 for any temperature should yield a straight line through the origin. Three such plots are shown in the inset to Fig. 7.11, where the two at lower temperatures are seen to behave as expected. The one at 70 K yields a straight line, but appears to miss the origin. That behavior may have to do with baseline subtraction, which is a bit less certain at 70 K.

The frequency dependence for these cases is satisfactory. A much more pressing question is the magnitude and temperature dependence of the indirect coupling that drives T_{1CR}, i.e. of $\langle \Delta\omega^2 \rangle_\beta$.

As we shall see, the crucial point here is the magnitude and origin of $\langle \Delta\omega^2 \rangle_\beta$. In a system with itinerant fermions, there is the possibility of indirect spin–spin coupling between unlike species. In NpO_2, however, the only known interaction is direct dipolar coupling. An estimate shows that the effective value of $\langle \Delta\omega^2 \rangle_\beta$ at low temperatures is *three orders of magnitude larger than simple dipolar coupling*. Thus, the T_{1CR} effect in NpO_2 stumbled onto a new source of indirect spin–spin coupling between the nucleus of an f–electron–bearing ion and a ligand neighbor. Casting about for possibilities, it seems likely that the following scenario is in effect here. The three ingredients are as follows: There is a modest coupling (i.e., classical dipolar, hybridization) between the f-ion moment and the ligand nuclear spin. The f-ion nucleus has a *strong* HF interaction with the f moment. Finally, the f*-ion* moment has a large susceptibility. We reason as follows: The f–moment nucleus produces an inverse HF field that induces a moment on the f-electron ion. Any such moment then couples with the ligand ion nucleus. It is easy to write down an estimate of this mechanism that comes out in the right ballpark for a greatly enhanced coupling.

We can test this hypothesis by fitting the second ("β") term in Eq. (7.4.1) to the field (i.e., ω_{17}) and temperature dependence of the data. Fitted results for the temperature dependence of $\langle \Delta^2\omega \rangle_\beta$ and the numerical values of T_{1Np} data are shown in Fig. 7.12. In Part (a) T_{1Np} is seen to give a flat temperature characteristic with a value around 40 nsec. In part (b) the driving matrix element is seen to scale very nicely with χ_0^2 as expected from the foregoing paragraph. We consider this to be a clear–cut validation of the simple model described above for the indirect coupling.

At T_0, the conventional T_1 mechanism for both ^{237}Np and ^{17}O collapses abruptly, since the Np^{4+} dipolar moment disappears with the onset of octupolar order. What is interesting is that T_{1CR} does not vanish altogether, but makes a transition into a new set of values according to the value of applied field H as T is lowered through the ordered state region, as is shown in Fig. 7.13. Although one can only guess at what is taking place microscopically, it is noteworthy that the conditions required to support T_{1CR} are sustained here. First, although the dipolar moments driving the paramagnetic state T_1 processes are gone, the molar susceptibility of the Np^{4+} ions decreases only slightly in the ordered state [305]. The HF couplings to ^{237}Np nuclear moments may not vary much, either. What evidently changes markedly is the fluctuation spectrum of the ^{237}Np nuclear moments. This was given in Eq. (7.4.1) as Lorentzian in character, suitable for fluctuations driven by T_{1Np}. That this spectrum is now rather different is evident from the sharp dropoff with field in Fig. 7.13. The value of $1/T_{1O}$ at 2.8 MHz drops about a factor of 10, but for the higher fields the drop is several orders of magnitude. What was Lorentzian in shape above T_0 is now much steeper, possibly Gaussian. This is no longer a T_1 spectrum, but may possibly be a "T_2

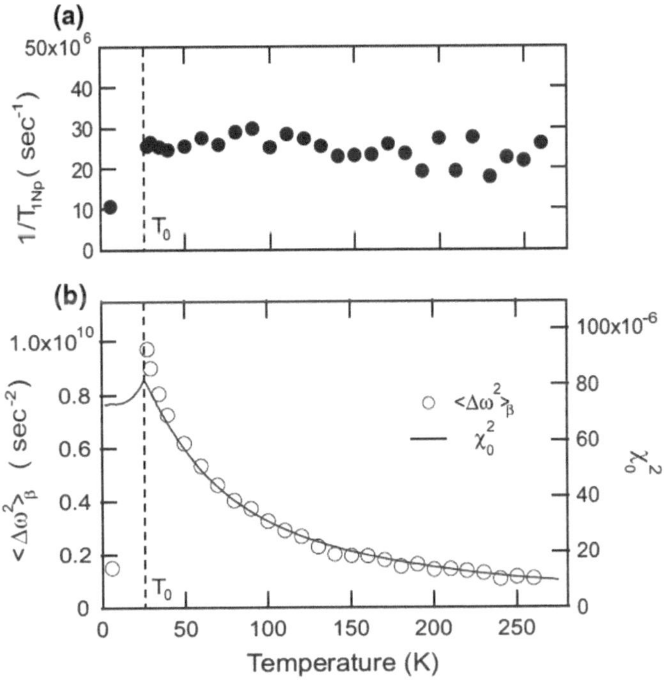

Fig. 7.12 **a** Values of T_{1Np} derived from fits to the β term of Eq. (7.4.1) at temperatures ranging from T_0 up to 270 K. The data are roughly constant at $T_{1Np} \simeq 40$ nsec. Such a number projects a spectral width of ~ 4 MHz to account for the distribution of T_{1CR} values. **b** The effective unlike–spin second moment measured by the data for $1/T_{1CR}$ as a function of temperature. Peak values are nearly three orders of magnitude greater than classical dipolar coupling. Scaled to the data (solid line) is a plot of the squared paramagnetic state susceptibility, corroborating the simple indirect–coupling mechanism described in the text

spectrum", i.e. driven by spin–spin couplings among the ^{237}Np. Further, the latter are greatly enhanced compared with classical dipolar interactions, yielding a spectrum that is MHz in width, but has much steeper sides than the $T1$ spectrum found in the paramagnetic state. What is truly fascinating is that these conditions lead to T_{1CR} for the ^{17}O that is asymptotically *independent of temperature* at temperatures low compared with T_0. This is strongly suggestive of T_2 fluctuations that satisfy the condition $\hbar/T_2 \ll k_B T$.

Seemingly, it only requires abundant nuclear spins in a Rare Earth or actinide compound for CR effects to appear. Thus, one would expect them to be found in many f–electron compounds. Up to now, the only other such compound for which they have been reported is $NpPd_5Al_2$ [314].

Fig. 7.13 Cross relaxation rates for ^{17}O in NpO_2 on a log–log scale to emphasize behavior at the lowest temperatures measured (i.e., to ~6 K). The CR rates are seen to drop with increasing H as in the paramagnetic state, but to do so much more steeply. This is suggestive of a T_2 spectrum of fluctuations as discussed in the text. Such an outcome would require strong indirect $^{237}Np-^{237}Np$ couplings analogous to the $^{237}Np-^{17}O$ couplings analyzed in the text

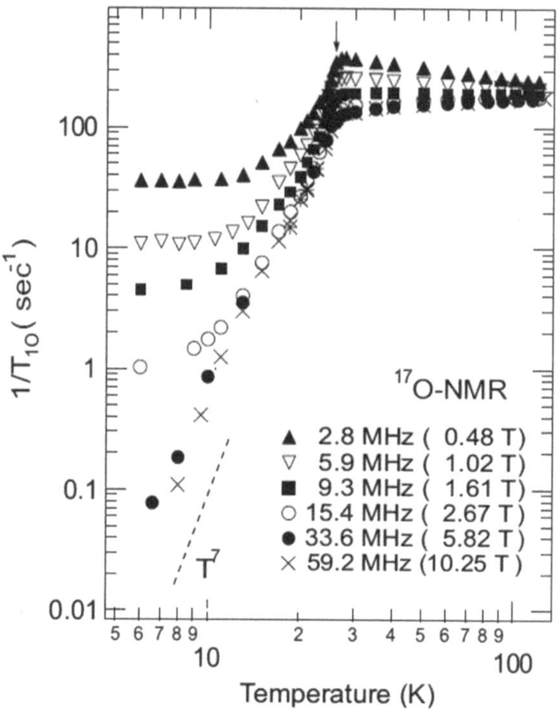

7.5 NMR of ^{239}Pu in PuO_2

The isotope ^{239}Pu has a nuclear spin I = 1/2, and up until 2012 its NMR signal in a solid phase had never been observed. Perhaps this was because the element Pu, either alone or when combined with oxygen, exists in multiple non-stoichiometric phases, so that the solid forms are difficult to manage. In any case, Yasuoka and co-workers [299] obtained a very nearly stiochiometric sample of PuO_2 of good crystallinity and were finally able to observe and identify the ^{239}Pu NMR line over a series of fields and frequencies. The problem with observing ^{239}Pu NMR is not that T_1 is unmanageably short, but that, on the contrary, the ground state of Pu^{4+} is a magnetically inert crystal field singlet, so that T_1 is almost unmeasurably long, and thus, the resonance is difficult to excite without saturating the line and rendering it unobservable. All this was known in advance, because neutron scattering had been used to verify the Γ_1 singlet ground state assignment [300].

The NMR line profile of ^{237}Np in NpO_2 (T = 4 K) at ν = 20.48 MHz is shown in Fig. 7.14. This NMR line is inhomogeneously broadened to a width of ~ 320 G, which is much greater than the simple dipolar broadening. The breadth is ascribed to crystalline disorder, but has an approximately Lorentzian shape. The reason for such a large inhomogeneous width is undoubtedly because the Van Vleck shift is estimated to be 24.8%. A great deal of the broadening would accrue from fluctuations in the shift

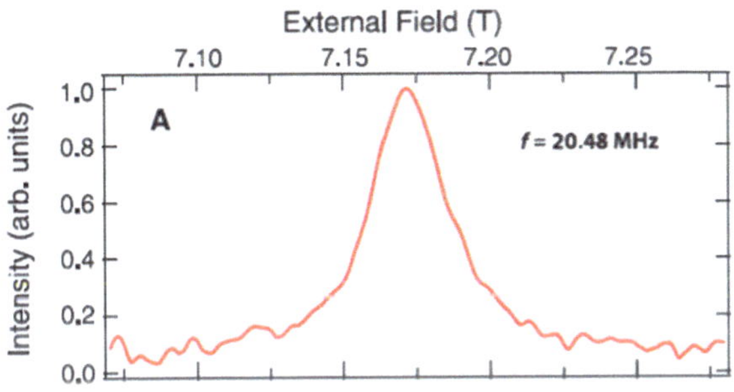

Fig. 7.14 A scan of the ^{237}Pu NMR line in NpO_2 is shown at a frequency of 20.48 MHz and a temperature of 4 K. The approximately Lorentzian line has FWHM ~320 G. This study is the first ever of ^{239}Np NMR in a solid–state host material

driven by crystalline disorder. The shift was estimated by multiplying the measured susceptibility ($\chi_0 = 5.36 \times 10^{-4}$ emu/mol) by the free–ion HF coupling constant (A_{hf} = 253 T/μ_B), a procedure that was checked against other actinide ions [299]. The effective gyromagnetic ratio was deduced from a series of line scans at 4 K and at different frequencies. The "bare" nuclear gyromagnetic ratio was thereby estimated to be $\gamma_{239}/2\pi$ = 2.29 MHz/T. At T = 4 K, T_1 could not actually be measured. It was estimated to be at least 100 s.

7.6 A Study of ^{17}O NMR in the Host Compound AmO_2

As one progresses across the actinide series, the half–life of available isotopes decreases by roughly two orders of magnitude per step. For the longest–lived isotope of americium, ^{243}Am, the half–life is 7,370 years. That may seem long, but for a crystalline sample of AmO_2 it is fast enough for massive damage to the crystal structure to take place in just a few days. The reason for this is that each α decay event causes recoil damage to many of the neighboring unit cells. These effects, along with the low-temperature state of this compound, have been investigated through ^{17}O NMR studies of freshly–synthesized samples of $^{243}AmO_2$ [315, 316].

With normal procedures for sample synthesis, retrieval and characterization, the delay for observation of the ^{17}O NMR spectrum is 13 days, after which the spectrum appears as in the top (a) tracing of Fig. 7.15. The spectrum is seen to consist of a very narrow spike on top of a much broader background spectrum. The spike has the approximate width of the paramagnetic state resonance, while the background resonance has a width that can only be explained if the system has entered into a highly disordered state with, however, magnetic ordering. The large difference between these two linewidths is explained by the fact that in a crystalline state the dipolar fields at any oxygen site cancels out because of the cubic symmetry; meanwhile, if the crystal becomes amorphous and the magnetic moments freeze, then a very large broadening effect is enabled. The same spectrum is shown in the bottom (b) half of Fig. 7.15 after 25 days following synthesis. Now there is no crystalline vestige

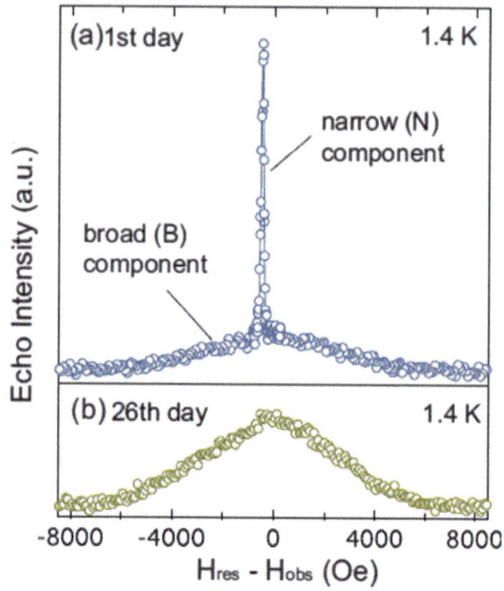

Fig. 7.15 **a** Top of figure shows the 17*O* NMR spectrum 13 days after the synthesis of the $^{243}AmO_2$ polycrystalline sample material. The narrow peak corresponds to the crystalline portion of the sample, and the broad peak to the part that has been decrystallized by α decay of the ^{243}Am. At the experimental sample temperature of 1.4 K, the latter portion has become a spin glass. See text for further discussion. **b** The ^{17}O NMR spectrum under the same conditions as **a** 25 days after sample synthesis

remaining, and only the spin glass NMR line survives. These spectra were taken at T = 1.4 K, where the spin glass transition temperature is estimated to be 8.9 K [316].

It is interesting to note that Fig. 7.15a shows no sign of magnetic ordering for the crystalline portion at T = 1.4 K. The nominal crystal field ground state of this compound is a Γ_7 doublet capable only of dipolar ordering. The Curie–Weiss character of the paramagnetic susceptibility suggests the possibility of ordering at some tens of degrees. Thus, the paramagnetic behavior at He temperatures seems rather anomalous. However, it has been pointed out more recently that through competition between spin–orbit coupling and Coulomb interactions it is possible that the ground state has Γ_8 character. In that event, quadrupolar and octupolar ground states are also possible [317]. It seems, however, that to resolve the ^{243}Am ground state question it will be necessary to take measurements on a sample within just a few days of its synthesis.

Chapter 8
The Kondo Effect and Heavy Fermion Phenomena

8.1 Introduction

Recent generations of condensed-matter physicists have devoted a great deal of attention to the metallic environment and, e.g., the behavior of d and f-electron ions doped into s–p band host metals. From the indirect exchange between nuclei in simple metals of Ruderman and Kittel [27] came spin density oscillations surrounding localized magnetic moments [318] and indirect exchange couplings between such moments [319]. There then followed the Anderson model on how such local moments form on impurities in simple metals [320]. There was a surge of theoretical activity on these and related questions, [321] and out of that surge came a remarkable disclosure (among many), namely that in the case of a localized moment that undergoes antiferromagnetic exchange with electron spins in a surrounding electron gas, the exchange scattering diverges at low temperatures. And thus, the local moment d-state itself becomes unstable and collapses. Named after its discoverer, this came to be known as the "Kondo effect" [322].

The Kondo effect was identified in a number of systems, where the simplest manifestation of the collapse of an embedded localized magnetic moment appeared to be a Curie–Weiss form for the susceptibility $\chi_K(T) = \chi_{max}T^*/(T + T^*)$, where the peak susceptibility χ_{max} occurs as $T \rightarrow 0$. Such a form for $\chi(T)$ is well known in the paramagnetic phase of dense, interacting magnetic systems, where the Curie–Weiss constant is generated by magnetic interactions with neighbors. In the Kondo effect, however, $\chi_K(T)$ is found to occur for impurities that are isolated from any magnetic neighbors. It is, thus, the Hallmark of a single magnetic site whose main interaction with its surroundings is an exchange coupling with quasiparticles.

It was a number of years before a complete theory of the Kondo effect was available, and in the meantime there was a controversy about just how the local moment was "quenched" by interactions with the surrounding electron gas. One school of thought held that a quasiparticle with equal and opposite spin to that of the local

© Springer-Verlag GmbH Germany 2018

R.E. Walstedt, *The NMR Probe of High-T_c Materials and Correlated Electron Systems*, Springer Tracts in Modern Physics 276,
https://doi.org/10.1007/978-3-662-55582-8_8

moment gradually became localized in the vicinity of the Kondo impurity, eventually cancelling out the original localized moment and leaving only an enhanced susceptibility. Such an effect would occasion a substantial rearrangement of spin and charge densities in the vicinity of the Kondo ion. Opposing theories suggested that the original local moment simply lost its character as an orientable moment at low temperatures without changing the near-field environment of the Kondo impurity. A very elegant NMR experiment on dilute Fe impurities in Cu metal—a well-known Kondo system—was performed by the Slichter group, [323, 324] clearly distinguishing between the competing physical pictures. We begin by describing the latter experiment in some detail.

8.2 The Isolated Kondo Impurity

Not surprisingly, the Kondo effect emerged at a time when the study of isolated d-ions in sp–band metals was fashionable. The question of whether such a d-ion impurity state would possess a localized magnetic moment gave us the Anderson model, [320] because there were cases such as AlMn where no local moment appears to be present at any temperature. The Kondo case is something in between, where a Curie–Weiss susceptibility makes the ion seem local-moment-like at high T, but then it simply melts away at $T < T_K$.

Before an accepted theory of the details of Kondo condensation had emerged, there was debate about how Kondo condensation took place in detail. It was known then that localized d-orbitals on impurity sites in sp-band metallic environments caused spin-density oscillations in their environment (often referred to as "Friedel oscillations" [318, 319]). The debate concerned how these oscillations would change with temperature as a local moment was gradually modified into just an enhanced local susceptibility at T_K and below. Boyce and Slichter [323] framed the issue very clearly when they noted that the controversy was over whether the induced spin polarization $\sigma(r)$ at distance r from a Kondo impurity was given by

$$\sigma(r, T)/H = \chi_s(T) f(r), \tag{8.2.1}$$

or by

$$\sigma(r, T)/H = \chi_s(T) g(r, T), \tag{8.2.2}$$

where $\sigma(r, T)$ is the induced spin polarization in the lattice surrounding the impurity, H is the applied field, and χ_s is the conduction electron spin susceptibility. In (8.2.1) $f(r)$ is the temperature-independent distribution of spin-density oscillations in the lattice surrounding the Kondo site, while in (8.2.2) g(r, T) is a similar distribution, except that it is assumed to depend explicitly on temperature as would be required if another quasiparticle were to be incorporated into the impurity state of the Kondo ion.

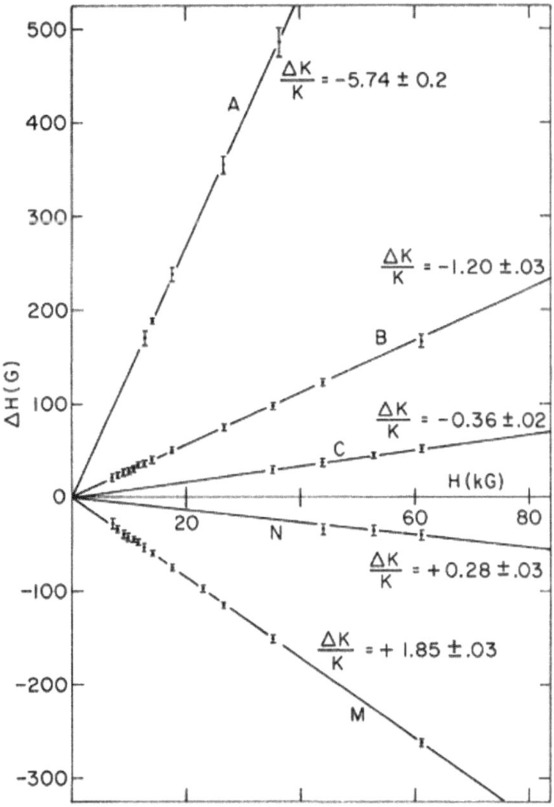

Fig. 8.1 Magnetic field dependence of ^{63}Cu NMR satellite separation from the main resonance line at 300 K. Plotted is shift ΔH, in Gauss, from the ^{63}Cu resonance versus applied field H. Each line of shift data is labeled with a value of $\Delta K/K$, where δK is the shift of that satellite, and K is the Knight shift of the host Cu metal ($K_s = 0.232\%$—Table 2.1). The fractional deviation $\Delta K/K$ for satellite A includes a direct dipolar contribution of $-0.5 \pm 0.1G$ so that the isotropic part of $\Delta K/K|_A = -5.24 \pm 0.3$. Figure reprinted from J.B. Boyce and C.P. Slichter, Phys. Rev. Lett. **32**, 61 (1974). Copyright 1974 American Physical Society

Since bulk measurements were not able to distinguish between the two foregoing possibilities, it seemed that only a microscopic probe of the host lattice would be able to provide the evidence needed to decide between the two possibilities posed by (8.2.1) and (8.2.2). A scan of the ^{63}Cu NMR line for satellites in the CuFe Kondo system provided just what was needed: Five satellite NMR lines whose shift away from the main ^{63}Cu resonance line is precisely proportional to the field [323]. Shift data taken at T = 300 K for satellite splittings (in Gauss) were plotted as shown in Fig. 8.1, where data taken in fields ranging from below 10 kG to above 60 kG are presented. Data were taken on specimens of CuFe with Fe concentrations c from as low as $c = 500$ ppm up to 5000 ppm. The satellite splittings were found to be independent of c.

The temperature dependence is, of course, a crucial point for the investigation. This aspect of the study was impacted by the fact that the satellite lines broaden rapidly at low temperatures. This was because even at these low concentrations, there apparently occurred in these powder samples a substantial concentration of "clusters", i.e., pairs of Fe impurities and possibly larger agglomerations for which local moment magnetism persisted at low T. These would be a major source of line broadening for all or nearly all Cu sites. Moreover, satellite sites near to the host Fe impurity would undergo dipolar broadening averaged over all orientations of the displacement vector **r**. As a consequence, line A in Fig. 8.1 could not be effectively observed below 77 K and was omitted from further study. Line broadening analysis of line A strongly suggests that this is the first neighbor. The identity of other satellites is commented on below.

Denoting the satellite splitting ΔK, plots of $1/\Delta K$ versus T showed that where it could be determined, ΔK varied with T as 1/(T+29) in agreement with the Fe ion susceptibility. In Fig. 8.2 the shift ratio $\Delta K/K$ is plotted versus const./(T+29) for the four satellites available for such a plot. The data are all consistent with straight line

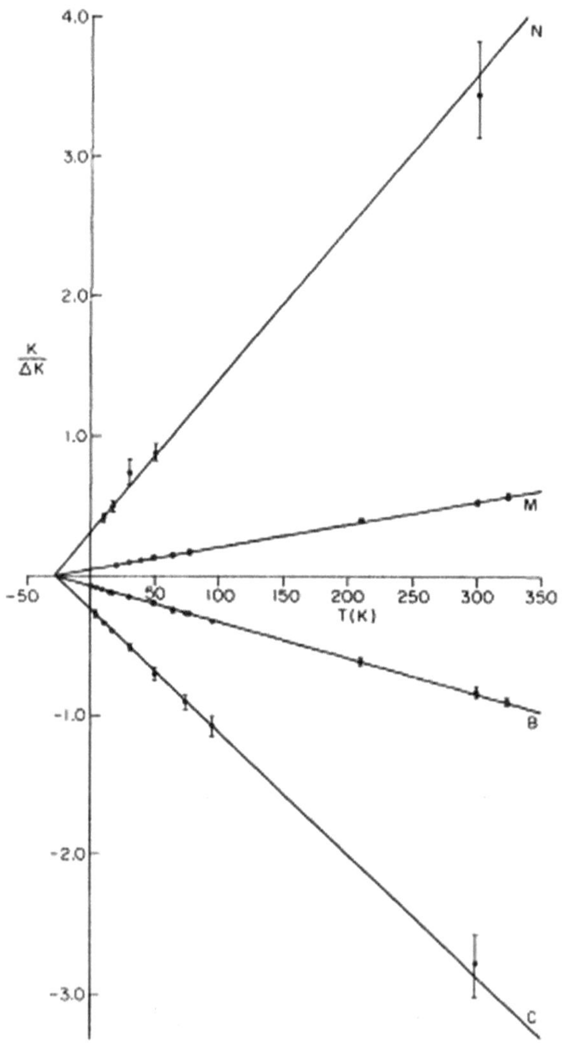

Fig. 8.2 The shift deviation ΔK as a fraction of the total shift K (i.e., $\Delta K/K$) is plotted as a function of $constant/(T+29)$, emphasizing the low-temperature points. Since the splittings from the main line were all proportional to H, it is convenient here and in Fig. 8.1 to define the quantity $\Delta K = \Delta H/H$ and to plot the ratio $\Delta K/K$ and its inverse as measures of the relative splittings. Figure reprinted from J.B. Boyce and C.P. Slichter, Phys. Rev. Lett. **32**, 61 (1974). Copyright 1974 American Physical Society

behavior, verifying that (8.2.1) is the equation that describes these results. Boyce and Slichter wrote a second paper [324] in which they summarize their results, but also confront a much larger body of data relevant to the choice between (8.2.1) and (8.2.2), from bulk measurements to Mössbauer, neutron scattering, other NMR studies, etc., concluding in the end that the bulk of known data, as well as theory, were consistent with the validity of (8.2.1). This was indeed a well-documented conclusion.

We offer just a brief note about theory. At the time, a theory consistent with (8.2.1) had been given by E. Muller–Hartmann [325], while the ultimately incorrect theory that supported (8.2.2) was that of Heeger et al. [326] Since then, complete theories of the physics of isolated Kondo "impurities" have been established [327] and interest has passed on to the case of densely populated Kondo compounds, i.e., the "Kondo lattice", which we consider next.

The Boyce and Slichter study of $\underline{\text{Cu}}$Fe was without doubt one of the most elegant NMR studies of the era. And one of the most remarkable things about it was that the work was conducted with "old fashioned" CW NMR spectrometers. It was conducted at a time when the vast majority of condensed-matter NMR studies employed spin echo techniques, as has been the case ever since. But it clearly demonstrates that the CW method functions on a par with spin echoes in cases where *extreme broadening of the resonance is not a problem.*

8.3 The "Kondo Lattice" and Heavy Fermions

While the single-site Kondo problem was ultimately manageable, the Kondo lattice has proved a good bit more difficult, so that as of this writing, no comprehensive theory is yet available. In an excellent historical sketch, Hewson recounts how it was realized that the heavy fermion phenomenon was simply a dense Kondo lattice [327]. Early interpretations of this nature were made, for example, by Mott [328] on the heavy fermion system $CeAl_3$. In fact, there are a number of heavy-fermion systems that are simply intermetallic compounds including either Ce or Yb, and thus a single 4f electron or a single 4f hole, with the consequent simplification of atomic 4f states in their makeup. Interest in these systems was stimulated early on by the discovery of superconductivity among heavy fermions in the compound $CeCu_2Si_2$ in 1979 by Steglich and co-workers [329].

In this section we present and discuss briefly the two-fluid phenomenological model by Nakatsuji, Pines and Fisk (NPF) [330]. This model, based on a few simple assumptions about the physical content of a heavy-fermion system, appears to provide a reliable framework for systematizing a number of the important properties of these systems. The original reference compound for this derivation is $Ce_{1-x}La_xCoIn_5$, "CeCoIn", although the results will be expected to describe a wide variety of heavy fermion systems. CeCoIn is very flexible in that for $x \ll 1$ one will have a heavy-fermion transition at some temperature T^* to a Fermi liquid with relatively massive quasiparticles, while for $(1 - x) \ll 1$ one will have dilute Kondo impurities whose behavior as a function of T is well known, and which do not form a heavy-fermion state. In NPF the focus is on physical behavior in these two limits, and most particularly on characterization of the heavy-fermion state.

The discussion begins with two-fluid representations of the specific heat and susceptibility of the system considered. These are written out as follows:

$$\begin{aligned} C_{MAG}(T)/T &= [1-f(T)]C_{KI}(T)/T + f(T)C_{HF}(T)/T \\ \chi_{MAG}(T) &= [1-f(T)]\chi_{KI}(T) + f(T)\chi_{HF}(T), \end{aligned} \tag{8.3.1}$$

where $(C, \chi)_{MAG}(T)$ are measured quantities on the system investigated, while $(C, \chi)_{KI}(T)$ are taken from "Kondo impurity" molar data on a dilute sample with $(1-x) \ll 1$. In (8.3.1) the parameter f(T) is used as an effective order parameter to describe the interplay between "HF" and "KI" terms for both $C(T)$ and $\chi(T)$, a step that must be justified. This point is argued in terms of the kind of plot, $C_{MAG}(T)/T$ versus $C_{KI}(T)/T$, shown in Fig. 8.3. In this kind of plot, the low temperature values are at the far right. There are several important points to note here. First, at low T, the variation of $C_{MAG}(T)/T$ with $C_{KI}(T)/T$ is linear for all compositions. Next, the slope is necessarily 1 for $x \to 1$, but converges to a value $\simeq$0.1 for all $x \leq 0.25$. Thus, for high concentrations of Ce, $f(T) \simeq 0.1$, meaning that in samples with highly concentrated Ce, roughly 10% of the specific heat behavior is that of a dilute Kondo impurity. Finally, it is to be emphasized that a plot of $\chi_{MAG}(T)$ versus $\chi_{KI}(T)$ given in [330] yields closely parallel behavior to that of the specific heat, so that the variation of f(T) with temperature and composition is essentially identical with what is seen in Fig. 8.3. Thus, using the identical format for C_{MAG} and χ_{MAG} in (8.3.1) is justified here by *experimental data*.

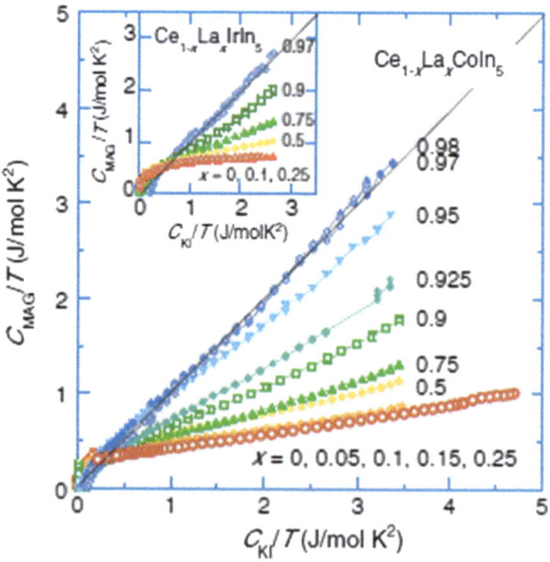

Fig. 8.3 The f-electron contribution to the specific heat divided by T is plotted versus its single impurity C_{KI}/T for $Ce_{1-x}La_xCoIn_5$ with temperature T as an implicit variable. Inset: The same plot for $Ce_{1-x}La_xIrIn_5$. The specific heat was measured by a relaxation method using high-quality single crystals grown by an In self-flux method [338, 339]. Figure reprinted from S. Nakatsuji, D. Pines and Z. Fisk, Phys. Rev. Lett. **92**, 016401 (2004). Copyright 2004 American Physical Society

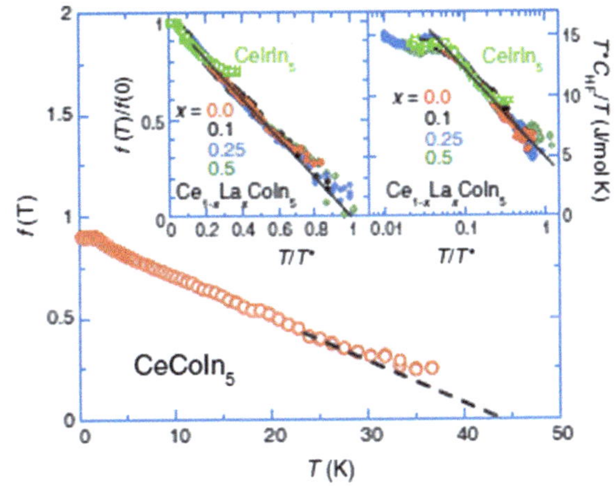

Fig. 8.4 The relative weight of the heavy fermion component f(T) for $CeCoIn_5$ as a function of T. The broken line shows a linear fit. Left inset: The scaling behavior with T^* of f(T)/f(0) for $Ce_{1-x}La_xCoIn_5$ and $CeIrIn_5$. The solid line represents a fit to the linear increase of f(T) that universally appears below $\sim T^*$. Right inset: $T^* C_{HF}/T$ as a function of $ln(T/T^*)$ for $Ce_{1-x}La_xCoIn_5$ and $CeIrIn_5$ under B = 0 T. Here, C_{HF}/T represents the heavy fermion component of C_{MAG}/T. The solid line is a fit to the $ln(T/T^*)$ behavior that universally appears for $0.05T^* < T < T^*$. Figure reprinted from S. Nakatsuji, D. Pines and Z. Fisk, Phys. Rev. Lett. **92**, 016401 (2004). Copyright 2004 American Physical Society

Having validated the use of (8.3.1), the remaining challenge is to use these equations to determine $C_{HF}(T)$, $\chi_{HF}(T)$ and f(T) with the available data. Since there are three unknowns, however, another equation relating these quantities is required. For that, the authors of [330] turned to the Wilson ratio. This is defined as $R_W = \alpha\chi/CT =$ where $\alpha = \pi^2 k_B^2/3\mu_B^2$. For a non-interacting electron gas one expects $R_W \simeq 1$. Interestingly, however, it has been shown (Ref. [327], p. 90), that for isolated Kondo impurities, on a molar basis, $R_W = 2$. For Ce-rich CeCoIn, the estimate $R_W \simeq 2$ has been given for the heavy fermion fluid[1] [330]. On this basis, the authors introduce their "local Wilson ansatz", namely that $R_W = 2$ at all temperatures. This ansatz then yields the relation

$$\chi_{HF}(T) = 2C_{HF}(T)/\alpha T, \tag{8.3.2}$$

which, combined with (8.3.1), allows the determination of f(T), $C_{HF}(T)$ and $\chi_{HF}(T)$ for all temperatures below T^*. The results are summarized in Fig. 8.4, where it is seen that f(T) is essentially linear with $T^* \simeq 45\,\text{K}$, and $C_{HF}(T)/T$ varies $\propto ln(T)$ over the low-temperature range of measurements.

[1]This was obtained by extrapolating the *low-temperature behavior* of C_{MAG}/T and $\chi(T)$ to $C_{KI}/T \to 0$ and $\chi_{KI} \to 0$, respectively [330]. This procedure yields the low-T estimates $f(T)C_{HF}/T \to 2.90 \times 10^6$ ergs/mol-Ce K^2 and $f(T)\chi_{HF}(T) \to 0.008$ emu/mol-Ce. Noting that f(T) cancels in the estimate of R_W, the latter values lead to $R_W = 2.01$.

The remarkable results of this two-fluid model show universal behavior, where even at the lowest temperatures $\simeq$10% of the behavior of the system is that of the non-interacting Kondo impurity state. There are many other features expounded by the authors [330]. How the validity of their main approximation, i.e., the Wilson ratio ansatz, will hold up vis-a-vis a first-principles theory remains to be seen. Since no such theory appears to be imminent, this two-fluid model remains the state-of-the-art for the time being.[2] It forms the basis for the discussion of NMR shift data on systems described in the next section.

8.4 A Universal NMR Shift Scaling Plot

In this section we are concerned with the implications of the NPF two-fluid model for NMR frequency shift effects in heavy fermion systems [330]. This topic was analyzed by Curro et al., where by using the two-fluid formulation of the susceptibility and corresponding NMR shift, a universal plot format was arrived at that exhibits scaling behavior for the heavy fermion component of susceptibility [331]. Constructing this plot is an important step in the characterization of these compounds. We begin by restating the derivation that leads to the universal plot of shift $K_{HF}(T)$ versus T, where K_{HF} is the heavy fermion component of shift. Examples of the scaling behavior are then presented along with examples of other effects that elucidate the behavior of these systems.

To analyze the NMR shift we first need to understand the susceptibility. We use a slightly different notation from that in [331] and comment on the difference as we go along. The magnetization operator may be written,

$$M_{sp} = g_{sp}\mu_B \sum_i S_i^c + g_f \mu_B \sum_j S_j^f = S^c + S^f, \tag{8.4.1}$$

where for simplicity we do not identify an axis of quantization, assuming it is the field axis. In (8.4.1) the sum on i is over a localized Wannier representation of the conduction electron band and that on j is over the f-electron sites. For completeness we add a density matrix scheme to derive the form of the two-part susceptibility. Thus we write for the magnetization,

$$\langle M_{sp}\rangle = C_N Tr\{[S^c + S^f]exp[-\mathcal{H}/kT]\}, \tag{8.4.2}$$

where $\mathcal{H} = \mathcal{H}_0 - H_0[S^c + S^f]$ and C_N is the normalization. Taking $\chi = d\langle M_{sp}\rangle/dH_0$, one finds easily that

$$\begin{aligned}\chi &= (C_N/kT)\, Tr\{[(S^c)^2 + 2S^cS^f + (S^f)^2]exp(-\mathcal{H}_0/kT)\}\\ &\approx \chi_{ff} + 2\chi_{cf}),\end{aligned} \tag{8.4.3}$$

[2]However, please see the report on LDA-DMFT calculations later in this chapter.

where the small χ^{cc} term is neglected, and where the correspondence is clear. In the same approximation, for the shift calculation below we state the results

$$\langle S^c \rangle = \chi_{cf} \quad \text{and} \quad \langle S^f \rangle = \chi_{cf} + \chi_{ff}. \tag{8.4.4}$$

For completeness we note the correspondence of the foregoing developments with the fundamental two-fluid equation of NPF for the susceptibility, namely $\chi = [1 - f(T)]\chi_{KI} + f(T)\chi_{HF}$. Comparing the latter with (8.4.3) and (8.4.4), we see that $[1 - f(T)]\chi_{KI} = \chi_{ff}$ and $f(T)\chi_{HF} = 2\chi_{cf}$.

8.4.1 The Knight Shift K in Heavy Fermion Materials

Here we deviate slightly from the style of [331] in taking the traditional hyperfine Hamiltonian,

$$\mathcal{H}_{hyp,\alpha} = A_\alpha \sum_l I_{\alpha l} S^c_\alpha(r_l) + B_\alpha \sum_{i,l} I_{\alpha l} S^f_\alpha(r_i), \tag{8.4.5}$$

where A_α is the on-site (i.e., contact) interaction and B_α is the transferred coupling, understood to be that between $I_{\alpha l}$ and neighboring f-electron spins at sites i. Here A_α and B_α are in units of ergs, and the constant B_α is assumed re-defined to be that of z identical neighbor f spins to nuclear site l for simplicity. The subscript α indicates a crystalline symmetry axis for possible anisotropy. Again, for simplicity, these subscripts will henceforth be omitted, but are understood to be included. Taking the thermal expectation value of (8.4.5) = $\langle \mathcal{H} \rangle$, using the magnetization operators of (8.4.1) and the definition of the NMR shift (2.32) $\langle \mathcal{H} \rangle = \gamma\hbar K H_0 \sum_l I(r_l)$, we find

$$K(T) = K_0 + (\alpha_A + \alpha_B)\chi_{cf}(T) + \alpha_B \chi_{ff}(T), \tag{8.4.6}$$

where α_A and α_B are the shift coefficients defined as, typically, $\alpha_A = A/(N_0\gamma\hbar g\mu_B)$, where the susceptibilities are taken to be emu/mol. Further, a possible T-independent shift K_0 is included.

Equation (8.4.6) illustrates how at high temperature ($T > T^*$) the NMR shift gives a straight line when plotted against χ, but below T^* when χ_{cf} is non-zero, it has a different slope and may not be linear at all. The question is how to sense the presence of $\chi_{cf}(T)$ when $T < T^*$ and even to estimate its magnitude. The strategy adopted by Curro et al. [331] was to (i) determine the value of α_B with a K-χ plot at $T > T^*$, then (ii) using (8.4.3), (8.4.4) and (8.4.6), plot

$$K(T) - K_0 - \alpha_B\chi(T) = K_{cf}(T) = (\alpha_A - \alpha_B)\chi_{cf}(T) = (\alpha_A - \alpha_B)f(T)\chi_{HF}(T)/2 \tag{8.4.7}$$

as a function of T to reveal the presence of the heavy fermion susceptibility $\chi_{HF}(T)$ below T^*. When you plot the left-hand side of (8.4.7) versus T, you reveal how a

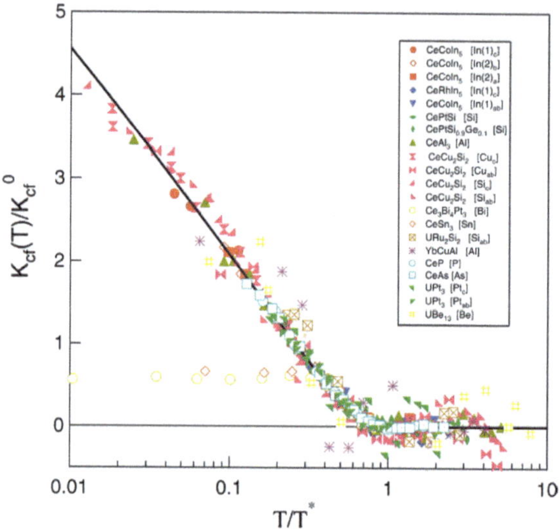

Fig. 8.5 $K_{cf}(T)/K^0_{cf}$ versus $ln(T/T^*)$ for a series of Kondo lattice systems, illustrating the scaling behavior of the Kondo liquid component of the susceptibility. The solid line is a plot of (8.4.8). Figure reprinted from N.J. Curro, B.-L. Young, J. Schmalian and D. Pines, Phys. Rev. B **70**, 235117 (2004). Copyright 2004 American Physical Society

quantity proportional to $f(T)\chi_{HF}(T)$ varies with T, but not, of course, $\chi_{HF}(T)$ itself. However, there are more effects to uncover using the results of the NPF model, as discussed in Sect. 8.4.2.

First, let us note that (8.4.7) can be cast as a universal plot. The temperature scale itself can be plotted in units of T^*, which can be reasonably well estimated by observing where the high-temperature K versus χ plot deviates from linearity (see below for examples). Further, the vertical scale of $f(T)\chi_{HF}(T) \times Const.$ can be gauged by its low-T intercept, so that a universal plot can be constructed of $f(T)\chi_{HF}(T)/f(0)\chi_{HF}(0)$ versus T/T^*, where the denominators of both scales are estimated experimentally. Such a plot, as the superposition of data from fourteen different Kondo lattice compounds, is shown in Fig. 8.5. There is clearly a universal scaling of $K_{cf}(T)$ for the systems concerned. The functional form of this behavior was clearly anticipated by the NPF paper [330]. Thus, the heavy fermion NMR shift is shown to be given by [331]

$$K_{cf}(T) = K^0_{cf}(1 - T/T^*)ln(T^*/T), \tag{8.4.8}$$

where the solid line in Fig. 8.5 is precisely the latter function.

The authors of [331] go beyond what is discussed here to analyze a series of compounds in terms of K-χ plots where the high-temperature data smoothly merge into (8.4.8). We illustrate this behavior with the K-χ plot for the In(2) site in $CeCoIn_5$. Recall that the In(2) site is located in the "wall" of the unit cell and thus has only orthorhombic, i.e., twofold, symmetry for field orientation in the basal plane. The K-χ plot for the In(2) site in Fig. 8.6 compares results for a and b orientation of the field. The full K-χ plots have solid lines drawn for the asymptotic high-temperature behavior. For higher T (lower χ) there is good linear behavior, then both orientations

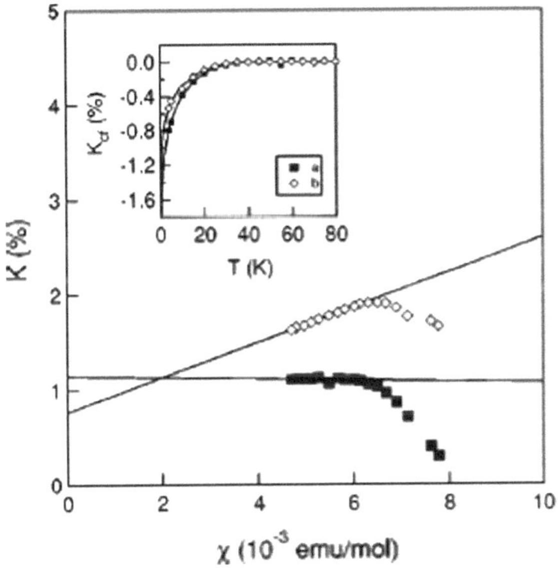

Fig. 8.6 Plot of the In(2)-site (^{115}In) NMR shift in $CeCoIn_5$ versus bulk susceptibility. Solid lines are fits to the high-temperature ($T > T^*$) data. Inset shows a plot of the heavy fermion shift component K_{cf}, determined as described in the text, as a function of T. $K_{cf}(T)$ curves for both a and b axes are fitted to (8.4.8) (solid lines). Figure reprinted from N.J. Curro, B.-L. Young, J. Schmalian and D. Pines, Phys. Rev. B **70**, 235117 (2004). Copyright 2004 American Physical Society

show $K(T)$ deviating downward starting at about $\chi = 0.006$ emu/mol. These deviations are the onset of $K_{cf}(T)$. By subtracting off the solid line (high-temp.) behavior, one obtains data for $K_{cf}(T)$ versus T. Those data are plotted in the inset to Fig. 8.6 for both the a and b field orientation. Next, the fitting function of the NPF model, shown in (8.4.8), can be adjusted to fit the $K_{cf}(T)$ data. The fit is made by choosing values of T^* and K^0_{cf}, the latter of which is simply the value of $K_{cf}(T = 0.259T^*)$ [331].

Many other heavy fermion systems are analyzed in a similar fashion in [331], obtaining uniformly high-quality results. These results serve to validate the NPF two-fluid picture for behavior of Kondo lattice systems in a very convincing way. This picture may be simply a phenomenological model, but it offers a very useful analytical tool for investigating these systems and certainly a foundation for first-principles theoretical work as well.

8.4.2 *Further Refinement of the NPF Model*

Yang and Pines have refined the two-fluid model introduced by NPF [330] to correct small discrepancies with the results of Curro et al. [331] and to incorporate the results of Hall measurements on $CeCoIn_5$ as well as the brilliant theoretical LDA + DMFT results by Shim et al. [333] into their phenomenological model [332]. Principally, they introduce a modified reduced density of states for the Kondo liquid (KL) contribution,

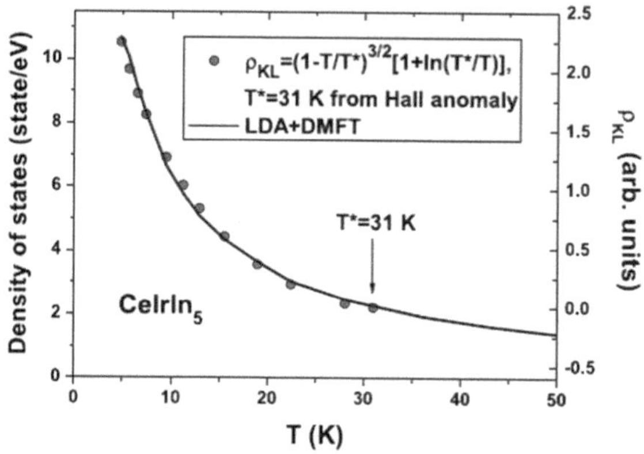

Fig. 8.7 The LDA + DMFT density of states of quasiparticles in *CeIrIn*$_5$ from [333] (solid line) is compared with the Yang–Pines KL (Kondo heavy fermion liquid) density of states ρ_{KL} (Eq. (7) from [332]—filled circles), showing near coincidence. Figure reprinted from Y.Y. Yang and D. Pines, Phys. Rev. Lett. **100**, 096404 (2008). Copyright 2008 American Physical Society

$$\rho_{KL} = \left(1 - \frac{T}{T^*}\right)^{3/2}\left(1 + ln\frac{T^*}{T}\right), \tag{8.4.9}$$

where the specification of T^* is discussed below. Here the dynamic part of the shift is given by $K - K_0 = $ Af(T)χ_{KL} + B[1 − f(T)]χ_{SL}, where χ_{KL} is the emergent "Kondo Liquid" term and χ_{SL} is the "Spin Liquid" or isolated Ce local moment term. Then the anomalous term $\chi_{anom} = $ (A − B)f(T)χ_{XL} (cf. (8.4.7)) is to be compared with (8.4.9).[3] An excellent correspondence is seen to be obtained [332]. These authors also analyze an anomalous Hall coefficient for *CeMIn*$_5$ compounds that not only follows (8.4.9) but also offers a reliable measure of T^*.

Finally, we would like to highlight the remarkable agreement that Yang and Pines find between their phenomenological characterization of the Kondo Liquid term given by (8.4.9) and the calculated density of states of heavy quasiparticles in *CeIrIn*$_5$ obtained using dynamical mean field theory combined with the local density approximation (DMFT + LDA) given by Shim et al. [333]. The fitted result is shown in Fig. 8.7, where an excellent fit is seen to hold. An interesting side point of this result is that it illustrates that T^* is just a parameter and that nothing special really happens at T^*. The quasiparticle density of states in Fig. 8.7 goes on with appreciable values to temperatures well above T^*. We shall discuss further comparisons between the SHK density of states "D(T)" and NMR data in the next section.

8.5 One-Component Description of Magnetic Excitations in *CeIrIn*$_5$

The evident success of the two-fluid model as described up to this point, plus the nonlinear behavior of the K-χ plot, give good evidence that below T^* there are

[3]Note the slightly different hyperfine coefficient notation compared with (8.4.7).

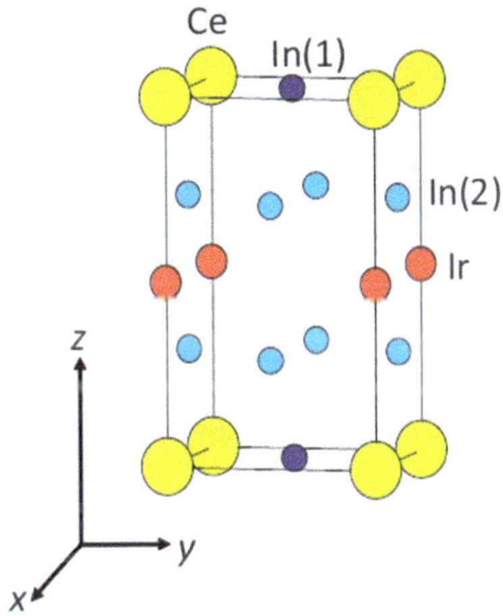

Fig. 8.8 The $HoCoGa_5$ ($I4/mmm$) crystal structure is illustrated for the case of heavy fermion compound $CeIrIn_5$, showing the location of its various constituents. In the text the NMR properties of the In(1) and In(2) nuclei are presented and discussed in some detail. Reprinted from S. Kambe et al., Phys. Rev. B**82**, 144503 (2010). Copyright 2010 American Physical Society

two distinct components to the susceptibility. Nonetheless, it is interesting to develop an interpretation of both NMR shift and T_1 for the ^{115}In isotope in $CeIrIn_5$ in terms of a single, temperature-dependent hyperfine coupling constant as well as a one-component dynamical susceptibility. We shall see that a surprisingly successful picture emerges from such an approach. Our discussion here follows the account of NMR data interpretation along these lines given by Kambe et al. [334].

At the outset of a discussion of NMR studies on Ce "115" compounds it is useful to keep in mind the associated crystal structure and the distinction between local symmetry and environment of the, e.g., In(1) and In(2) sites for NMR-active ^{115}In nuclear spins. The 115 structure is shown in Fig. 8.8, where the tetragonal In(1) site symmetry is seen to be rather higher than that of In(2) (orthrhombic), which is twofold in the basal plane. NMR studies are, of course, typically limited to the $\pm 1/2$ transitions.

We first examine the behavior of the K-χ plot for the In(1) and In(2) sites (e.g., Fig. 2.8). In Fig. 8.9 we see typical shift behavior for a heavy fermion compound. The plots are linear for temperatures above some limit, here $T \sim 60\,\mathrm{K}$. Then, as the heavy fermion susceptibility term χ_α^{hf} rises up, substantial curvature appears in the K-χ plots. However, one can account for this in terms of the DMFT-LDA density of states D(T) [333] shown in Fig. 8.7, and thus indirectly by (8.4.9). We therefore write

$$K_{i,\alpha}^{spin} = K_{i,\alpha} - K_{i,\alpha}^0 = A_{i,\alpha}^{hf}\chi_\alpha + C_{i,\alpha}\chi_\alpha^{hf}, \tag{8.5.1}$$

where χ_α is the total measured susceptibility and $\chi_\alpha^{hf} = f(T)\chi_\alpha^{KL} \propto D(T)$. The constant shift term $K_{i,\alpha}^0$, presumably of orbital origin, is subtracted off as irrelevant to the behavior of $K_{i,\alpha}^{spin}$. Estimates of $K_{i,\alpha}^0$ are derived from extrapolation of high-temperature data to $\chi_\alpha \to 0$ (see, e.g., dashed line in Fig. 8.9). The $C_{i,\alpha}$ terms in (8.5.1) are added to account for the deviations from linearity at $T < T^*$ exhibited by the data plots in Fig. 8.9, using $\chi_\alpha^{hf} \propto$ D(T). A satisfactory account of the data

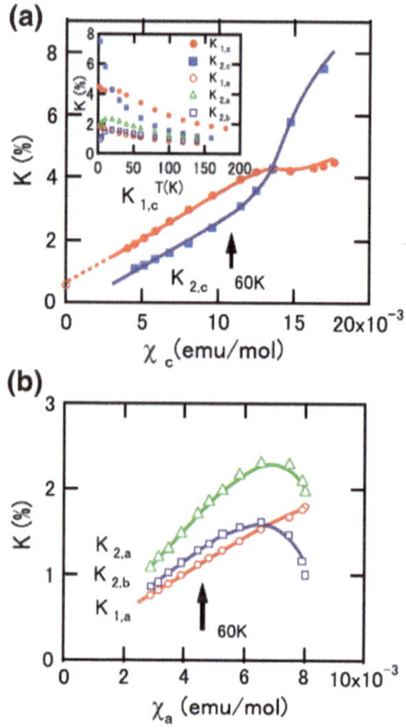

Fig. 8.9 Knight shift $K_{\alpha,i}$ plotted versus static susceptibility χ_α for the In(1) (i = 1) and the In(2) (i = 2) sites in *CeIrIn*$_5$. **a** Results for $H \parallel$ c axis ($\alpha =$ c) and **b** results for $H \parallel$ a axis ($\alpha =$ a, b). The nonlinear behavior looks different for these two orientations, since χ_c increases more rapidly with declining T than χ_a. In (**a**), as an example of $K^{ori}_{i,\alpha}$ (see text), $K^{ori}_{1,c}$ is shown as an extrapolation (dotted line) to $\chi_c = 0$. Solid lines drawn are based on (8.5.1) with $C_{i,\alpha} = 0$ (see text for discussion). The sizes of symbols represent experimental errors. *Inset* T-dependence of $K_{i,\alpha}$. Figure reprinted from S. Kambe et al., Phys. Rev. B**81**, 140405(R) (2010). Copyright 2010, American Physical Society

can be achieved in this fashion. In fact, this fitting procedure validates T^* as a universal parameter for all sites and all field orientations. Such a result supports the density-of-states interpretation by Yang and Pines [332].

Pursuing the density-of-states model for χ^{hf}_α, we have

$$\chi^{hf}_\alpha = 0.5 g^2_\alpha \mu^2_B \lambda^* D(T), \tag{8.5.2}$$

where $g_a = g_b = 1.31$ and $g_c = 1.93$, [334] and where the mass enhancement factor $\lambda^* \simeq 10$.[4] In (8.5.2), g_α is determined by the condition that this expression is equal to the measured χ_α at $T \to 0$ where the "spin liquid" term $\chi_{SL}(T)$ is expected to vanish.

[4] $\lambda^* = \gamma_{el}/\gamma_{cal}$, where $\gamma_{cal} = \pi^2 k_B^2 D(T)/3 = 0.073\,\text{J/molK}^2$ is the calculated Sommerfeld coefficient for D(T) = 12 (states/eV).

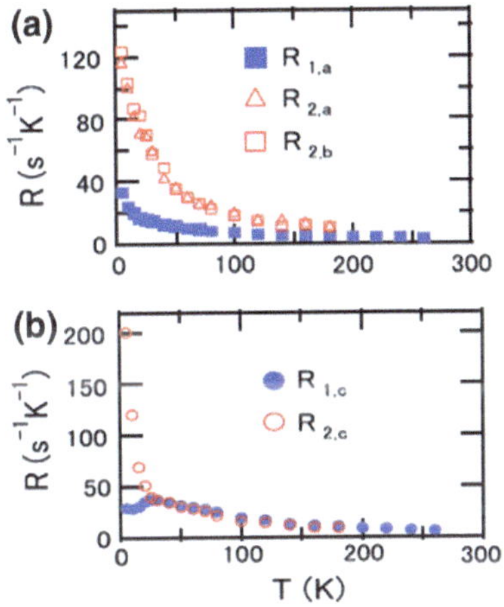

Fig. 8.10 **a** T-dependence of relaxation rates along the a axis: $R_{1,a}$, $R_{2,a}$ and b axis: $R_{2,b}$. **b** T-dependence of relaxation rates along the c axis: $R_{1,c}$, $R_{2,c}$. Figure reprinted from S. Kambe et al., Phys. Rev. B**81**, 140405(R) (2010). Copyright 2010, American Physical Society

We turn our attention now to the spin-lattice relaxation rates measured for $CeIrIn_5$. At each In site there are contributions from nn Ce sites, of which there are four for the In-1 and two for the In-2. The relaxation can be parameterized in terms of contributions from fluctuations along certain axes. Thus, for the In-1 site we have $1/(T_1T)_{H\|a} = R_{1,a} + R_{1,c}$ and $1/(T_1T)_{H\|c} = 2R_{1,a}$. At the In-2 site, $1/(T_1T)_{H\perp \mathbf{n}_{zz}} = R_{2,b} + R_{2,c}$; $1/(T_1T)_{H\|\mathbf{n}_{zz}} = R_{2,c} + R_{2,a}$; $1/(T_1T)_{H\|c} = R_{2,a} + R_{2,b}$. By measuring each of those relaxation rates all of the $R_{i,\alpha}$ subcomponent rates can be evaluated. The results are plotted in Fig. 8.10. It is seen that these rates rise monotonically starting often above 200 K. There is one peculiarity, however, namely the disparity between $R_{1,c}$ and $R_{2,c}$ at $T < T^*$. These rates are made up of the c-axis fluctuations of moments on the Ce sites, yet how could the fluctuations at the In-2 site rise dramatically with lowering temperature, while the In-1 site shows no increase whatever? This point was raised in [334], and a suggestion of correlated fluctuations was mentioned but not discussed in detail.

8.5.1 *Treatment of Spin-Spin Correlations in T_1 for ^{115}In*

There is always, of course, the possibility of short-range correlations among the fluctuations of neighboring Ce spins in a system such as $CeIrIn_5$. Such correlations can weaken or strengthen the net effect of two or more fluctuating spins in a T_1 process. This effect has been treated in detail in Sect. 3.5, and we shall have recourse to those results presently. First, it is important to discuss possible correlation scenarios that might lead to the observed behavior. In the present case these seem at first to be a bit puzzling. First, a pair of nn sites that contribute equally to fluctuations at an In-2 site are producing a very large effect ($R_{2,c}$), while at an In-1 site the combined effects of four nn sites in a square appears to cancel out for the most part ($R_{1,c}$). We propose an arrangement of nn correlations that could bring about such an effect.

While it is probably not the only possibility, the following fluctuating AFM SRO scenario would almost certainly lead to the observed result. We presume the largest spin-spin exchange coupling to occur between *second neighbors*, while that between first neighbors is either weak or even slightly ferromagnetic. Such an arrangement of couplings may seem counter-intuitive, but see below for comments on its plausibility. Such an arrangement would lead to strongest AFM SRO between second neighbor spins, i.e., along diagonals of the quadratic array. There are two planar sublattices of such couplings that do not interact with each other. If the nn coupling is weak in comparison, then the observed T_1 behavior would almost certainly result. For In-1 nuclear spins, the four nn Ce spins would consist of two strongly correlated diagonal pairs, i.e., one pair from each sublattice. Thus, in terms of the T_1 formulation developed in Sect. 3.5 and applied, e.g., to the cuprate LSCO in Sect. 4.1, the relaxation rate can be expressed as

$$\frac{1}{T_{In-1}} = \frac{\gamma_{115}^2}{2} C^2(1 + \mathcal{K}_{01})\tau_e, \tag{8.5.3}$$

where C is the hyperfine constant coupling diagonally opposed Ce nn's, $\tau_e = (k_B T/\mu_B^2)\sum_q^{(N)} \chi''_{ab}(\mathbf{q}, \omega_0)/\omega_0$ and $\mathcal{K}_{02} = \sum_q^{(N)} cos(\sqrt{2}q_{diag}a)\chi''_{ab}(\mathbf{q}, \omega_0)$. Here, $\mathcal{K}_{02}$ is the correlation factor for diagonally opposite Ce spins in the unit cell and is a number close to -1 in value if correlation is strong. In Fig. 8.10a $R_{1,c}$ is thus greatly reduced by the factor $(1 + \mathcal{K}_{02})$, and its attenuating effect grows as T declines. In this way we account for the flatness of $R_{1,c}$ across the HF region.

On the other hand, $R_{2,c}$ grows very rapidly below T^*. The two Ce neighbors whose combined effect relaxes the In(2) nuclear spin belong to separate AFM SRO sublattices. If, as we suggest, their (nn) exchange coupling is rather weaker than the AFM coupling in the diagonal sublattices or is possibly slightly ferromagnetic in character, then there is no expected correlation between their fluctuations and maybe even a slight mutual enhancement. Thus, a similar expression holds as in (8.5.3) ($\times 1/2$), except that $\mathcal{K}_{02}$ in this case is either zero or slightly positive. The In(2) T_1 process is unaffected or slightly enhanced. Thus, as HF fluctuations increase with decreasing T, the large rise in $R_{2,c}$ is not unexpected.

The exchange scenario described here is undoubtedly one of many possibilities. For the compound *CeRhIn*$_5$ [1] there are neutron scattering data showing AFM ordering in the basal plane with an ordering vector $\mathbf{q} = [0.5, 0.5]$, which does not correspond to the mechanism described. For this case one can only argue in general terms using the long-established sources on RKKY exchange in conductors [27] and superexchange [336, 337] in systems where ligand species are embedded with magnetic ions. In the latter case, for example, there is a general finding of antiferromagnetic coupling along straight ("180°") bonds with a tendency for ferromagnetic links in the case of right-angle exchange paths. How RKKY couplings would factor into this is unknown. Nonetheless, such a scenario would support the above picture leading to the observed T_1 behavior. We await further clarification of these matters.

8.6 A Glimpse at Future Additions

There has been a considerable body of work on superconductivity, antiferromagnetic ordering and quantum critical points in recent years, to which NMR studies have made significant contributions. This area appears to provide fruitful topics for a future chapter or chapters of this volume. Not only is there substantive and interesting material to review, but further investigation on themes of this nature seems very likely to be forthcoming.

Appendix A
Appendices

A.1 Some Basic Properties of Spin Echoes

We present in this Appendix some basic derivations of spin echo properties, along with discussion of a number of special topics on spin echo spectroscopy deemed to be of use to the practitioner. The reader is referred to the original paper by Prof. E.L. Hahn [23] for a unique and insightful perspective on the problem. An article by the present author gave a basic derivation, but with emphasis on topics that were somewhat inaccessible in the early days, such as machine calculations of the echo response dependence on rf pulse shapes and pulse phasing techniques to mitigate spurious electronic signals resulting from pulse–induced vibrations of rf coils [24]. While the former topic is not especially useful, the latter will be presented here in some detail because of its practical importance.

The coverage of spin echo–related topics has been expanded in the second edition of this monograph. While there are many aspects of spin echo behavior that are uniquely quantum mechanical in nature, there are also many basic phenomena that can be studied using classical models without loss of generality. We begin, then, with a basic classical derivation of the two–pulse spin echo response with its dependence on the phase and pulse–angle of the radio frequency (rf) excitation pulses. The density–matrix based (quantum) derivation will be included as before, with its ready application to such topics as spin–echo double resonance and pulse–induced echo relaxation through reversal of neigboring spin moments, the latter having been pioneered by Pennington and Slichter [31]. On the topic of spin echo decay processes, a basic formal treatment has been given by Alloul and Froidevaux [294], which is also applicable to oscillatory spin echo decay [340]. A basic discussion of these results is given below. Other sources may also be useful [26, 284]. Finally, a phase–stepping method is described here to eliminate signal interference from pickup coil oscillations in a magnetic field [24].

© Springer-Verlag GmbH Germany 2018

R.E. Walstedt, *The NMR Probe of High-T_c Materials and Correlated Electron Systems*, Springer Tracts in Modern Physics 276,
https://doi.org/10.1007/978-3-662-55582-8

A.1.1 Classical Derivation of the Two-Pulse Spin Echo

This derivation was presented in Ref. [24], where a comprehensive expression is given for both the in–phase (absorption) and quadrature (dispersion) components of the echo response.[1] In the classical picture, we study the motion of a magnetic moment $\boldsymbol{m}$ under the influence of a steady, uniform applied field $\boldsymbol{H}_0$, applied along the z axis. The equation of motion for $\boldsymbol{m}$ in field $\boldsymbol{H}_0$ is $d\boldsymbol{m}/dt = \gamma \boldsymbol{m} \times \boldsymbol{H}_0$. The resulting motion is known as *Larmor precession*, namely that if $\gamma > 0$, $\boldsymbol{m}$ will precess in a clockwise direction around $\boldsymbol{H}_0$. The thermal equilibrium position of $\boldsymbol{m}$ is along the (z-axis) field direction. We can simulate a study of the precessional motion of $\boldsymbol{m}$ that is the essence of an NMR experiment by applying an oscillating "rf" field along a chosen axis in the plane $\perp$ $\boldsymbol{H}_0$. Such an oscillating field of amplitude $2\boldsymbol{H}_1$ can be viewed as two counter-rotating circularly polarized $\boldsymbol{H}_1$ components, only one of which will be at or near the Larmor precession (NMR) frequency of the nuclei under study, depending on the sign of γ. The other component will be "off resonance" by roughly twice the Larmor frequency $\omega_0 = \gamma H_0$ and can safely be ignored.

To visualize the mechanism of rf pulses to excite NMR signals and ultimately spin echoes, we picture their mechanics in a reference frame rotating at the frequency of the applied circularly rotating field $\boldsymbol{H}_1$. In the rotating frame, the transformed field along the z axis is $\Delta H = H_0 - \omega/\gamma$. The equations of motion can be written as

$$\begin{aligned} dm_x/dt &= \Delta\omega m_y - \omega_1 cos\phi\, m_z \\ dm_y/dt &= -\Delta\omega m_x + \omega_1 sin\phi\, m_z \\ dm_z/dt &= \omega_1(m_x cos\phi - m_y sin\phi), \end{aligned} \tag{A.1.1}$$

where $\omega_1 = \gamma H_1$, and $\Delta\omega = \gamma \Delta H$. In the rotating frame diagram of Fig. A.1, the motion according to these equations can be pictured as precession around the vector resultant $\boldsymbol{\omega}_e$ of ΔH along the z axis and $\boldsymbol{\omega}_1$ along a direction in the xy plane at an angle ϕ relative to the y axis, where ϕ is the phase of the rf pulse. $\boldsymbol{\omega}_e$ is γ times the "effective field" in the rotating frame, which makes an angle θ relative to the z axis. If $\boldsymbol{m}$ is at equilibrium along the z axis at $t = 0$, it will precess clockwise in a cone at an angle θ from $\boldsymbol{\omega}_e$, beginning toward the negative x axis. Note that $\omega_e = \gamma[sin\theta(\boldsymbol{i} sin\phi + \boldsymbol{j} cos\phi)H_1 + \boldsymbol{k} cos\theta \Delta H]$.

For the first excitation pulse, we choose $\phi = 0$, which will be the phase reference for the second pulse as well as for precessing components of magnetization that occur. The above equations then simplify to become

$$\begin{aligned} dm_x/dt &= \Delta\omega\, m_y - \omega_1 m_z \\ dm_y/dt &= -\Delta\omega\, m_x \\ dm_z/dt &= \omega_1 m_x, \end{aligned} \tag{A.1.2}$$

[1] Some unfortunate errors in the out–of–phase component echo expression in [24] have come to the author's attention. These are corrected in the current version.

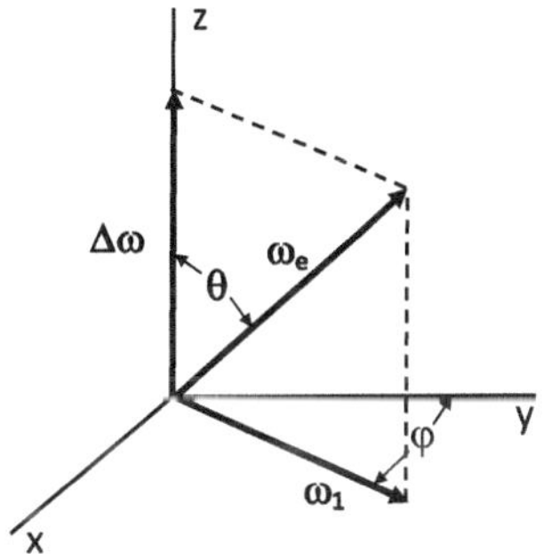

Fig. A.1 Effective magnetic fields that appear in a reference frame rotating around the z axis at the frequency ω of an applied rf pulse

These equations are straightforwardly solved under the initial conditions that $m_x(0) = m_y(0) = 0$ and $m_z(0) = m_0$. From this point forward we keep track only of magnetic components in the xy plane, because only they contribute to the spin echo signal. At the end of the first pulse at time t_{w1} we have

$$\begin{aligned} m_x(t_{w1}) &= -S\theta\, m_0 sin(\omega_e t_{w1}) \\ m_y(t_{w1}) &= S\theta\, C\theta\, m_0[1 - cos(\omega_e t_{w1})], \end{aligned} \tag{A.1.3}$$

where $S\theta = sin\theta = \omega_1/\omega_e$ and $C\theta = cos\theta = \Delta\omega/\omega_e$.

The pulses that generate a spin echo consist of the following time sequence. (i) First pulse from t = 0 to t = t_{w1}; (ii) dephasing interval from t = t_{w1} to t = τ; (iii) second pulse from t = τ to t = $\tau + t_{w2}$; (iv) rephasing interval from t = $\tau + t_{w2}$ to the vicinity of the echo, which is centered at $t_E = 2\tau - t_{w1} + t_{w2}$. Here we shall represent the status of the nuclear magnetization with a three vector $\boldsymbol{m}(t)$. The development of the magnetization vector during any of the time intervals listed is represented by a 3 × 3 matrix operating on the magnetization vector at the beginning of the interval. Thus, for the first pulse result above, we have $\boldsymbol{m}(t_{w1}) = \begin{bmatrix} m_x(t_{w1}) \\ m_y(t_{w1}) \\ 0 \end{bmatrix} = \begin{bmatrix} 0 & 0 & R_{13} \\ 0 & 0 & R_{23} \\ 0 & 0 & 0 \end{bmatrix} \begin{bmatrix} 0 \\ 0 \\ m_0 \end{bmatrix}$, where $R_{13} = -\ S\theta\, m_0\, sin(\omega_e t_{w1})$ and $R_{23} = C\theta\, m_0[1 -\ cos(\omega_e t_{w1})]$.

The next step is the dephasing interval from $t = t_{w1}$ to $t = \tau$, during which time there is no rf field, so that Eq. (A.1.1) become

$$\begin{aligned} dm_x/dt &= \Delta\omega m_y \\ dm_y/dt &= -\Delta\omega m_x \\ dm_z/dt &= 0. \end{aligned} \tag{A.1.4}$$

Both $m_x(t)$ and $m_y(t)$ are subject to dephasing to the extent ($\Delta\omega$) to which they are detuned from the applied rf frequency. The solution to Eq. (A.1.4) is easily found to be represented by $\begin{bmatrix} m_x(\tau) \\ m_y(\tau) \\ 0 \end{bmatrix} = \begin{bmatrix} c_1 & s_1 & 0 \\ -s_1 & c_1 & 0 \\ 0 & 0 & 0 \end{bmatrix} \begin{bmatrix} m_x(t_{w1}) \\ m_y(t_{w1}) \\ 0 \end{bmatrix}$, where $c_1 = cos(\Delta\omega(\tau - t_{w1}))$ and $s_1 = sin(\Delta\omega(\tau - t_{w1}))$.

For the second "rephasing" pulse, we need to use Eq. (A.1.1) as is, because we need a pulse with a variable phase ϕ. To simplify the solution of these equations, they are transformed from variables $m_{x,y}$ to variables $m_{1,2}$ as follows:

$$\begin{aligned} m_1 &= m_x\, s\phi + m_y\, c\phi \quad \text{and inverse} \quad & m_x &= m_1\, s\phi + m_2\, c\phi \\ m_2 &= m_x\, c\phi - m_y\, s\phi & m_y &= m_1\, c\phi - m_2\, s\phi, \end{aligned} \tag{A.1.5}$$

where $c\phi = cos\phi$ and $s\phi = \sin\phi$. In terms of $m_{1,2}(t)$, Eq. (A.1.1) become

$$\begin{aligned} dm_1/dt &= -\Delta\omega\, m_2 \\ dm_2/dt &= \Delta\omega\, m_1 - \omega_1\, m_z \\ dm_z/dt &= \omega_1\, m_2. \end{aligned} \tag{A.1.6}$$

It is easy to show that $d^2m_2/dt^2 + \omega_e^2 m_2 = 0$, so that $m_2(t') = A\, cos(\omega_e t') + B\, sin(\omega_e t')$, where the rephasing pulse goes from $t' = 0$ to $t' = t_{w2}$. But the initial condition for $m_2(t' = 0)$ involves $m_{x,y}(\tau)$. Thus A = $m_2(t' = 0) = m_x(\tau)\, c\phi$ - $m_y(\tau)\, s\phi$. Next, taking $dm_2/dt' = \Delta\omega m_1$ at $t' = 0$, thus dropping an $m_z(\tau)$ term that is unrelated to the echo, we find B = $C\theta\, [m_x(\tau)\, s\phi + m_y(\tau)\, c\phi]$. Using this determination of $m_2(t')$, we can integrate the first of Eq. (A.1.6) using similar methods to determine $m_1(t')$. For completeness we state the solutions for $m_1(t')$ and $m_2(t')$ at $t' = t_{w2}$:

$$\begin{aligned} m_1(t_{w2}) &= -C\theta\, S_2[m_x(\tau)\, c\phi - m_y(\tau)\, s\phi)] - C\theta^2(1 - C_2)[m_x(\tau)\, s\phi + m_y(\tau)\, c\phi] \\ &\quad + m_x(\tau)\, s\phi + m_y(\tau)\, c\phi \\ m_2(t_{w2}) &= C_2[m_x(\tau)\, c\phi - m_y(\tau)\, s\phi] + C\theta\, S_2[m_x(\tau)\, s\phi + m_y(\tau)\, c\phi], \end{aligned} \tag{A.1.7}$$

where $S_2 = sin(\omega_e t_{w2})$ and $C_2 = cos(\omega_e t_{w2})$. Substituting (A.1.7) into m_x and m_y from (A.1.5) gives us the R_{ij} matrix for the rephasing pulse, which may be expressed as the matrix equation $\begin{bmatrix} m_x(\tau + t_{w2}) \\ m_y(\tau + t_{w2}) \end{bmatrix} = \begin{bmatrix} R_{11} & R_{12} \\ R_{21} & R_{22} \end{bmatrix} \begin{bmatrix} m_x(\tau) \\ m_y(\tau) \end{bmatrix}$, where we have the values

$$\begin{aligned} R_{11} &= s\phi^2 S\theta^2(1 - C_2) + C_2 \\ R_{12} &= c\phi\, s\phi\, S\theta^2(1 - C_2) + C\theta\, S_2 \\ R_{21} &= c\phi\, s\phi\, S\theta^2(1 - C_2) - C\theta\, S_2 \\ R_{22} &= c\phi^2 S\theta^2(1 - C_2) + C_2. \end{aligned} \tag{A.1.8}$$

We shall see that in the echo expression the R_{ij} only appear in the following combinations, which we quote for convenience:

$$\begin{aligned} R_{22} - R_{11} &= (c\phi^2 - s\phi^2)\, S\theta^2(1 - C_2) = cos(2\phi)\, S\theta^2(1 - C_2) \\ R_{12} + R_{21} &= 2\, s\phi\, c\phi\, S\theta^2(1 - C_2) = sin(2\phi)\, S\theta^2(1 - C_2). \end{aligned} \tag{A.1.9}$$

Following the second pulse is the "rephasing" period leading to the formation of the echo. The rephasing effect is represented by a similar matrix to the dephasing matrix following (A.1.4) with $s_2 = sin(\Delta\omega(t-\tau-t_{w2}))$ and $c_2 = cos(\Delta\omega(t-\tau-t_{w2}))$. The formal expression for echo formation may then be written $\begin{bmatrix} m_x(t) - t_E \\ m_y(t) - t_E \end{bmatrix} = \begin{bmatrix} c_2 & s_2 \\ -s_2 & c_2 \end{bmatrix} \begin{bmatrix} R_{11} & R_{12} \\ R_{21} & R_{22} \end{bmatrix} \begin{bmatrix} c_1 & s_1 \\ -s_1 & c_1 \end{bmatrix} \begin{bmatrix} m_x(t_{w1}) \\ m_y(t_{w1}) \end{bmatrix}$, where $m_{x,y}$ at t = t_{w1} are given in (A.1.3). The product of the three 2×2 matrices is then a matrix with components T_{ij} which can be written

$$\begin{aligned}
T_{11} &= R_{11}c_2c_1 - R_{12}c_2s_1 + R_{21}s_2c_1 - R_{22}s_2s_1 \\
T_{12} &= R_{11}c_2s_1 + R_{12}c_2c_1 + R_{21}s_2s_1 + R_{22}s_2c_1 \\
T_{21} &= -R_{11}s_2c_1 + R_{12}s_2s_1 + R_{21}c_2c_1 - R_{22}c_2s_1 \\
T_{22} &= -R_{11}s_2s_1 - R_{12}s_2c_1 + R_{21}c_2s_1 + R_{22}c_2c_1 \,.
\end{aligned} \tag{A.1.10}$$

Each of the products c_2c_1, c_2s_1, s_2c_1 and s_2s_1 contains a term that yields an echo response and one that doesn't. Thus, $c_2c_1 = [cos(\Delta\omega(t - t_E)) + cos(\Delta\omega(t - t_{w1} - t_{w2}))]/2$, where $t_E = 2\tau + t_{w2} - t_{w1}$ is the center point of the spin echo. At $t = t_E$, the first term will assume the value "1" regardless of its position in the line, leading to macroscopic signal (echo) formation. The second term will vary randomly over the spins in the system for any value of $t > \tau$ and can be discarded. The sine–cosine products give a different result; thus, $c_2s_1 = (1/2) \times [sin(\Delta\omega(t - t_{w1} - t_{w2})) - sin(\Delta\omega(t - t_E))]$. Here the first term leads to no result, while the second one gives an antisymmetric echo form that is zero at precisely $t = t_E$. See below for more discussion of this term. For the present discussion we introduce the notation $c_E \equiv cos(\Delta\omega(t - t_E))$ and $s_E \equiv sin(\Delta\omega(t - t_E))$. For evaluation of the $T'_{ij}s$, then, we have both c_2c_1 and $s_2s_1 \to c_E/2$, $s_2c_1 \to s_E/2$ and $c_2s_1 \to -s_E/2$.

Insertion of these results into (A.1.10) then leads to

$$\begin{aligned}
T_{11} &= \frac{1}{2}[-(R_{22} - R_{11})c_E + (R_{12} + R_{21})s_E] = \frac{S\theta^2}{2}(1 - C2)(-cos(2\phi)c_E + sin(2\phi)s_E) \\
T_{12} &= \frac{1}{2}[(R_{22} - R_{11})s_E + (R_{12} + R_{21})c_E] = \frac{S\theta^2}{2}(1 - C2)(cos(2\phi)s_E + sin(2\phi)c_E) \\
T_{21} &= \frac{1}{2}[(R_{22} - R_{11})s_E + (R_{12} + R_{21})c_E] = \frac{S\theta^2}{2}(1 - C2)(cos(2\phi)s_E + sin(2\phi)c_E) \\
T_{22} &= \frac{1}{2}[(R_{22} - R_{11})c_E - (R_{12} + R_{21})s_E] = \frac{S\theta^2}{2}(1 - C2)(cos(2\phi)c_E - sin(2\phi)s_E).
\end{aligned} \tag{A.1.11}$$

Combining the sines and cosines, this becomes

$$T_{11} = -\frac{S\theta^2}{2}(1 - C2)\,cos(2\phi + \Delta\omega(t - t_E))$$

$$T_{12} = \frac{S\theta^2}{2}(1 - C2)\,sin(2\phi + \Delta\omega(t - t_E))$$
$$T_{21} = \frac{S\theta^2}{2}(1 - C2)\,sin(2\phi + \Delta\omega(t - t_E))$$
$$T_{22} = \frac{S\theta^2}{2}(1 - C2)\,cos(2\phi + \Delta\omega(t - t_E)),$$

where we see quite generally that $T_{11} = -\ T_{22}$ and $T_{12} = T_{21}$. Substituting these expressions into the foregoing matrix form for $m_{x,y}(t)$ we obtain the general answer for spin echo response of a single frequency component of the NMR spectrum,

$$\begin{aligned} m_x(t) &= \frac{m_0\omega_1^3}{2\omega_e^3}(1 - cos(\omega_e t_{w2}))\Big\{sin(\omega_e t_{w1})cos(2\phi + \Delta\omega(t - t_E)) \\ &+ \frac{\Delta\omega}{\omega_e}(1 - cos(\omega_e t_{w1}))sin(2\phi + \Delta\omega(t - t_E))\Big\} \\ m_y(t) &= \frac{m_0\omega_1^3}{2\omega_e^3}(1 - cos(\omega_e t_{w2}))\Big\{-sin(\omega_e t_{w1})sin(2\phi + \Delta\omega(t - t_E)) \\ &+ \frac{\Delta\omega}{\omega_e}(1 - cos(\omega_e t_{w1}))cos(2\phi + \Delta\omega(t - t_E))\Big\}, \end{aligned} \tag{A.1.12}$$

where we see that m_x and m_y both have both a symmetrical term (in $t - t_E$) and an antisymmetrical one. In addition we see that a positive rotation of the phase of the rf field $\boldsymbol{H}_1$ by ϕ leads to a rotation of the phase of the echo signal in the rotating frame by 2ϕ. This is an important and useful result, as we shall see below.

We may now write down the expression for the spin echo response from irradiating an NMR "line" that we take to be centered at frequency $\omega_0 = \gamma H_0$. The lineshape function $g(\omega')$ will be centered at ω_0, which will be taken to be the zero of ω' and thus, $g(0)$. Then the macroscopic echo response can be written,

$$\begin{aligned} M_x(t) &= \frac{M_0}{2}\int_{-\infty}^{\infty} d\omega' g(\omega')(1 - cos(\omega_e t_{w2}))\Big\{\frac{\omega_1^3}{\omega_e^3}sin(\omega_e t_{w1})cos(\Delta\omega(t - t_E)) \\ &+ \frac{\omega_1^3\Delta\omega}{\omega_e^4}(1 - cos(\omega_e t_{w1}))sin(\Delta\omega(t - t_E))\Big\} \\ M_y(t) &= \frac{M_0}{2}\int_{-\infty}^{\infty} d\omega' g(\omega')(1 - cos(\omega_e t_{w2}))\Big\{-\frac{\omega_1^3}{\omega_e^3}sin(\omega_e t_{w1})sin(\Delta\omega(t - t_E)) \\ &+ \frac{\omega_1^3\Delta\omega}{\omega_e^4}(1 - cos(\omega_e t_{w1}))cos(\Delta\omega(t - t_E))\Big\}, \end{aligned} \tag{A.1.13}$$

where we have specialized to the $\phi = 0$ case for simplicity, and where we emphasize that $\Delta\omega = \omega - \omega' - \omega_0$ and ω_e are both variables along with ω' in (A.1.13), so that the integrands vary widely when the integrals are performed. The two principal assumptions made to arrive at (A.1.13) are, first, the rf excitation pulses are "square", i.e. H_1 is assumed to be constant during the pulse, and second, the duration of the echo signal is assumed to be short compared with T_2. The classical picture used here is also equivalent to the assumption that the nuclear resonance addressed involves a single pair of magnetic energy levels.

A.1.2 The "Area Theorem" and Special Cases

While the foregoing formulas for the echo response are relatively simple, they have a few complications if one examines both the x-axis "absorption" and the y-axis "dispersion"–type signals. However, simplifications and interesting relationships emerge if one studies the integrated area under the echo signal, i.e. A = $\int M_{x,y}(t)dt$, where the integral needs to extend over the entire region where appreciable signal is observed. To integrate the area under both $M_x(t)$ and $M_y(t)$, we use the Fourier integral theorem, which says that $\int sin(\Delta\omega(t-t_E))dt = 0$ and $\int cos(\Delta\omega(t-t_E))dt = 2\pi\delta(\Delta\omega)$. Applying this to $\int M_x(t)dt$, one finds that the only surviving term is

$$\begin{aligned}\int M_x(t) &= \frac{M_0}{2}\int_{-\infty}^{\infty} d\omega' g(\omega')(1-cos(\omega_e t_{w2}))\frac{\omega_1^3}{\omega_e^3} sin(\omega_e t_{w1})\delta(\Delta\omega) \\ &= \frac{M_0}{2} g(\omega-\omega_0)(1-cos(\omega_1 t_{w2}))sin(\omega_1 t_{w1}),\end{aligned} \tag{A.1.14}$$

where on the second line, the $\int d\omega'$ is evaluated using $\delta(\Delta\omega) = \delta(\omega'-\omega+\omega_0)$ The useful result here is that by recording the area under the echo while ω_0 (i.e., H_0) is scanned, one can plot out the profile $g(\omega-\omega_0)$ of the NMR line. There is no *a priori* assumption about the lineshape or about the magnitude of H_1 relative to the linewidth of the profile $g(\omega')$. Note that the comparable integral for $M_y(t)$ vanishes, so the measured lineshape will not be distorted if the phase of the detector is not set to precisely select $M_x(t)$. The foregoing interpretation of the area integral is an exact result which requires only that any effects of transverse relaxation T_2 are uniform over the line scan.

A related but simpler method can be useful in the case that $H_1 \gg \delta H$, were δH is the measured linewidth of the NMR spectrum studied. In such a case the expression for $M_x(t)$ in (A.1.13) simplifies to

$$M_x(t) \approx \frac{M_0}{2}\int_{-\infty}^{\infty} d\omega' g(\omega')sin(\omega_1 t_{w1})(1-cos(\omega_1 t_{w2}))cos(\omega'(t-t_E)), \tag{A.1.15}$$

where the field and frequency are tuned to the resonance center ($\omega \approx \omega_0$) for this measurement. Then the cosine transform of this result can be written

$$\begin{aligned}CosTr\{M_x(t)\} &\approx \frac{M_0}{2} sin(\omega_1 t_{w1})(1-cos(\omega_1 t_{w2}))\int_{-\infty}^{\infty} d\omega' g(\omega')\int_{-\infty}^{\infty} dt cos(\omega_{sc}(t-t_E)cos(\omega'(t-t_E)), \\ &\approx \pi M_0\, sin(\omega_1 t_{w1})\,(1-cos(\omega_1 t_{w2}))g(\omega_{sc}),\end{aligned} \tag{A.1.16}$$

and thus one can extract the entire NMR lineshape function $g(\omega_{sc})$ from a single echo waveform $M_x(t)$. Note that there is no assumption about the shape or symmetry of the line. The integration limits need to go over the entire time interval where there is appreciable echo signal amplitude. An experimental example where this method has been used to good effect is an NMR study of ^{29}Si in URu_2Si_2 [343].

A.1.3 Density Matrix Derivation for the Spin Echo

It is useful to derive the spin echo in the quantum language of the density matrix, since nuclear spins are, after all, essentially quantum objects. Such a formulation is also useful for the purpose of illustrating spin–echo decay processes [294], and in particular spin–echo decay *oscillations*. Finally, the same calculations can illustrate the closely–related basis for spin–echo double resonance. These effects have not been utilized widely in NMR studies relating to condensed matter physics, but they have an interesting potential in that regard.

The nuclear polarization can be represented by a density matrix operator $\rho(T_s)$, where T_s is the effective spin temperature. Here we shall employ the density matrix formulation given by Abragam [5], in which the state of the spin system at thermal equilibrium may be represented by the quantum operator

$$\rho(T_s) = exp(-\mathcal{H}/k_B T_s) \approx \mathbf{1} - \mathcal{H}/k_B T_s = \mathbf{1} - \Sigma_i(\gamma H_0 + \Delta_i) I_{zi}/k_B T_s, \quad \text{(A.1.17)}$$

where $\mathcal{H}$ is taken to be a Zeeman Hamiltonian with an inhomogeneous broadening parameter Δ_i ($\gamma H_0 \gg \Delta_i$). Some degree of static broadening is necessary for an echo to be observable. We shall take the Δ_i's to be static and independent of field. Since the energy scale of individual nuclear spins in the present context is typically very small compared with $k_B T_s$, the single–term expansion in (A.1.17) will always be valid. In the manipulations to follow, transient values of nuclear polarization will not always lie parallel to actual physical magnetic fields. Following (A.1.17), we shall always represent the nuclear polarization with $\rho(t) = \alpha(t) I_\delta$, where δ represents some direction in space. The polarization along the δ axis is then given by $\langle I_\delta \rangle = Tr\{\rho(t) \sum_i I_{\delta i}\}$.

The Interaction Representation and rf Pulses.
The equation of motion for $\rho(t)$ is [5]

$$i\frac{\mathrm{d}\rho}{\mathrm{d}t} = [\mathcal{H}, \rho] \, . \quad \text{(A.1.18)}$$

When $\mathcal{H}$ is a time–independent operator such as the Zeeman Hamiltonian in (A.1.17), then the solution is given by $\rho(t) = exp(-i\mathcal{H}_Z t)\rho(0)exp(i\mathcal{H}_Z t)$. However, the echo excitation pulses are represented by a time–dependent Hamiltonian term

$$\mathcal{H}_1(t) = -\omega_1[I_x cos(\omega t) - I_y sin(\omega t)], \quad \text{(A.1.19)}$$

where $\omega_1 = \gamma H_1$. Thus, (A.1.19) represents magnetic coupling with a rotating magnetic field $\boldsymbol{H}_1(t)$ that is in resonance or nearly so with the Larmor precession of the nuclear spins. To describe the effects of such rf pulses, it is necessary to transform the equation of motion into coordinates rotating in sync with $\boldsymbol{H_1(t)}$. This is an "interaction representation" for describing the problem, but is also simply the rotating frame for the density matrix (see Fig. A.1). It is important to view echo formation as a coherent process, where the rf waveform described by (A.1.19) is the reference system required to describe that coherence.[2] Second, in this representation the equation of motion for the density matrix incorporating $\mathcal{H}_1$ has no explicit time dependence and can be solved easily as shown above.

The required transformation may be written $\rho^*(t) = exp(-i\omega t I_z)\rho(t)exp(i\omega t I_z)$. Then the equation of motion for $\rho^*(t)$ becomes

$$i\frac{d\rho^*}{dt} = (\omega - \omega_0)[I_z, \rho^*] + [\mathcal{H}_1^*, \rho^*], \tag{A.1.20}$$

where $\mathcal{H}_1^* = exp(-i\omega t I_z)\mathcal{H}_1 exp(i\omega t I_z)$. We substitute from (A.1.19) and use the relation

$$exp(-i\phi I_\alpha)I_\beta exp(i\phi I_\alpha) = I_\beta cos(\phi) \pm I_\gamma sin(\phi), \tag{A.1.21}$$

where $(\alpha\beta\gamma)$ is a permutation of (xyz). Thus, "+" is used if $(\alpha\beta\gamma)$ is a positive permutation and "–" is used otherwise. We find

$$i\frac{d\rho^*}{dt} = [\mathcal{H}^*, \rho^*], \quad \text{where} \quad \mathcal{H}^* = -(\Delta\omega I_z + \omega_1 I_x), \tag{A.1.22}$$

which is equivalent to the classical finding earlier of precession around an effective field in the rotating frame of $\boldsymbol{\omega}_e = \omega_1 \boldsymbol{j} + \Delta\omega \boldsymbol{k}$.

It is a straightforward matter to calculate the mechanics of the rf pulses with (A.1.22) in terms of the pulse phase ϕ and the detuning parameter θ (i.e., $\Delta\omega$ and ω_1), etc., as was carried out above for the classical case. Here, the discussion will be simplified by assuming very short rf pulses, i.e. that $\omega_1 \gg \Delta\omega$ for tuning within a linewidth or two of the resonance peak.[3] Thus, we assume a $\pi/2$–π pulse excitation for all the nuclei with negligible width in time and focus our attention on the echo formation process which proceeds with the use of (A.1.22) with $\omega_1 = 0$. The initial condition will be $\rho^*(0_+) = -\alpha I_x$. The spins will dephase during the time interval $(0_+, \tau)$, after which a π pulse of negligible width will be applied with rotation around the rotating–frame y axis. After the π pulse, rephasing will occur, generating an echo in the vicinity of $t = 2\tau$.

Solving (A.1.22) during the dephasing interval then gives, at $t = \tau$,

---

[2]This is because the rf waveform that drives the rotating frame also drives the phase detector that is used to observed the echo.

[3]The linewidth here will be roughly given by the RMS inhomogeneous broadening parameter $\langle\Delta_i^2\rangle^{1/2}$.

$$\rho^*(\tau) = -\alpha \Sigma_i \mathbf{T}_{dp,i} I_{xi} \mathbf{T}^{\dagger}_{dp,i}, \tag{A.1.23}$$

where the dephasing operator for spin i is $\mathbf{T}_{dp,i} = exp[i(\Delta\omega - \Delta_i)\tau I_{zi}]$. Of course, to achieve the result in (A.1.23) one has to apply the full dephasing operator $\mathbf{T}_{dp} = exp[i\Sigma_i(\Delta\omega - \Delta_i)\tau I_{zi}]$, where for each term i in the sum in (A.1.23), only operators pertaining to spin i will survive. This procedure holds for each of the other two steps as well. In this fashion the π–pulse operator $\mathbf{T}_{pi,i} = exp(-i\pi I_{yi})$ at $t = \tau$, followed by the rephasing pulse $\mathbf{T}_{rp,i} = exp[i(\Delta\omega - \Delta_i)(t - \tau)I_{zi}]$, are applied, yielding for the complete echo sequence,

$$\rho^*(t > \tau) = -\alpha \Sigma_i \mathbf{T}_{rp,i} \mathbf{T}_{pi,i} \mathbf{T}_{dp,i} I_{xi} \mathbf{T}^{\dagger}_{dp,i} \mathbf{T}^{\dagger}_{pi,i} \mathbf{T}^{\dagger}_{rp,i}. \tag{A.1.24}$$

On expanding out the rotation operators and combining the resulting terms using (A.1.21), the latter result simplifies to

$$\rho^*(t > \tau) = -\alpha \Sigma_i I_{xi} cos((\Delta\omega - \Delta_i)(t - 2\tau)) - \alpha \Sigma_i I_{yi} sin((\Delta\omega - \Delta_i)(t - 2\tau)). \tag{A.1.25}$$

Recalling from (A.1.3) that $\alpha < 0$, we then find the magnetization

$$\begin{aligned} M_x(t) &= Tr\{\Sigma_i I_{xi} \rho^*(t)\} \\ &= (M_0/2\pi N)\Sigma_i cos((\Delta_i - \Delta\omega)(t - 2\tau)); \\ M_y(t) &= Tr\{\Sigma_i I_{yi} \rho^*(t)\} \\ &= (M_0/2\pi N)\Sigma_i sin((\Delta_i - \Delta\omega)(t - 2\tau)). \end{aligned} \tag{A.1.26}$$

The summation is over the distribution of the Δ_i, which we represent as a continuum described by the lineshape function $g(\Delta')$, where $\int g(\Delta')d\Delta' = 1$. Thus,

$$\begin{aligned} M_x(t) &= (M_0/2\pi) \int d\Delta' g(\Delta') cos((\Delta' - \Delta\omega)(t - 2\tau)); \\ M_y(t) &= (M_0/2\pi) \int d\Delta' g(\Delta') sin((\Delta' - \Delta\omega)(t - 2\tau)). \end{aligned} \tag{A.1.27}$$

We examine these waveforms just briefly. A clear–cut picture emerges if one supposes that $g(\Delta')$ is symmetrical and centered at zero. Then expanding the cosine in $M_x(t)$ and the sine in $M_y(t)$, we find

$$\begin{aligned} M_x(t) &= (M_0/2\pi) cos((\Delta\omega)(t - 2\tau)) \int d\Delta' g(\Delta') cos((\Delta')(t - 2\tau)); \\ M_y(t) &= -(M_0/2\pi) sin((\Delta\omega)(t - 2\tau)) \int d\Delta' g(\Delta') cos((\Delta')(t - 2\tau)), \end{aligned}$$

where the two sine transform terms of $g(\Delta')$ vanish, because the Δ' integrand is antisymmetric. The cosine transform of $g(\Delta')$ is the waveform of the free–induction signal, with a decay time generally labeled T_2^*. In the case of the echo, however, the time variable can be negative as well as positive, so the shape function is two free–induction signals back-to-back. When the frequency is off resonance, the cosine transform waveform is further modulated by $cos(\Delta\omega(t - 2\tau))$ in the case of $M_x(t)$, maintaining the symmetry of the echo waveform around $t = 2\tau$, while $M_y(t)$ is modulated by $sin(\Delta\omega(t - 2\tau))$, which is antisymmetric around $t = 2\tau$. Thus, the area under $M_y(t)$ is zero, as discussed below.

Next, we integrate the area under the echo envelope, i.e. $\int M_{x,y}(t)dt$. Again using the Fourier integral theorem gives $\int cos((\Delta' - \Delta\omega)(t - 2\tau))dt = 2\pi\,\delta(\Delta' - \Delta\omega)$ and $\int sin((\Delta' - \Delta\omega)(t - 2\tau))dt = 0$. With these results, there is no area under $M_y(t)$ as noted above, and the area under $M_x(t)$ is

$$\int M_x(t)dt = M_0 \int d\Delta' g(\delta')\delta(\Delta' - \Delta\omega) = M_0 g(\omega - \omega_0). \tag{A.1.28}$$

This is the same result as (A.1.14) when $\omega_1 t_{w1} = \pi/2$ and $\omega_1 t_{w2} = \pi$. Thus, one can scan the profile of the NMR line by scanning the area under the echo versus ω or ω_0 (i.e., the applied field). In the case of linewidth $\ll \omega_1$ one can simply Fourier transform the echo envelope to get the NMR lineshape.

The density matrix method is seen to give equivalent results to the classical derivation discussed earlier. It is a bit surprising how simply and quickly the density matrix result emerges from the formalism. Of course, that was helped by the simplifying assumptions of a $\pi/2$–π pulse sequence and that $\omega_1 \gg \Delta\omega$ within the resonance line. Comparing the expressions (A.1.27) with the equivalent classical expressions for the spin echo response in (A.1.15), the present result appears simpler, first, because $sin(\omega_e t_{w1}$ and $(1-cos(\omega_e t_{w2}))/2$ in (A.1.15) are replaced here by 1, the ratio $(\omega_1/\omega_e)^3$ is also unity here, and the terms $\propto \Delta\omega/\omega_e$ do not appear in the limit of large ω_1. The remaining terms are the same as in (A.1.27). However, it may be useful to note that solving the density matrix expressions for the general case of arbitrary pulse angles, as well as of more complex broadening mechanisms gives rise to many operators in the formal solutions that do not commute with one another, remains a mathematical challenge.

A.1.4 Spin Echo Decay Phenomena

In the following we shall consider mainly the effects of coupling with other nuclear spins located within a few tens of Angstroms of the relaxing nuclear spin. The relevant nuclear spin–spin coupling can be written

$$\mathcal{H}_{ss} = \frac{1}{2}\Sigma_{i\neq j}\{a_{ij}I_{zi}I_{zj} + b_{ij}\boldsymbol{I}_i \cdot \boldsymbol{I}_j\}. \tag{A.1.29}$$

These couplings appear in addition to the distribution of local inhomogeneous broadening frequencies Δ_i. There are in general three sources for the $\mathcal{H}_{ss}$ couplings in a metallic host. These are the classical dipolar spin–spin coupling and the Bloembergen–Rowland pseudodipolar coupling [28], each contributing to both a_{ij} and b_{ij}, and finally, the Ruderman–Kittel indirect exchange coupling [27], that contributes only to b_{ij}.

There are a number of markedly different cases that arise in the context of (A.1.29), depending on whether some or all of the neighboring spins are a different nuclear species, as well as on the detailed nature of the broadening parameters Δ_i. Such a general case has been discussed at length and a variety of T_2 effects calculated for the high–Tc materials [26]. Here we limit the scope of the discussion mainly to like–spin cases with no T_1 effects. First, however, we make a brief comment on the "unlike spin" case. If unlike spins are fluctuating among themselves, whether through T_1 flips or exchange–driven flip–flops, they will relax the like–spin echo through the like–unlike spin coupling terms. On the other hand, if neighboring unlike spins are static on the time scale of the echo decay, they will have no effect on T_2.

In contrast, like–spin neighbors will have an effect on T_2 whether they are fluctuating or not. The nature of the effect for these two cases will, however, be somewhat different. If the spins are fluctuating "normally", i.e. on the time scale of the larger coefficients in $\mathcal{H}$ in (A.1.29), then T_2 will be affected in the same way it is with fluctuating unlike neighbor spins. The more interesting case is when the spins are static.

For the discussion of like–spin–generated echo decay, the crucial point is whether the broadening by the Δ_i is long–range, so that terms such as $I_{i\pm}I_{j\mp}$ are allowed to cause 'flip–flop' fluctuations among the resonant nuclei. If such is the case, then there will be echo decay that is independent of the pulse angle of the refocusing pulse. The opposite is the case where the Δ_i broadening sufficiently detunes neighboring like–spins so that they cannot undergo mutual flip–flop fluctuations. This is an interesting case that does arise in practice, in which the echo decay often has an oscillatory component. As we shall see below, there are cases of oscillatory decay where simple theory gives an exact closed–form decay function [294].

Let us set up the general echo formation process including the spin–spin coupling perturbation given by (A.1.29). The dephasing operator used in (A.1.23) would then be rewritten as

$$\mathbf{T}_{dp,r} \rightarrow exp\{i\tau[\Sigma_i(\Delta\omega - \Delta_i)I_{zi} + \mathcal{H}_{ss}]\}, \tag{A.1.30}$$

with a similar modification for $\mathbf{T}_{rp,r}$. And while we can still write down the solution to the equation of motion (A.1.22), the form must be modified from that of (A.1.24), because the terms of $\mathcal{H}_{ss}$ do not commute with each other or with I_{zi} by itself. Thus, (A.1.24) must be rewritten as

$$\rho^*_{rlx}(t > \tau) = -\alpha\mathbf{T}_{dp,r}\mathbf{T}_{pi,r}\mathbf{T}_{rp,r}\Sigma_i I_{xi}\mathbf{T}^\dagger_{dp,r}\mathbf{T}^\dagger_{pi,r}\mathbf{T}^\dagger_{rp,r}, \tag{A.1.31}$$

where the rotation operator exponents must each be summed over all sites, because of non-commutation of the relaxation terms. (See $\mathbf{T}_{dp}$ following (A.1.23).) The form (A.1.31) represents a more general form of echo formation and decay, for which there is no general solution available, and is very similar to the free–induction decay problem. In the latter case, series expansions in powers of t are the only available recourse [4, 341].

Exact Solution in the Static Spin Case

Here, we restrict our attention to a different case that arises in alloys [294] and in f-electron compounds, namely where the flip–flop terms in (A.1.29) are rendered inactive by local inhomogeneities. Thus, terms of the form $I_{i\pm}I_{j\mp}$ become non-secular and may be dropped. The broadening implied by $\mathcal{H}_{ss}$ then becomes entirely static. In such a static environment, the echo relaxation process is derived solely from spin reorientation driven by the refocusing pulse. Moreover, it is then possible to write out an exact evaluation of (A.1.31).

Thus, in the static case, we consider the evaluation of the echo decay process via $E(2\tau) = C\,Tr[I_x\rho^*_{rlx}(2\tau)]$, using (A.1.31). C is adjusted so that $E(0) = 1$. In the static limit all of the terms in $\mathcal{H}_{ss}$ commute with each other and with I_{zi}. It is, then, a straightforward matter to evaluate $E(2\tau)$ for the case of a single crystal specimen, giving

$$E(2\tau)_I = C\Sigma_\lambda\Pi_{j\neq 0}\Sigma^{+I}_{m=-I}exp(2i\tau m\tilde{J}_{0j}), \qquad \text{(A.1.32)}$$

where $\tilde{J}_{0j} = J_{0j} - (\gamma^2\hbar/r^3_{0j} + b_{0j})(3cos^2\theta_{jl} - 1)$, and where the Σ_λ is a sum over all possible configurations of neighboring like spins in a lattice with less than 100% occupation of the resonant nuclear species. For echo formation with 100% occupation by the resonant nuclear isotope, the latter summation may be omitted. The derivation of (A.1.32) with the assumptions given is straightforward. The details may be found in Sec. IIC of [294].

There is further simplification of the above expression for $E(2\tau)$ if the resonant nucleus has spin quantum number $I = 1/2$. In that case the sums on m may be carried out explicitly to give

$$E(2\tau)_{1/2} = C'\Sigma_\lambda\Pi_{j\neq 0}cos(\tilde{J}_{0j}\tau), \qquad \text{(A.1.33)}$$

where the product of cosines should extend over neighbor sites having appreciable coupling with the nucleus at the origin. Within the assumption of static spins, both (A.1.32) and (A.1.33) are exact results that can be evaluated to high precision if the range function of $\tilde{J}_{0j}$ is known to good accuracy. That, unfortunately, is rarely the case. If the combination of structure and probability of site occupancy gives rise to dominance by nuclei with one or two nearest neighbor spins, then it is clear from (A.1.33) that there will be a significant oscillatory component to the T_2 decay process. This is the basis for oscillatory echo decay observed in doped samples of platinum and lead [294]. A recent example is described here briefly.

These results have been applied to ^{29}Si NMR in the compound $YbRh_2Si_2$ [340], which is a nearly ideal system for the application of (A.1.33). This is an f–electron compound with exotic properties such as heavy fermions and a quantum critical point [342]. The structure is well suited to application of (A.1.33) in that the Si sites are organized in pairs separated by only 2.39Å. Each Si site has a single close neighbor, then 12 more neighboring Si sites at distances ranging from 4 to 5Å. The latter sites provide a very nearly Gaussian background decay structure, while the first neighbors (nn) impose a clear sinusoidal oscillation on the decay. As a result, the entire waveform can be fitted to (A.1.33) in such a way that both the RKKY and pseudodipolar coupling constants for the nearest neighbors can be measured experimentally over a wide range of temperatures, providing important physical input for our understanding of this system [339].

An important question, then, arises as to how to decide whether the spin system under study is static or not. A useful test on this point is whether the echo decay process depends on the pulse angle of the refocusing pulse. We may envision two limits for this effect. If the spin system is fluctuating freely, it won't make much difference if the refocusing pulse inverts all the spins halfway through the echo sequence. Thus, T_2 won't change as the refocusing pulse angle is varied. At the other limit, T_2 will change in a systematic way with pulse angle if the spins are static, becoming very long for small pulse angles. Such a check provides a straightforward method to determine whether the spins are actually static in any case of interest.

A More General Case of Static Spins

The exact expression for spin echo decay (A.1.32) was derived under the assumption of a π–pulse for the rephasing step (i.e., θ_2). It is a bit more trouble to generalize this result to an arbitrary value of θ_2. Such a result, however, is called for by experiments that evaluate the indirect "$I_{zi}I_{zj}$" spin–spin coupling measurements pioneered by Pennington and Slichter [31], where it is important to be able to calibrate echo decay as a function of θ_2. We revert, then, to (A.1.31) and expand the $M_x(t)$ echo expression for arbitrary values of θ_2. Again, $\mathcal{H}_{ss}$ (A.1.29) is simplified in the dropping of any $I_{i\pm}I_{j\mp}$ terms, so that all perturbation terms in the relaxation (A.1.31) commute, but they don't commute with $\mathbf{T}_{pi,r} \rightarrow exp(i\theta_2 I_y)$. To deal with this, we expand the general expression for $\rho^*(t > \tau)$ using (A.1.21), keeping only the terms $\propto I_{xi}$, as they represent the principal echo response. The y–axis rotation operator is worked through in this process, giving for $t > \tau$ the expression

$$E(t > \tau)/E_0 = \frac{1}{N}\Sigma_i \langle cos\theta_2\, cos(\beta_i(t-\tau))\, cos(\beta_i'\tau) - sin(\beta_i(t-\tau))\, sin(\beta_i'\tau)\rangle_i, \tag{A.1.34}$$

where $\beta_i = \Delta_i + \Sigma_j \tilde{J}_{ij} m_j$ and $\beta_i' = \Delta_i + \Sigma_j \tilde{J}_{ij} m_j'$. The prime indicates that the quantum number m_j' must show the effect of operator I_{zj} being modified by the θ_2 pulse. $N^{-1}\Sigma_i$ is a normalized sum on the resonating spins and Σ_j a sum on neighbor spins to spin site i with enough coupling to affect its behavior. To show the echo behavior of (A.1.34) we rewrite it as

$$\begin{aligned} E(t > \tau)/E_0 = \frac{1}{2N}\{ &\Sigma_i \langle cos(\beta_i(t-\tau) - \beta_i'\tau)(cos\theta_2 - 1)\rangle_i \\ &+ \Sigma_i \langle cos(\beta_i(t-\tau) + \beta_i'\tau)(cos\theta_2 + 1)\rangle_i \}. \end{aligned} \tag{A.1.35}$$

If we examine the t–dependence of the Δ_i terms in the cosine arguments in (A.1.35), we find in the first sum $\Delta_i(t - 2\tau)$ and in the second, $\Delta_i t$. These are the major terms that determine overall behavior. Thus, the first term gives an echo response at $t = 2\tau$, while the second term shows no echo response at all. Dropping the second term and evaluating the first one at $t = 2\tau$, (A.1.35) simplifies to

$$E(2\tau)/E_0 = -\frac{1}{2}(1 - cos\theta_2)\langle cos[2\tau\Sigma_j \tilde{J}_{ij}(m_j - m'_j)]\rangle_i, \quad \text{(A.1.36)}$$

where on dropping the Σ_i, the angle brackets are, in effect, an average over all possible distributions of like–spin neighbors in the lattice surrounding the site of a resonant nucleus.

Pressing on, we first note that (A.1.36) closely resembles echo expressions given earlier, where the essential quantity in $\langle\ \rangle_i$ is unity at $\tau = 0$ and decays to lower values as τ increases. It is an approximate formulation of the decay process, where effects of the off–diagonal terms from the θ_2 rephasing pulse are neglected. To continue, we expand the cosine in (A.1.36) to order τ^2, in what is a Gaussian approximation to the decay process [26, 31].

$$E(2\tau)/E_0 \approx (1 - cos\,\theta_2)\left\langle 1 - \frac{\tau^2}{2}\left[\sum_j \tilde{J}_{ij}(m'_j - m_j)\right]^2\right\rangle_i . \quad \text{(A.1.37)}$$

In squaring the sum $\sum_j \tilde{J}_{ij}(m'_j - m_j)$ the cross terms will average to zero. In the diagonal terms the quantity $(m'_j - m_j)^2$ will equal 1 with a probability $sin^2(\theta_2/2)$ and will be zero otherwise.[4] (A.1.37) becomes [26]

$$E(2\tau)/E_0 \simeq (1 - cos\,\theta_2)\left\langle 1 - \frac{\tau^2}{2} sin^2(\theta_2/2)\sum_j{}' \tilde{J}_{ij}^2\right\rangle_i , \quad \text{(A.1.38)}$$

where the $\sum'_j$ indicates a sum only over $^{63}Cu(2)$ neighbors which are affected by the θ_2 pulse. This means, typically, couplings with the same isotope, and m_j-values belonging to the resonance transition observed. In high-field NMR for $I = 3/2$, it applies to half of the neighbor spins. For NQR with $I = 3/2$, one sums all of the $^{63}Cu(2)$ neighbor spins, because they are all part of the observed resonance. The difficulty in applying this is, of course, that often the line is broad and the θ_2 pulse is not uniform over all of the spins. The factor $sin^2(\theta_2/2)$ in such a case is an effective mean value.

[4] This result is one of the most basic in all of spin resonance. It's derivation is discussed at some length in Ch. II of [5]. Noting that $sin^2(\theta_2/2) = (1 - cos\theta_2)/2$, it is seen, e.g., to occur in many formulations of the spin echo as the probability of flipping the resonant spins with the rephasing pulse.

If we suppose that (A.1.38) gives the first two terms of the Gaussian $exp[-(2\tau)^2/2T_{2g}^2]$, then we have for the final result for a high-field NMR transition for ^{63}Cu(2),

$$\frac{1}{T_{2g}^2} = sin^2(\theta_2/2)\frac{c_{63}}{8}\sum_j \tilde{J}_{ij}^2 , \tag{A.1.39}$$

where the final sum over all neighbor sites is, of course, independent of i. In naturally abundant ^{63}Cu, $c_{63} \simeq 0.69$.

A.1.5 Indirect Coupling with a Gaussian Susceptibility

The indirect coupling range function used in the text is discussed here for the simple case of a $\boldsymbol{q}$-independent, isotropic HF tensor and a Gaussian susceptibility $\chi'(\boldsymbol{q})$. After deriving the range function from (6.1.7), results for the second moment $(\sum_{j(\neq i)} a_{ij}^2)$ obtained in several ways are discussed.

The general result for the indirect coupling coefficient of $I_{zi}I_{zj}$ is

$$a_\alpha(i,j) = \frac{\hbar}{N_A g_\alpha^2 \mu_B^2}\sum_{\boldsymbol{q}}{}' e^{i\boldsymbol{q}\cdot(\boldsymbol{R}_j-\boldsymbol{R}_i)}\chi'_\alpha(\boldsymbol{q})F_{\alpha\alpha}(\boldsymbol{q})^2 , \tag{A.1.40}$$

where the prefactor is suited to a molar susceptibility and an HF tensor in units of s^{-1}. Here we take $F_{\alpha\alpha}(\boldsymbol{q}) = A_0$, and we use the Gaussian susceptibility form employed by Auler et al. [245], which is

$$\chi'_\alpha(\boldsymbol{q}) = \chi'_{AF} e^{-(\boldsymbol{q}-\boldsymbol{Q})^2\xi^2/4} . \tag{A.1.41}$$

Since $\chi'_\alpha(\boldsymbol{q})$ is symmetric about $\boldsymbol{Q}$, we move the origin to that point with $\boldsymbol{q} = \boldsymbol{p}+(\pi,\pi)$. The imaginary sine term is then seen to vanish by symmetry. For simplicity we drop the prefactor $(\hbar\chi'_{AF}A_0^2/N_A g_\alpha^2\mu_B^2)$ and restore it at the end. The sums may then be taken as integrals, giving,

$$a(i,j) = \frac{(-1)^{(n_x+n_y)}}{4\pi^2}\int_{-\pi}^{\pi} dp_x \int_{-\pi}^{\pi} dp_y e^{-(p_x^2+p_y^2)\xi^2/4} cos(n_x p_x + n_y p_y) . \tag{A.1.42}$$

The cosine factor can be written $cos(n_x p_x)\,cos(n_y p_y)$, since the sine terms vanish. The integrals can now be written $g(n_x)g(n_y)$, where

$$g(n) = \frac{(-1)^n}{\pi}\int_0^{\pi} dp e^{-p^2\xi^2/4} cos(np) \simeq \frac{(-1)^n}{\sqrt{\pi}\,\xi} e^{-n^2/\xi^2} , \tag{A.1.43}$$

where the last step approximates the upper limit of the integral as ∞. This is a reasonable approximation, since the exponential factor $\simeq 0.015$ at $p = \pi$. We shall check this approximation below, as well. The final result for our dimensionless range function is

$$a(i, j) = \frac{(-1)^{(n_x+n_y)}}{\pi\xi^2} \mathrm{e}^{-(n_x^2+n_y^2)/\xi^2} . \tag{A.1.44}$$

We now evaluate the second moment sum $\sum_{j(\neq i)} a(i, j)^2$ in three separate ways, in order to test the approximation in (A.1.43). First, we may do this directly with (A.1.44), giving the expression

$$\sum_{j(\neq i)} a(i, j)^2 = \frac{1}{\pi^2\xi^4} \left[\sum_{n_x,n_y} \mathrm{e}^{-2(n_x^2+n_y^2)/\xi^2} - 1 \right], \tag{A.1.45}$$

where the summation is over all integer values of n_x, n_y which would give an appreciable contribution. In what may seem a crude approximation for $\xi \simeq 1.3$ (see text), we evaluate the sums in (A.1.45) as an integral in a continuum approximation. This is straightforwardly shown to yield the value $\pi\xi^2/2$, so that we find

$$\sum_{j(\neq i)} a(i, j)^2 = \frac{1}{\pi^2\xi^4} \left[\frac{\pi\xi^2}{2} - 1 \right]. \tag{A.1.46}$$

The foregoing result gives the value $\sum_{j(\neq i)} a(i, j)^2 = 0.0587$ for $\xi = 1.3$, which is the minimum value we shall use it for. This sum has also been evaluated by computer for the same value of ξ, giving 0.0588 for the result. It seems then that the continuum approximation to the (n_x, n_y) sums is accurate to a fraction of a percent. However, the range function itself is an approximate result from (A.1.43).

We make a check on the latter result by evaluating the second moment sum using the $\boldsymbol{q}$-space result

$$\sum_{j(\neq i)} a(i, j)^2 = \sum_q \sum_p \mathrm{e}^{-(q^2+p^2)\xi^2/4} - \left[\sum_q \mathrm{e}^{-q^2\xi^2/4} \right]. \tag{A.1.47}$$

When the first sum is converted to an integral in polar coordinates, the integrand is proportional to $\exp(-q^2\xi^2/2)$, which, at $q = \pi$ falls to the value $\sim 2.4 \times 10^{-4}$. While this could be accounted for in approximate fashion, we neglect errors of this magnitude for simplicity here. This term then becomes $1/(2\pi\xi^2)$. The second term in (A.1.47) can be cast into a very similar form as that of (A.1.43), where we neglect a correction term of the order of 1.5%. Since this is squared, neglecting these correction means that the second term in (A.1.48) is small by about 3%. The two terms together give $\sum_{j(\neq i)} a(i, j)^2 = 1/(2\pi\xi^2) - 1/(\pi^2\xi^4)$, which is identical to the summing result in (A.1.46). We now see that for $\xi = 1.3$, this result probably errs on the high side by an amount of order 1%. For larger values of ξ the error will diminish rapidly.

Restoring the prefactor, and multiplying by the factor $c_{63}/8$ to give the physically correct second moment (or $(1/T_{2g})^2$), which is

$$\left(\frac{1}{T_{2g}}\right)^2 = \frac{c_{63}}{8}\sum_{j(\neq i)} a(i,j)^2 = \frac{c_{63}}{8}\frac{\hbar\chi'_{AF}A_0^2}{N_A g_\alpha^2\mu_B^2)}\frac{1}{\pi^2\xi^4}\left[\frac{\pi\xi^2}{2}-1\right]. \quad \text{(A.1.48)}$$

A.2 Spin Echo Decay via Spin-Lattice Relaxation

Spin-lattice relaxation of a spin echo was first treated to explain the large disparity between conventional T_1 values and the enhanced exponential decay rates $1/T_{1E}$ of spin echoes for nuclear spins with large I values [284]. The relaxation rate for the $(-\frac{1}{2} \leftrightarrow \frac{1}{2})$ transition was shown to be $1/T_{1E} = 2\,W_0(I+\frac{1}{2})^2$, where $1/T_1 = 2\,W_0$. This result was modified for the case of an anistropic T_1 process by Narath [286], finding

$$\frac{1}{T_{1Ez}} = (W_x + W_y)[I(I+1) - 1/4] + W_z\,, \quad \text{(A.2.1)}$$

where W_α is the contribution of fluctuations along the α axis to the relaxation rate for a quantization axis perperdicular to that axis. Thus, $1/T_{1z} = W_x + W_y$. The latter results have been derived using the "fictitious spin $\frac{1}{2}$" technique. They have been derived using other methods as well [139, 285].

Since the W_α's are not measured directly, it is also useful to express the result (A.2.1) in terms of the anisotropic T_1 values $T_{1x,y,z}$:

$$\frac{1}{T_{1Ez}} = \frac{1}{T_{1z}}[I(I+1) - 3/4] + \frac{1}{2\,T_{1x}} + \frac{1}{2\,T_{1y}}\,. \quad \text{(A.2.2)}$$

We also treat the case of arbitrary direction for the quantization axis, represented by the Euler angles (θ, ϕ). We let that be the z' axis with corresponding x' axis in the xy plane and mutually perpendicular y' axis. The associated rate parameters for the (x', y', z') axes are

$$W_{x'} = W_x sin^2\phi + W_y cos^2\phi \quad \text{(A.2.3a)}$$

$$W_{y'} = W_x cos^2\theta\, cos^2\phi + W_y cos^2\theta\, sin^2\phi + W_z sin^2\theta \quad \text{(A.2.3b)}$$

$$W_{z'} = W_x sin^2\theta\, cos^2\phi + W_y sin^2\theta\, sin^2\phi + W_z cos^2\theta\,, \quad \text{(A.2.3c)}$$

whereupon $1/T_{1Ez'}$ is given in terms of the $W_{\alpha'}$'s by (A.2.1). This result was used to correct the planar ^{17}O spin echo decay data in [26] for T_1 effects.

A.3 Using rf Phase Toggling to Eliminate Transient NMR Interference

Spin Echo phase toggling to eliminate spurious signals. As any NMR practitioner well knows, there are many sources of spurious rf signals that can find their way into the best of spectrometers. Simple rf pickup from induction heaters and other electrical equipment can usually be eliminated by metal enclosures and other shielding techniques. One of the most difficult interference problems to deal with is transient rf signals generated by the high–power rf excitation pulses used to excite spin echoes. One version of this is transient mechanical vibrations generated in the rf coil of a typical single–coil spectrometer when high–power rf pulses are applied to the coil while in a high magnetic field. The result is transient mechanical vibrations in the coil wire which, in turn, induce spurious electrical signals in the spectrometer receiver system. Their generation is so closely related to how NMR signals, themselves, are generated, that no type of shielding can be put to use in mitigating such effects. One is occasionally called upon to conduct measurements at relatively low frequencies, which is where these coil vibration transients are at their worst.

We describe here a scheme using rf phase toggling that is capable of reducing spurious pulse transients by at least three orders of magnitude. While it is worthwhile to employ strategies such as winding rf coils with bundles of very fine wire to suppress mechanical vibrations, or possibly using an old–fashioned two–coil NMR head, ultimately the phase toggling scheme can be counted on to do the "heavy lifting" of transient cancellation. The central principles of rf phase toggling are (i) that the phase of the echo signal can be reversed by either reversing the phase of pulse 1 or by changing the phase of pulse 2 by $\pi/2$. And (ii) a digital signal averager can be set to digitize and alternately add and subtract incoming spin echo signals. In this way, a four–step sequence of phase toggling can be combined with an add–subtract–add–subtract... phase reversal of the signal averager to diminish any accumulation of spurious transients by orders of magnitude. We illustrate the phase toggling scheme in Table A.1, and illustrate the results below with data taken at Bell Laboratories in Murray Hill, NJ, in the 1980's.

The four–step sequence of pulse phases is presented in Table A.1, where Pulse one is at phase 0 for two steps and then at π for two steps. Pulse two toggles between 0 and $\pi/2$ for two steps at each of these, but one step out of phase with Pulse one. This sequence produces a series of echo responses that alternate indefinitely between "+" and "−". The digital signal averager simply toggles between "add" and "subtract", so that while pulse–driven transients are cancelled on alternate steps, the echo signal accumulates continuously.

Application of this scheme is illustrated in Figs. A.2 and A.3. In Fig. A.2 the accumulated result is shown from 64,000 passes of a two-pulse echo sequence with no form of transient cancellation in use. As the echo should be in the 50 μs region, clearly, any echo signal has been completely swamped by a large oscillating transient that extends out beyond 100 μs. In Fig. A.3, an identical sequence is shown with the four–step transient cancellation scheme described above and in Table A.1 activated

Table A.1 Table giving pulse phasing sequence used to cancel rf pulse–induced electrical transients

STEP	Pulse 1	Pulse 2	Echo	Averager
1	0	0	+	Add
2	0	$\pi/2$	−	Subtract
3	π	$\pi/2$	+	Add
4	π	0	−	Subtract

for the digital averaging process. There we see a cleanly resolved echo signal on a baseline with a noise level giving a signal–to–noise ratio of better than 10. Note that in the second figure, the amplitude scale is ten times smaller than in the first one. The transient cancellation scheme has reduced the random noise level by a factor ~ 250. Since there is no discernible sign of the transient in the final result, the transient has been reduced by a similar or greater factor.

The process of transient cancellation can be analyzed from the pulse toggle sequences in Table A.1. Both pulses are applied in sequences of identical pairs, while the digitizer subtracts the receiver transients from the two pulses of any adjacent pair. For example, transients from the two identical "Pulse 1" pulses in Steps 1 and 2 are cancelled by the "Add/Subtract" toggle of the digitizer for those steps. The same is

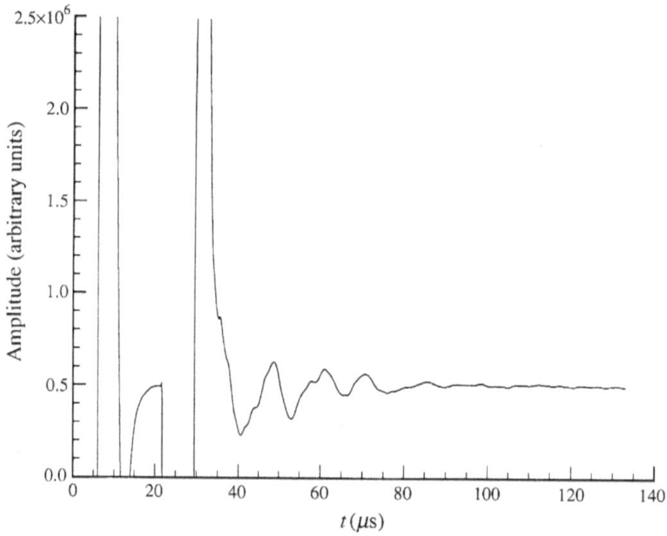

Fig. A.2 Digitally recorded sum of 64,000 ^{63}Cu spin echo signals taken without employing any procedure to cancel and oscillatory electrical transient. This transient is barely visible on a single trace. See text for a description of the sample and other experimental details. The rf pulses begin at $t = 0$ and at $22\,\mu s$ and are $3\,\mu s$ and $5\,\mu s$ in length, respectively

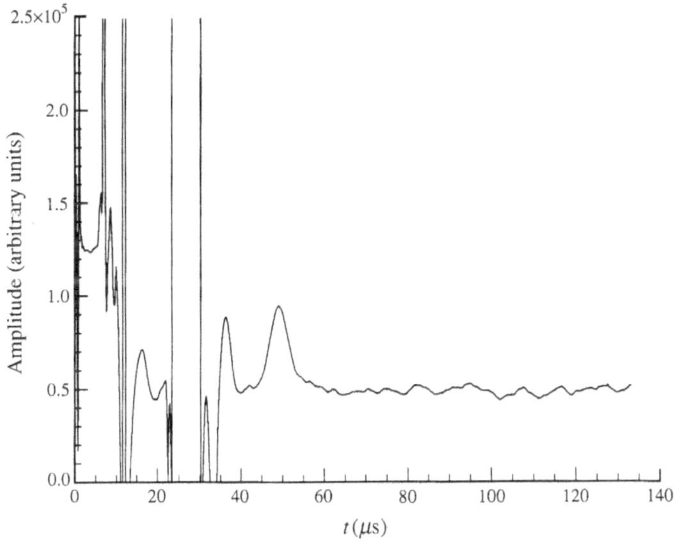

Fig. A.3 Digitally recorded spin echo signal with identical conditions to Fig. A.2, except that the "add-subtract" method for cancelling the electrical transient was employed as described in the text. Note that the vertical scale is an order of magnitude smaller than in Fig. A.2. Resolution of the echo signal is now only limited by random noise, and could be improved by further averaging

true of Pulse 2 in steps 2 and 3. There is evidently a very nearly linear superposition of transients from the two pulses in any echo sequence. At least if the second pulse modifies the transient from the first one in any degree, it does so in a fashion that is completely independent of its phase. Thus, the combined transients from any pulse pair are both cancelled almost perfectly. In any case, the cancellation error of this scheme is well below the resolution of the experimental example presented. The scheme described has been in operation for many years with entirely satisfactory results.

References

1. *Physical Properties of High Temperature Superconductors I*, ed. D.M. Ginsburg (World Scientific, Singapore, 1989); *Physical Properties of High Temperature Superconductors II*, ed. D.M. Ginsburg (World Scientific, Singpore, 1990)
2. T. Graf, G. Triscone, J. Muller, J. Less-Common Met. **159**, 349 (1990)
3. T. Tanamoto, H. Kohno, H. Fukuyama, J. Phys. Soc. Japan **62**, 717 (1993)
4. M. Horvatić, T. Auler, C. Berthier, Y. Berthier et al., Phys. Rev. B **47**, 3461 (1993)
5. A. Abragam, *Principles of Nuclear Magnetism* (Oxford University Press, New York, 1983)
6. C.P. Slichter, *Principles of Magnetic Resonance*, 3rd edn. (Springer, New York, 1992)
7. A. Narath: *Hyperfine Interactions*, ed. A.J. Freeman, R. B. Frankel (Academic Press, New York, 1967), p. 287
8. J. Winter, *Magnetic Resonance in Metals* (Clarendon Press, Oxford, 1971)
9. D.E. MacLaughlin: Magnetic Resonance in the Superconducting State, in *Solid State Physics*, vol. 31, ed. H. Ehrenreich, F. Seitz, D. Turnbull (Academic Press, New York, 1976), p. 1
10. F. Bloch, Phys. Rev. **70**, 460 (1946)
11. N. Bloembergen, E.M. Purcell, R.V. Pound, Phys. Rev. **73**, 679 (1948)
12. G.E. Pake, E.M. Purcell, Phys. Rev. **74**, 1184 (1948)
13. C.H. Pennington, C. P. Slichter, *Physical Properties of High Temperature Superconductors*, vol. II, ed. D.M. Ginsberg (World Scientific, Singapore, 1990), p. XX
14. R.E. Walstedt, W.W. Warren Jr., Science **248**, 1082 (1990)
15. C.P. Slichter, *Strongly Correlated Electron Systems*, ed. K.S. Bedell et al. (Addison–Wesley, Reading, MA, 1993), p. 427
16. D. Brinkmann, M. Mali, *NMR Basic Principles and Progress*, vol. 31 (Springer, Berlin, 1994), p. 171
17. C. Berthier, M.H. Julien, M. Horvatić, Y. Berthier, J. Phys. I France **6**, 2205 (1996)
18. M. Tinkham, *Introduction to Superconductivity* (McGraw-Hill, New York, 1975)
19. J.R. Schrieffer, *Theory of Superconductivity* (Westview Press, Boulder, 1999)
20. One diagnostic of this condition is that the effective γ for transitions between the $\pm 1/2$ levels (i.e. for *rf* pulse conditions) is larger that the tabulated value by a factor $(I + 1/2)$. See Appendix A for details
21. It is possible to separately excite, e.g. the (1/2,3/2) and the (−1/2,−3/2) transitions, using circularly polarized *rf* pulses (M.J. Weber and E.L. Hahn, Phys. Rev. **120**, 365 (1960)). However, for most measurements, there is no advantage in doing so

© Springer-Verlag GmbH Germany 2018

R.E. Walstedt, *The NMR Probe of High-T_c Materials and Correlated Electron Systems*, Springer Tracts in Modern Physics 276,
https://doi.org/10.1007/978-3-662-55582-8

22. There are certain advantages to *cw* technique, however. If T_2 is shorter than a few μs, spin echoes may become unobservable, whereupon *cw* methods become the only way to observe an NMR signal
23. E.L. Hahn, Phys. Rev. **80**, 580 (1950)
24. R.E. Walstedt, The two-pulse spin echo revisited, in *Pulsed Magnetic Resonance: NMR, ESR and Optics*, ed. D.M.S. Bagguley (Clarendon Press, Oxford, 1992), p. 242
25. H. Yasuoka, T. Imai, T. Shimizu, Strong Correlations and Superconductivity, ed H. Fukuyama, S. Maekawa, A. P. Malozemoff (Springer, Berlin, 1989). See also H. Yasuoka, S. Kambe, et al. Physica **199&200**, 278 (1994)
26. R.E. Walstedt, S.-W. Cheong, Phys. Rev. B **51**, 3163 (1995)
27. M.A. Ruderman, C. Kittel, Phys. Rev. 96, 99 (1954); T. Kasuya, Prog. Theor. Phys. 16; K. Yosida. Phys. Rev. **106**, 893 (1957)
28. N. Bloembergen, T.J. Rowland, Phys. Rev. **97**, 1679 (1955)
29. H. Suhl, Phys. Rev. 109, 606 (1958); T. Nakamura. Prog. Theor. Phys. **20**, 542 (1958)
30. J. Van Kranendonk, Physica **20**, 781 (1954)
31. C.H. Pennington, C.P. Slichter, Phys. Rev. Lett. **66**, 381 (1991)
32. W.D. Knight, Phys. Rev. 76, 1259 (1949); C.H. Townes, C. Herring, W.D. Knight. Phys. Rev. **77**, 852 (1950)
33. D.L. Randles, Proc. Roy. Soc. London, Ser. A **331**, 85 (1972)
34. R.E. Walstedt, Y. Yafet, A.I.P. Conference Proceedings **24**, 522 (1974)
35. V. Heine, Phys. Rev. **107**, 1002 (1957)
36. R. Kubo, Y. Obata, J. Phys. Soc. Japan **11**, 547 (1956)
37. A.M. Clogston, A.C. Gossard, V. Jaccarino, Y. Yafet, Phys. Rev. Lett. **9**, 262 (1962)
38. A.M. Clogston, V. Jaccarino, Phys. Rev. **121**, 1357 (1961)
39. J. Bardeen, L.N. Cooper, J.R. Schrieffer, Phys. Rev. **108**, 1175 (1957)
40. Oriented powder samples are samples with anisotropic susceptibilities which have been induced to form partially-to-fully ordered dominant magnetic axes in a magnetic field. This technique was of great importance for NMR and susceptibility studies of high-T_c materials. See Sect. 2.1 for a more extensive discussion
41. J. Korringa, Physica **16**, 601 (1950)
42. This discussion only applies to nuclei with odd half-integer spin quantum numbers. Many nuclei of interest answer to this description, and, in fact, there will be no nuclear spins of interest in this volume which have integer values of I
43. Inability to invert polarization can result from fast spectral diffusion, but it is also caused by an inhomogeneous *rf* field H_1 so that the pulse condition varies over the sample volume
44. The author is indebted to W.G. Clark for this observation
45. W.A. Hines, W.D. Knight, Phys. Rev. Lett. 18, 341 (1967). Phys. Rev. **4**, 893 (1971)
46. F. Wright, W.A. Hines, W.D. Knight, Phys. Rev. Lett. **18**, 115 (1967)
47. R.H. Hammond, G.M. Kelly, Rev. Mod. Phys. **36**, 185 (1964)
48. F. Reif: Phys. Rev. **102**, 1417 (1956); **106**, 208 (1957)
49. K. Yosida, Phys. Rev. **110**, 769 (1958)
50. See Ref. 5 for a more detailed account of these developments
51. R.A. Ferrell, Phys. Rev. Lett. **3**, 262 (1959)
52. P.W. Anderson, Phys. Ref. Lett. **3**, 325 (1959)
53. A.A. Abrikosov, L.P. Gorkov, Zh. Eksp. Teor. Fiz. **42**, 1088 (1962); [Sov. Phys. –JETP **15**, 752 (1962)]
54. P.W. Anderson, J. Phys. Chem. Solids **11**, 26 (1959)
55. R.H. Hammond, G.M. Kelly, Phys. Rev. Lett. **18**, 156 (1967)
56. H.L. Fine, M. Lipsicas, M. Strongin, Phys. Lett. **29A**, 366 (1969)
57. D. Pines, Solid State Physics, vol. I, ed. F. Seitz, D. Turnbull (Academic Press, New York, 1954), p. 1
58. T. Moriya, J. Phys. Soc. Jpn. **18**, 516 (1963)
59. G.M. Androes, W.D. Knight, Phys. Rev. **121**, 779 (1961)

60. A brief discussion defining the coherence length and penetration depth parameters is given in Sec. 1.2.4
61. A plot showing $\mathcal{K}(\alpha)$ vs. α is given in [6]
62. Experimental values of the Korringa product can deviate sharply from that in Eq. (2.2.8) for a number of reasons. We give three examples: First, if the s-character of Fermi surface orbitals is unusually weak, then p-electron contributions which have been neglected here may play a large role. Second, if the degree of s-character is strongly non-uniform over the Fermi surface, then one of the principal assumptions of the calculation breaks down. Third, if there are multiple bands or multiple neighbors (in the case of ligand nuclei), the experimental Korringa product will change drastically. For example, if a ligand nucleus has N independent but equivalent sources of HF field and fluctuations, then the shift and relaxation rate will both be N times that from a single source. It follows that the Korringa product will be the value from a single source divided by N. The latter case is encountered frequently for ligand nuclei in intermetallic compounds
63. For simplicity, detailed balance considerations are omitted here
64. L.C. Hebel, C.P. Slichter, Phys. Rev. **113**, 1504 (1957)
65. Y. Masuda, A.C. Redfield, Phys. Rev. **125**, 159 (1962)
66. This broadening mechanism was suggested by [65], who noted that it gave a good account of the variation of T_1 below T_c. The uniform energy gap parameter $\Delta(T)$ in BCS theory is, in fact, an approximation, where the actual gap parameter is properly a function of $\boldsymbol{k}$. See the book by Tinkham [17], Sect. 2–6, for more discussion of this point
67. H. Bommel, Phys. Rev. **96**, 220 (1954)
68. C.R. Leavens, J.P. Carbotte, Solid State Commun. **9**, 75 (1971); Ann. Phys. (*N.Y.*) **70**, 338 (1972)
69. Y. Masuda, Phys. Rev. **126**, 1271 (1962)
70. J.D. Williamson, D.A. MacLaughlin, Phys. Rev. B **8**, 125 (1973)
71. J.R. Clem, Phys. Rev. **148**, 392 (1966)
72. R.E. Watson, A.J. Freeman, *Hyperfine Interactions*, ed. A.J. Freeman, R.B. Frankel (Academic Press, New York, 1967), p. 53
73. E.D. Jones, K.B. Jefferts, Phys. Rev. 135, A1277 (1964); H. Yasuoka, et al. Phys. Rev. **177**, 667 (1969)
74. J.H. Van Vleck, *Electric and Magnetic Susceptibilities*
75. In discussions of experimental susceptibilities, it will be more convenient to use molar units, whereupon the shift coefficient α_s will be written $8\pi\langle|u_{\boldsymbol{k}}(0)|^2\rangle/3N_A$, where N_A is Avogadro's number
76. A.M. Clogston, V. Jaccarino, Y. Yafet, Phys. Rev. **134**, A650 (1964)
77. J.H. Wood, G.W. Pratt Jr., Phys. Rev. **107**, 995 (1957)
78. A.J. Freeman, R.E. Watson, *Treatise on Magnetism*, vol. IIA, ed. G.T. Rado, H. Suhl (Academic Press, New York, 1965)
79. From the standpoint of the microscopic physics of solids, we regard the susceptibilities per atom χ_{at} (or per formula unit χ_{fu}) as fundamental, representing the magnetic polarization which occupies one Wigner-Seitz (or unit) cell. Their companion molar susceptibilities are $\chi_M = N_A\chi_{at}$ and $\chi_{Mfu} = N_A\chi_{fu}$. If Ω is the atomic (or Ω_c the unit cell) volume, then the dimensionless volume susceptibility fundamental to electromagnetic theory is $\chi_V = \chi_{at}/\Omega$ (χ_{fu}/Ω_c), and the gram susceptibility is $\chi_{gm} = \chi_{at}/\Omega\rho$, where ρ is the volume density. It is clear that the actual dimensions of χ_M are cm^3 and of the α shift coefficients are cm^{-3}, although these are usually cited as $emu/mole$ and $(emu/mole)^{-1}$
80. F.E. Hoare, J.C. Matthews, Proc. Roy. Soc. (London) **A212**,137 (1952); D.W. Budworth et al., ibid. **A257**, 250 (1960)
81. Y. Obata, J. Phys. Soc. Jpn. **18**, 1020 (1963)
82. Y. Yafet, V. Jaccarino, Phys. Rev. **133**, A1630 (1964)
83. Explain d-functions in cubic space group
84. L.P. Kadanoff, Phys. Rev. **132**, 2073 (1963)

85. J. Butterworth, Phys. Rev. Lett. **5**, 305 (1960); J. Butterworth, Proc. Phy. Soc. London **85**, 735 (1965)
86. P. Pincus, A.C. Gossard, V. Jaccarino, J.H. Wernick, Phys. Lett. **13**, 21 (1964)
87. A.G. Redfield, Phys. Rev. **162**, 367 (1967)
88. D. Rossier, D.E. MacLaughlin, Phys. Kondens. Mater. **11**, 66 (1970)
89. J.-M. Delrieu, J.-M. Winter, Solid State Commun. **4**, 545 (1966)
90. G. Eilenberger, Phys. Rev. **153**, 584 (1967)
91. P.R. Radebaugh, P. Keesom, Phys. Rev. **149**, 209 (1966); **149**, 217 (1966)
92. A. A. Abrikosov, Zh. Eksperim. i Teor. Fiz. **32**,1442 (1957) [English transl.: Soviet Phys.-JETP **5**, 1174 (1957)
93. W.H. Kleiner, L.M. Roth, S.H. Autler, Phys. Rev. **133**, A1226 (1964)
94. T.J. Greytak, J.H. Wernick, J. Phys. Chem. Solids **25**, 535 (1964)
95. U. Brandt, W. Pesch, L. Tewordt, Z. Physik **201**, 209 (1967)
96. U. Brandt, Phys. Lett. **27A**, 645 (1968)
97. W. Pesch, Phys. Lett. **28A**, 71 (1968)
98. M. Cyrot, J. Phys. (Paris) **27**, 283 (1966)
99. N. Okuda, Y. Masuda, Phys. Rev. Lett. **20**, 1475 (1968)
100. Y. Masuda, N. Okuda, J. Phys. Soc. Japan **26**, 309 (1969)
101. W. Fite, A.G. Redfield, Phys. Rev. **162**, 358 (1967)
102. B.G. Silbernagel, M. Weger, J.H. Wernick, Phys. Rev. Lett. **17**, 384 (1966)
103. B.G. Silbernagel, M. Weger, W.G. Clark, J.H. Wernick, Phys. Rev. **153**, 535 (1967)
104. B.W. Batterman, C.S. Barrett, Phys. Rev. Lett. **13**, 390 (1964)
105. A.C. Gossard, Phys. Rev. **149**, 246 (1966)
106. D. Rossier, D.E. MacLaughlin, Phys. Rev. Lett. **22**, 1300 (1969)
107. C. Caroli, P.G. deGennes. J. Matricon Phys. Lett. **9**, 307 (1964)
108. A.G. Redfield, W.N. Yu, Phys. Rev. **169**, 443 (1968); **177**, 1018 (1969)
109. A.Z. Genack, A.G. Redfield, Phys. Rev. Lett. **31**, 1204 (1973)
110. G.B. Furman, S.D. Goren, Phys. Rev. B **68**, 064402 (2003)
111. H. Monien, D. Pines, Phys. Rev. B 41, 6297 (1990); N. Bulut, D. Scalapino. Phys. Rev. **45**, 2371 (1992)
112. J.G. Bednorz, K.A. Mueller, Z. Phys. **64**, 189 (1986)
113. R.J. Cava, R.B. Van Dover, B. Batlogg, E.A. Rietman, Phys. Rev. Lett. **58**, 408 (1987)
114. M.K. Wu, J.R. Ashburn et al., Phys. Rev. Lett. **58**, 908 (1987)
115. J.D. Jorgensen et al., Phys. Rev. Lett. **58**, 1024 (1987)
116. M. Onoda, S. Shamoto et al., *Novel Superconductivity*, ed. S.A. Wolf, V.Z. Kresin (Plenum, New York, 1987) p. 919
117. D.E. Farrell et al., Phys. Rev. B **36**, 4025 (1987)
118. F. Mila, T.M. Rice, Physica C **157**, 561 (1989)
119. R.E. Walstedt, S.-W. Cheong, Phys. Rev. B **64**, 014404 (2001)
120. P. Hüsser, H.U. Suter, E.P. Stoll, P.F. Meier, Phys. Rev. B **61**, 1567 (2000)
121. J. Owen, J.H.M. Thornley, Rep. Prog. Phys. **29**, 675 (1966)
122. N.F. Mott, Proc. Phys. Soc. London Sect. A **62**, 416 (1949)
123. P.W. Anderson, Science **235**, 1196 (1987)
124. P.W. Anderson, *Frontiers and Borderlines in Many-Particle Physics, International School of Physics "Enrico Fermi", Course CIV, edited by R* (A. Broglia, J. R, Schrieffer (North-Holland, Amsterdam, 1988)
125. F.C. Zhang, T.M. Rice, Phys. Rev. B **37**, 3759 (1988)
126. S. Trugman, Phys. Rev. 37, 1597 (1988); B.J. Shraiman, E.D. Siggia, Phys. Rev. Lett. 60, 740 (1988); S. Sachdev, Phys. Rev. 39, 12232 (1989); C.K. Lane, P.A. Lee, N. Read. Phys. Rev. **39**, 6880 (1989)
127. J.T. Markert, T.W. Noh, S.E. Russek, R.M. Cotts, Sol. St. Comm. **63**, 847 (1987)
128. D. Brinkmann, L. Pauli, J. Roos, H. Zimmermann, Phys. Lett. A **124**, 112 (1987)
129. W.W. Warren Jr., R.E. Walstedt et al., Phys. Rev. Lett. **59**, 1860 (1987)
130. R.E. Walstedt, W.W. Warren Jr. et al., Phys. Rev. B **36**, 5727 (1987)

131. H. Lütgemeier, M.W. Pieper, Sol. St. Comm. **64**, 267 (1987)
132. I. Furo, A. Jánossy, L. Mihály et al., Phys. Rev. B **36**, 5690 (1987)
133. C.H. Pennington, D.J. Durand et al., Phys. Rev. B **37**, 7944 (1988)
134. T. Shimizu, H. Yasuoka, T. Imai et al., J. Phys. Soc. Japan **57**, 2494 (1988)
135. M. Takigawa, P.C. Hammel, R.H. Heffner, Z. Fisk, Phys. Rev. B **38**, 2832 (1988)
136. H. Alloul, P. Mendels et al., Phys. Rev. Lett. **61**, 746 (1988)
137. R.E. Walstedt, W.W. Warren Jr. et al., Phys. Rev. B **38**, 9299 (1988)
138. R.E. Walstedt, W.W. Warren Jr., R. Tycko et al., Phys. Rev. B **38**, 9303 (1988)
139. C.H. Pennington, D.J. Durand, C.P. Slichter, Phys. Rev. B **39**, 274 (1989)
140. M. Takigawa, P.C. Hammel et al., Phys. Rev. B **39**, 300 (1989)
141. C.H. Pennington, D.J. Durand, C.P. Slichter et al., Phys. Rev. B **39**, 2902 (1989)
142. F. Adrian, Phys. Rev. B **38**, 2426 (1988)
143. R.E. Walstedt, W.W. Warren Jr. et al., Phys. Rev. B **40**, 2572 (1989)
144. B. Bleaney, K.D. Bowers, M.H.L. Pryce, Proc. R. Soc. London A **228**, 166 (1955)
145. A. Abragam, B. Bleaney, *Electron Paramagnetic Resonance of Transition Ions* (Clarendon Press, Oxford, 1970), p. 455
146. B.S. Shastry, Phys. Rev. Lett. **63**, 1288 (1989)
147. S.E. Barrett, D.J. Durand et al., Phys. Rev. B **41**, 6283 (1990)
148. R.E. Walstedt, W.W. Warren Jr. et al., Phys. Rev. B **41**, 9574 (1990)
149. W.C. Lee, D.C. Johnston, Phys. Rev. B **41**, 1904 (1990)
150. A.K. McMahan, R.M. Martin, S. Satpathy, Phys. Rev. B **38**, 6650 (1988)
151. F. Mila, Phys. Rev. B **38**, 11358 (1988)
152. T. Moriya, J. Phys. Soc. Japan **19**, 681 (1964)
153. J.A. Martindale, S.E. Barrett et al., Phys. Rev. Lett. **68**, 702 (1992)
154. G. Balakrishnan, R. Dupree et al., J. Phys. C Solid State Phys. **21**, L847 (1988)
155. V. Jaccarino, *in Physics of Solids, International School of Physics "Enrico Fermi", Course XVIII* (Academic, New York, 1967)
156. P.C. Hammel, M. Takigawa et al., Phys. Rev. Lett. **63**, 1992 (1989)
157. F. Mila, T.M. Rice, Phys. Rev. B **40**, 11382 (1989)
158. N. Bulut, D.W. Hone et al., Phys. Rev. 41, 1797 (1990). Phys. Rev. Lett. **64**, 2723 (1990)
159. C.M. Varma, P. Littlewood et al., Phys. Rev. Lett. **63**, 1996 (1989)
160. H. Monien, D. Pines, Phys. Rev. B **41**, 6297 (1990)
161. A.J. Millis, H. Monien, D. Pines, Phys. Rev. B **42**, 167 (1990)
162. A.J. Millis, H. Monien, Phys. Rev. B **45**, 3059 (1992)
163. T. Moriya, Prog. Theor. Phys. **28**, 371 (1962)
164. H. Monien, D. Pines, C.P. Slichter, Phys. Rev. B **41**, 11120 (1990)
165. F. Mila, T.M. Rice, Phys. Rev. B **40**, 11382 (1989)
166. H. Alloul, T. Ohno, P. Mendels, Phys. Rev. Lett. **63**, 1700 (1989)
167. V.J. Emery, G. Reiter, Phys. Rev. B **38**, 4547 (1988)
168. M. Takigawa, P.C. Hammel et al., Phys. Rev. Lett. **63**, 1865 (1989)
169. M. Horvatić, Y. Berthier et al., Physica C **159**, 689 (1989)
170. C. Coretsopoulos et al., Phys. Rev. B **39**, 781 (1989)
171. R. Sternheimer, Phys. Rev. **84**, 244 (1951); **86**, 316 (1952); **95**, 736 (1954)
172. J. Kirtley, C.C. Tsuei et al., Nature (London) **373**, 225 (1995)
173. D.H. Wollman, D.J. van Harlingen et al., Phys. Rev. Lett. **74**, 797 (1995)
174. H. Monien, A. Zawadowski, Phys. Rev. Lett. **63**, 911 (1989)
175. C. Gros, R. Joynt, T.M. Rice, Z. Phys. B **68**, 425 (1987)
176. F.C. Zhang, C. Gros, T.M. Rice, H. Shiba, Supercond. Sci. Technol. **1**, 36 (1988)
177. G. Kotliar, J. Liu, Phys. Rev. B **38**, 5142 (1988)
178. M. Takigawa, P.C. Hammel, R.H. Heffner, Z. Fisk, Phys. Rev. B **39**, 7371 (1989)
179. R.E. Walstedt et al., Phys. Rev. B **45**, 8074 (1992)
180. T. Imai, T. Shimizu, H. Yasuoka, J. Phys. Soc. Jpn. **57**, 2280 (1988)
181. G.-Q. Zheng et al., Phys. Rev. Lett. **86**, 4664 (2001)
182. N.J. Curro et al., Nature **434**, 622 (2005)

183. H. Sakai, Y. Tokunaga et al., J. Phys. Soc. Japan **74**, 1710 (2005)
184. T. Fujimoto, G.-Q. Zheng et al., Phys. Ref. Lett. **92**, 047004 (2004)
185. W.W. Warren Jr., R.E. Walstedt et al., Phys. Rev. Lett. **62**, 1193 (1989)
186. M. Horvatić, P. Ségransan, C. Berthier, Y. Berthier et al., Phys. Rev. B **39**, 7332 (1989)
187. J. Friedel, Physica C **153–155**, 1610 (1988)
188. P.A. Lee, N. Nagaosa, X.-G. Wen, Revs. Mod. Phys. **78**, 17 (2006)
189. D.C. Johnston, Phys. Rev. Lett. **62**, 957 (1989)
190. M. Takigawa, A.P. Reyes, P.C. Hammel, Phys. Rev. B **43**, 247 (1991)
191. M. Takigawa, W.L. Hults, J.L. Smith, Phys. Rev. Lett. **71**, 2650 (1993)
192. H. Monien, D. Pines, M. Takigawa, Phys. Rev. B **43**, 258 (1991)
193. J. Zaanen, M.L. Horbach, W. van Saarloos, Phys. Rev. B **53**, 8671 (1996)
194. J. Zaanen, W. van Saarloos, Physica C **282–287**, 178 (1997)
195. A. Uldry, P.F. Meier, Phys. Rev. B **72**, 094508 (2005)
196. M. Randeria, N. Trivedi, A. Moreo, R.T. Scalettar, Phys. Rev. Lett. **69**, 2001 (1992)
197. P.M. Gehring, J.M. Tranquada et al., Phys. Rev. B **44**, 2811 (1991)
198. J. Rossat-Mignod et al., Physica B **169**, 58 (1991)
199. J.M. Tranquada et al., Phys. Rev. B **46**, 5561 (1992)
200. H. Alloul, A. Mahajan, H. Casalta, O. Klein, Phys. Rev. Lett. **70**, 1171 (1993)
201. C. Stock, W.J.L. Buyers, R. Liang et al., Phys. Rev. B **69**, 014502 (2004)
202. R.J. Birgenau, R.W. Erwin, P.M. Gehring et al., Z. Physik B Condensed matter **87**, 15 (1992)
203. H. Chou, J.M. Tranquada, G. Shirane et al., Phys. Rev. B **43**, 5554 (1991)
204. R.E. Walstedt, T.E. Mason, G. Aeppli, S.M. Hayden, H.A. Mook, Phys. Rev. B **84**, 024530 (2011)
205. J.A. Martindale, P.C. Hammel et al., Phys. Rev. B **57**, 11769 (1998)
206. V. Nandor et al., Phys. Rev. B **60**, 6907 (1999)
207. L.D. Rotter et al., Phys. Rev. Lett. **67**, 2741 (1991)
208. M. Randeria: *Precursor Pairing Correlations and Pseudogaps*, [Varenna Lectures 1997]; cond-mat/9710223
209. J.W. Loram et al., Phys. Rev. Lett. **71**, 1740 (1993)
210. J.W. Loram et al., Physica C **135–140**, 134 (1994)
211. J.M. Harris, Z.-X. Shen et al., Phys. Rev. B **54**, R15665 (1996)
212. A.G. Loeser et al., Science **273**, 325 (1996)
213. H. Ding et al., Nature (London) **382**, 51 (1996)
214. R.M. Hazen, *Physical Properties of High Temperature Superconductors*, ed D. M. Ginsburg (World Scientific, Singapore, 1990), vol. II, p. 121
215. T. Machi, I. Tomeno et al., Physica C **173**, 32 (1991)
216. I. Tomeno, T. Machi et al., Phys. Rev. B **49**, 15327 (1994)
217. H. Yasuoka, Hyperf. Int. **105**, 27 (1997)
218. M. Bankay, M. Mali, J. Roos, D. Brinkmann, Phys. Rev. B **50**, 6416 (1994)
219. G.V.M. Williams et al., Phys. Rev. Lett. **80**, 377 (1998)
220. J. P. Franck, *Physical Properties of High Temperature Superconductors*, ed. D. M. Ginsburg (World Scientific, Singapore, 1994), vol. IV, p. 189
221. F. Raffa, T. Ohno et al., Phys. Rev. Lett. **81**, 5912 (1998)
222. C.M. Varma, Phys. Rev. B **73**, 155113 (2006)
223. B. Fauqué et al., Phys. Rev. Lett. **97**, 197001 (2006)
224. G. Xiao, M.Z. Cieplak et al., J. Phys. Soc. Japan **62**, 2591 (1993)
225. A.V. Mahajan, H. Alloul, G. Collin, J.F. Marucco, Phys. Rev. Lett. **72**, 3100 (1994)
226. J. Bobroff, W.H. MacFarlane, H. Alloul, P. Mendels et al., Phys. Rev. Lett. **83**, 4381 (1999)
227. W.H. MacFarlane, J. Bobroff, H. Alloul, P. Mendels, N. Blanchard, Phys. Rev. Lett. **85**, 1108 (2000)
228. G.-Q. Zheng, T. Odaguchi, T. Mito, Y. Kitaoka et al., J. Phys. Soc. Japan **62**, 2591 (1993)
229. H. Zimmerman et al., Physica C **159**, 681 (1989)
230. W.Y. Liang, J.W. Loram, Physica C **263**, 277 (1996)
231. N. Nagaosa, P.A. Lee, Phys. Rev. B **45**, 966 (1992)

232. T. Nakano et al., Phys. Rev. B **49**, 16000 (1994)
233. R. Yoshizaki et al., Physica C **166**, 417 (1990)
234. S. Ohsugi, Y. Kitaoka, K. Ishida, G.-Q. Zheng, K. Asayama, J. Phys. Soc. Japan **63**, 700 (1994)
235. K. Ishida et al., J. Phys. Soc. Jpn. **60**, 2351 (1991)
236. G.-Q. Zheng et al., Physica C **208**, 339 (1993)
237. K. Ishida et al., J. Phys. Soc. Jpn. **60**, 3516 (1991)
238. K. Yoshimura, T. Imai et al., J. Phys. Soc. Jpn. **58**, 3057 (1989)
239. R.E. Walstedt, B.S. Shastry, S.-W. Cheong, Phys. Rev. Lett. **72**, 3610 (1994)
240. V. Barzykin, D. Pines, Phys. Rev. B **52**, 13585 (1995)
241. D. Pines, C.P. Slichter, Phys. Rev. **100**, 1014 (1955)
242. N. Curro, T. Imai, C.P. Slichter, B. Dabrowski, Phys. Rev. B **56**, 877 (1997)
243. M. Takigawa, P. C. Hammel, et al., Physica (Amsterdam) **162–164C**, 853 (1989)
244. C.M. Varma, Phys. Rev. Lett. **77**, 3431 (1996)
245. T. Auler, M. Horvatic et al., Physica C **313**, 255 (1999)
246. L.P. Regnault et al., Physica C 235–240, 59 (1994). Physica B **213–214**, 48 (1995)
247. J. Rossat-Mignod et al., Physica B **192**, 109 (1993)
248. B.J. Sternlieb et al., Phys. Rev. 47, 5320 (1993). Phys. Rev. **50**, 12915 (1994)
249. M. Horvatić, C. Berthier, Y. Berthier et al., Phys. Rev. B **48**, 13848 (1993)
250. M. Horvatić, C. Berthier, Y. Berthier et al., Phys. Rev. B **48**, 13848 (1993)
251. N. Bulut, D.J. Scalapino, Phys. Rev. Lett. **68**, 706 (1992)
252. N. Bulut, D.J. Scalapino, Phys. Rev. B **45**, 2371 (1992)
253. Q. Si, J.H. Kim, J.P. Lu, K. Levin, Phys. Rev. B **42**, 1033 (1990)
254. J.P. Lu, Q. Si, J.H. Kim, K. Levin, Phys. Rev. Lett. **65**, 2466 (1990)
255. Q. Si, K. Levin, Phys. Rev. B **44**, 4727 (1991)
256. Q. Si, Y. Zha, K. Levin, J.P. Lu, Phys. Rev. B **47**, 9055 (1993)
257. Q. Si, J.P. Lu, K. Levin, Phys. Rev. B **45**, 4930 (1992)
258. J.A. Martindale et al., Phys. Rev. B **68**, 702 (1992)
259. M. Takigawa, J.L. Smith, W.L. Hults, Phys. Rev. B **44**, 7764 (1991)
260. M. Takigawa, P. C. Hammel, et al., Physica (Amsterdam) **162–164C**, 853 (1989)
261. A. Uldry, M. Mali, J. Roos, P.F. Meier, J. Phys. Condens. Matt. **17**, L499 (2005)
262. Y. Yoshinari, H. Yasuoka, Y. Ueda, J. Phys. Soc. Jpn. **61**, 770 (1992)
263. J.A. Martindale, S.E. Barrett et al., Phys. Rev. B **47**, R9155 (1993)
264. S.E. Barrett, J.A. Martindale et al., Phys. Rev. Lett. **66**, 108 (1991)
265. M. Bankay, M. Mali et al., Phys. Rev. B **46**, R11228 (1992)
266. M. Takigawa, Phys. Rev. B **49**, 4158 (1994)
267. D. Thelen, D. Pines, Phys. Rev. B **49**, 3528 (1994)
268. T. Imai, C.P. Slichter et al., Phys. Rev. B **47**, 9158 (1993)
269. D. Kaplan, E.L. Hahn, J. Phys. Radium **19**, 821 (1958)
270. R.E. Walstedt, J.H. Wernick, Phys. Rev. Lett. **20**, 856 (1968)
271. N. Bulut, D.J. Scalapino, Phys. Rev. Lett. **67**, 2898 (1991)
272. Y. Itoh, H. Yasuoka, J. Phys. Soc. Jpn. **61**, 1287 (1992)
273. A. Sokol, D. Pines, Phys. Rev. Lett. **71**, 2813 (1993)
274. S.-W. Cheong, G. Aeppli, T.E. Mason et al., Phys. Rev. Lett. **67**, 1791 (1991)
275. T.E. Mason, G. Aeppli, H. Mook, Phys. Rev. Lett. **68**, 1414 (1992)
276. T.R. Thurston, P.M. Gehring, G. Shirane et al., Phys. Rev. B **46**, 9128 (1992)
277. S.M. Hayden, G. Aeppli, H. Mook et al., Phys. Rev. Lett. **76**, 1344 (1996)
278. G. Aeppli, T.E. Mason et al., Science **278**, 1432 (1997)
279. S. Ohsugi, Y. Kitaoka, K. Ishida, G-q. Zheng, K. Asayama. J. Phys. Soc. Jpn. **63**, 700 (1994)
280. B. Batlogg, *Physics of High-Temperature Superconductors*, ed. S. Maekawa, M. Sato, Springer Series in Solid-State Sciences, vol. 106 (Springer, Berlin, 1992), p. 219
281. D.C. Johnston, J. Magn. Magn. Mater. **100**, 219 (1991)
282. T. Imai, C.P. Slichter et al., Phys. Rev. Lett. **70**, 1002 (1993)
283. T. Imai et al., J. Phys. Soc. Jpn. **59**, 3846 (1990)

284. R.E. Walstedt: Phys. Rev. Lett. **19**, 146 (1967); **19**, 816 (1967)
285. T. Imai et al., Phys. Rev. Lett. **71**, 1254 (1993)
286. A. Narath, Phys. Rev. B **13**, 3724 (1976)
287. T. Moriya, Y. Takahashi, K. Ueda, Phys. Rev. Lett. **59**, 2905 (1990)
288. R.E. Walstedt, S.-W. Cheong, Phys. Rev. B **53**, R6030 (1996)
289. A. Keren, H. Alloul, P. Mendels, Y. Yoshinara, Phys. Rev. Lett. **78**, 3547 (1997)
290. R.L. Corey, N. Curro et al., Phys. Rev. B **53**, 5907 (1996)
291. A. Sokol, R.P. Singh, N. Elstner, Phys. Rev. Lett. **46**, 4416 (1990)
292. Z.P. Han, R. Dupree et al., Phys. Rev. Lett. **69**, 1256 (1992)
293. H.Y. Hwang, B. Batlogg, H. Takagi et al., Phys. Rev. Lett. **72**, 2636 (1994)
294. H. Alloul, C. Froidevaux, Phys. Rev. **163**, 324 (1967)
295. C. Froidevaux, M. Weger, Phys. Rev. Lett. **12**, 123 (1964)
296. K. Ikushima, H. Yasuoka, S. Tsutsui, M. Saeki, S. Nasu, M. Date, J. Phys. Soc. Japan **67**, 65 (1998)
297. K. Ikushima, S. Tsutsui, Y. Haga, H. Yasuoka, R.E. Walstedt, N.M. Masaki, A. Nakamura, S. Nasu, Y. Onuki, Phys. Rev. B **63**, 104404 (2001)
298. R.E. Walstedt, Y. Tokunaga, S. Kambe, Comptes Rendus Physique **15**, 563 (2014)
299. H. Yasuoka, G. Koutroulakis, H. Chudo, S. Richmond, D.K. Veirs, A.I. Smith, E.D. Bauer, J.D. Thompson, G.D. Yarvinen, D.L. Clark, Science **336**, 901 (2012)
300. S. Kern, R. A. Robinson, H. Nakotte, G. H. Lander, B. Cort, R. Watson, F. A, Vigil, Phys. Rev B **59**, 104 (1999)
301. S.B. Wilkins, R. Caciuffo, C. Detlefs, E. Colineau, F. Wastin, G.H. Lander, Phys. Rev. B **73**, 060406(R) (2006)
302. P. Pincus, J. Winter, Phys. Rev. Lett. **7**, 269 (1961)
303. H. Abe, H. Yasuoka, A. Hirai, J. Phys. Soc. Japan **21**, 77 (1966)
304. D.W. Osborne, E.F. Westrum Jr., J. Chem. Phys. **21**, 1884 (1953)
305. A. Arrott, J.E. Goldman, Phys. Rev. 108, 948 (1957); J. W. Ross, D. J. Lam J. Appl. Phys. **38**, 1451 (1967)
306. J.M. Friedt, F.J. Litterst, J. Rebizant, Phys. Rev. B **32**, 257 (1985)
307. R. Caciuffo, G.H. Lander, J.C. Spirlet, J.M. Fournier, W.F. Kuhs, Solid State Commun. **64**, 149 (1987)
308. P. Santini, G. Amoretti, Phys. Rev. Lett. **85**, 2188 (2000)
309. W. Kopmann et al., J. Alloys Compd. **271–273**, 463 (1998)
310. J.A. Paixão, C. Detlefs, R. Caciuffo, P. Santini, N. Bernhoeft, J. Rebizant, G. Lander, Phys. Rev. Lett. **89**, 187202-1,4 (2002)
311. Y. Tokunaga, Y. Homma, S. Kambe, Phys. Rev. Lett. 94, 137209 (2005); Y. Tokunaga, D. Aoki, Y. Homma et al. et al., Phys. Rev. Lett. **97**, 257601 (2006)
312. O. Sakai, R. Shiina, H. Shiba, J. Phys. Soc. Jpn. **74**, 457 (2005)
313. T. Suzuki, N. Magnani, P.M. Oppeneer, Phys. Rev. B **82**, 241103(R) (2010)
314. H. Chudo, Y. Tokunaga, S. Kambe et al., Phys. Rev. B **84**, 094402 (2011)
315. Y. Tokunaga, T. Nishi, S. Kambe, M. Nakada, A. Itoh, Y. Homma, H. Sakai, H. Chudo, J. Phys. Soc. Jpn. **79**, 0537051 (2010)
316. Y. Tokunaga, T. Nishi, M. Nakada, A. Itoh, H. Sakai, S. Kambe, Y. Homma, F. Honda, D. Aoki, R.E. Walstedt, Phys. Rev. **89**, 214416 k(2014)
317. T. Hotta, Phys. Rev. B **80**, 024408 (2009)
318. J. Friedel, Advances in Physics (Taylor and Francis Ltd. London **3**, 447 (1954)
319. T. Kasuya, Prog. Theoret. Phys. 16; K. Yosida. Phys. Rev. **106**, 893 (1957)
320. P. W. Anderson, Phys. Rev. **124**, 41 (1961)
321. *Magnetism*, ed. H. Suhl (Academic Press, New York, 1973), vol. 5
322. J. Kondo, *Solid State Physics*, ed. H. Ehrenreich, F. Seitz, D. Turnbull (Academic Press, New York, 1969), vol. 23
323. J.B. Boyce, C.P. Slichter, Phys. Rev. Lett. **32**, 61 (1974)
324. J.B. Boyce, C.P. Slichter, Phys. Rev. **13**, 379 (1976)
325. E. Muller-Hartmann, Z. Phys. **223**, 277 (1969)

326. A.J. Heeger, L.B. Welsh, M.A. Jensen, G. Gladstone, Phys. Rev. **172**, 303 (1968)
327. A.C. Hewson, *The Kondo Problem to Heavy Fermions* (Cambridge Univerity Press, Cambridge, England, 1993)
328. N.F. Mott, Philos. Mag. (30), 403 (1974)
329. F. Steglich, J. Aarts, C. Bredl, W. Lieke, D. Meschede, W. Franz, H. Schafer, Phys. Rev. Lett. 43, 1892 (1979); F. Steglich, C. Geibel, K. Gloos, G. Olesch, C. Schank, C. Wassilew, A. Loidl, A. Krimmel, G. Stewart. J. Low Temp. Phys. **95**, 3 (1994)
330. S. Nakatsuji, D. Pines, Z. Fisk, Phys. Rev. Lett. **92**, 016401 (2004)
331. N.J. Curro, B.-L. Young, J. Schmalian, D. Pines, Phys. Rev. B **70**, 235117 (2004)
332. Y. Yang, D. Pines, Phys. Rev. Lett. **100**, 096404 (2008)
333. J.H. Shim, K. Haule, G. Kotliar, Science **318**, 1615 (2007)
334. S. Kambe, Y. Tokunaga, H. Sakai, H. Chudo, Y. Haga, T.D. Matsuda, R.E. Walstedt, Phys. Rev. B **81**, 140405(R) (2010)
335. W. Bao, G. Aeppli, J. W. Lynn, P. G. Pagliuso, J. L. Sarrao, M. F. Hundley, J. D. Thompson, Z. Fisk, PRB 65, 100505R (2002)
336. J.B. Goodenough, J. Phys. Chem. Solids (6), 287 (1958)
337. J. Kanamori, J. Phys. Chem. Solids **10**, 87 (1959)
338. C. Petrovic, P.G. Pagliuso, M.F. Hundley, R. Movshovich, J.L. Sarrao, J.D. Thompson, Z. Fisk, P. Monthoux, J. Phys. Condens. Matter **13**, L337 (2001)
339. S. Nakatsuji, S. Yeo, L. Balican, Z. Fisk, P. Schlottman, P.G. Pagliuso, N.O. Moreno, J.L. Sarrao, J.D. Thompson, Phys. Rev. Lett. **89**, 106402 (2002)
340. S. Kambe, H. Sakai, Y. Tokunaga, T. Hattori, G. Lapertot, T.D. Matsuda, G. Knebel, J. Flouquet, R.E. Walstedt, Phys. Rev. B **95**, 195121 (2017)
341. I.J. lowe, R.E. Norberg, Phys. Rev. **107**, 46 (1957)
342. P. Gegenwart, J. Custer, C. Geibel, K. Neumaier, T. Tayama, K. Tenya, O. Trovarelli, F. Steglich, Phys. Rev. Lett. **89**, 056402 (2002)
343. S. Kambe, Y. Tokunaga, H. Sakai, T.D. Matsuda, Y. Haga, Z. Fisk, R.E. Walstedt, Phys. Rev. Lett. **110**, 246406 (2013)

Index

© Springer-Verlag GmbH Germany 2018

R.E. Walstedt, *The NMR Probe of High-T_c Materials and Correlated Electron Systems*, Springer Tracts in Modern Physics 276,
https://doi.org/10.1007/978-3-662-55582-8

W

Y

Z

MIX
Papier | Fördert gute Waldnutzung
FSC® C083411

Kolibri 360 GmbH
Ferdinand-Jühlke-Straße 7
99095 Erfurt, Deutschland
produktsicherheit@kolibri360.de